Nonlinear Partial Differential Equations and Related Topics

Dedicated to
Nina N. Uraltseva

American Mathematical Society

TRANSLATIONS

Series 2 • Volume 229

Advances in the Mathematical Sciences—64

(*Formerly Advances in Soviet Mathematics*)

Nonlinear Partial Differential Equations and Related Topics

Dedicated to Nina N. Uraltseva

Arina A. Arkhipova
Alexander I. Nazarov
Editors

American Mathematical Society
Providence, Rhode Island

2000 *Mathematics Subject Classification*. Primary 35–06, 35F20, 35G20, 35J60, 35K55, 35Q30, 35J15, 35J20, 35J50.

Library of Congress Card Number 91-640741
ISBN-13: 978-0-8218-4997-2
ISBN-10: 0-8218-4997-2
ISSN 0065-9290

∞ The paper used in this book is acid-free and falls within the guidelines
established to ensure permanence and durability.
Visit the AMS home page at `http://www.ams.org/`

10 9 8 7 6 5 4 3 2 1 15 14 13 12 11 10

(Courtesy of Vladimir F. Demyanov)

Nina Nikolaevna Uraltseva

(Courtesy of Vladimir F. Demyanov)

Participants of the St. Petersburg PDE seminar.
Special Session dedicated to N.N. Uraltseva's anniversary,
June 2009

Dedicated to Nina N. Uraltseva on her 75th birthday

Contents

Preface

World-reknown mathematician Nina Nikolaevna Uraltseva was born on May 24, 1934, in Leningrad, and with this volume we celebrate her 75th birthday.

Nina Nikolaevna obtained fundamental results in the theory of elliptic and parabolic equations, with her works becoming classical for many generations of mathematicians. Since her first student years in Leningrad State University (LGU), she was supervised by Olga Aleksandrovna Ladyzhenskaya, with whom she remained in close creative contact for decades to follow.

Their joint papers are devoted to investigations of the second order quasilinear equations of elliptic and parabolic types and related higher-dimensional problems of the calculus of variations. These results are the main content of the two celebrated monographs on the theory of quasilinear equations of elliptic and parabolic types (the latter written jointly with V. A. Solonnikov).

These monographs present a rather complete theory for equations of divergence form: both classical solvability of the boundary value problems under natural growth conditions and analysis of smoothness of the generalized solutions are introduced there. Classes of quasilinear diagonal systems are also studied in these books.

Furthermore, the monographs present deep results on the theory of quasilinear equations of nondivergence form. At the beginning of 1980's, the authors developed the techniques introduced by N. Krylov and M. Safonov and essentially sharpened the results on such equations.

Warm personal relations with O. A. Ladyzhenskaya continued until the beginning of 2004, when Olga Aleksandrovna passed away.

The other series of N. N. Uraltseva's works is devoted to equations with various degeneracies of ellipticity with respect to the gradient. In particular, she is the author of the famous result on the $C^{1+\alpha}$-regularity of p-harmonic functions.

Nina Nikolaevna is a virtuoso of analytical technique. It has allowed her to obtain remarkable results on the smoothness for solutions of variational inequalities, in particular, for the Signorini problem. In recent years, she has successfully studied the smoothness of free boundaries in elliptic and parabolic obstacle-type problems.

A more detailed description of her scientific and personal biography can be found in the article "Nina Nikolaevna Uraltseva. To the 70-th anniversary of her birthday" *J. Math. Sci. (NY)*, **132** (2006), no. 3, 249–254.

Since 1959, Nina Nikolaevna has been teaching at the Department of Mathematical Physics of LGU (now St. Petersburg State University). She is the Head of this Department since 1977.

In 1960's, Nina Nikolaevna was awarded the Chebyshev Prize of the Academy of Science of the USSR and the State Prize of the USSR. Now she is an Honored

Scientist of Russian Federation, and an Honorary Professor of St. Petersburg University. She was also awarded the medal "For Labor Merit" for her scientific and pedagogical service.

In 2006, she received the prestigious Humboldt Research Award and became an Honorary doctor at the Royal Institute of Technology (Sweden) for her outstanding contribution to mathematics.

Colleagues, students, and friends admire her intelligence and charm, and congratulate her on this her 75th birthday, and wish her many more creative years ahead.

Editors

Amer. Math. Soc. Transl.
(2) Vol. **229**, 2010

Regularity below the C^2 Threshold for a Torsion Problem, Based on Regularity for Hamilton-Jacobi Equations

John Andersson, Henrik Shahgholian, and Georg S. Weiss

Dedicated to Nina Nikolaevna Uraltseva on the occasion of her 75th birthday

Abstract. We consider the regularity of minimizers of

$$\int_{B_1} |\nabla u|^2 dx$$

in the set $\{u \in W^{1,2}(B_1) \mid h^- \leq u \leq h^+\}$, where $h^\pm$ are solutions to the Hölder continuous Hamilton-Jacobi equations:

$$F^+(x, h^+, \nabla h^+) = g^+(x),$$
$$F^-(x, h^-, -\nabla h^-) = g^-(x),$$

$F^\pm, g^\pm \in C^\alpha$.

When h^+ and h^- solve

$$|\nabla(\pm h^\pm)|^2 = 1,$$

then this is the classical torsion problem for an elastic-plastic bar. When $F^\pm, g^\pm \notin C^2$, the classical techniques to show regularity fail. Here we develop a new approach to prove regularity based on blowup. Our main result is that the minimizer $u \in C^{1,\alpha/2}$, which is optimal.

As a corollary we get that the C^1 solutions to C^α Hamilton-Jacobi equations are $C^{1,\alpha/2}$.

1. Introduction

A, by now, classical problem in continuum mechanics is the torsion problem

$$\text{minimize} \int_\Omega \big(|\nabla u|^2 - u\big) dx \tag{1.1}$$

2000 *Mathematics Subject Classification.* Primary 35B65, 49J40, 70H20.

Key words and phrases. Regularity, Hamilton-Jacobi Equations, Torsion Problems.

H. Shahgholian has been supported in part by the Swedish Research Council.

G. S. Weiss has been partially supported by the Grant-in-Aid 18740086 of the Japanese Ministry of Education, Culture, Sports, Science and Technology. He also thanks the Knut och Alice Wallenberg foundation for a visiting appointment to KTH.

Both J. Andersson and G.S. Weiss thank the Göran Gustafsson Foundation for visiting appointments to KTH.

The present result is part of the ESF-program GLOBAL. It was completed while the first two authors were visiting the Petroleum Institute in Abu Dhabi.

among all functions in $\tilde{L} \equiv \{|\nabla u|^2 \leq 1 \mid u = f \text{ on } \partial\Omega\}$. In classical applications, $f \equiv 0$, but we do not need that restriction. We will assume that $\tilde{L} \neq \emptyset$, which in particular implies that $|\nabla f| \leq 1$. Here we are interested in local regularity of minimizers, so for our purposes it is enough to consider $\Omega = B_1$, with nonzero boundary data. For some classical results on the torsion problem, see [**4**], [**8**], [**9**] and [**10**].

Interestingly, it can be shown that minimizing (1.1) in $\tilde{L}$ is equivalent to minimizing (1.1) in $L = \{u \in W^{1,2}(\Omega) \mid h^- \leq u \leq h^+\}$, where

$$\begin{aligned} &|\nabla h^+|^2 = 1 && \text{in } \Omega, \\ &|\nabla(-h^-)|^2 = 1 && \text{in } \Omega, \\ &h^+ = h^- = f && \text{on } \partial\Omega; \end{aligned}$$

by this we mean that h^+ is a viscosity solution and that $-h^-$ is a viscosity solution to $|\nabla h|^2 = 1$ (see [**3**]).

The notation might strike someone not familiar with viscosity solutions as somewhat odd. We want to point out that the minus sign in the equation for h^- is not redundant and cannot be canceled. The minus sign tells us that h^- is the minimal solution to $|\nabla h| \leq 1$, with the prescribed boundary data. Correspondingly h^+ is the maximal solution.

Several people have generalized the torsion problem to more general constraints, that is, minimizing over

$$\tilde{L} = \{u \in W^{1,p} \mid F(x, u, \nabla u) \leq 1\}, \text{ or}$$
$$L = \{u \in W^{1,p} \mid h^- \leq u \leq h^+\}, \text{ where } F(x, \pm h^\pm, \pm\nabla h^\pm) = 1,$$

more general functionals

$$\int_\Omega G(x, \nabla u) dx$$

or both (for two such generalizations, see [**6**] and [**5**]). There seems to be a technical threshold that has not yet been crossed. All the previous results depend in one way or another on the C^2 differentiability of F. In this paper, we will announce and discuss some regularity proofs below the C^2 threshold. These results will be published elsewhere (see [**2**]).

In particular, we are interested in minimizers of the Dirichlet energy

$$\int_{B_1} |\nabla u|^2 \tag{1.2}$$

in the set

$$L = \{u \in W^{1,2} \mid h^- \leq u \leq h^+\}, \tag{1.3}$$

where

$$F(x, \pm h^\pm, \pm\nabla h^\pm) = 1$$

and $F \in C^\alpha$ (for some $\alpha \in (0, 1]$), together with some structural assumptions and the boundary conditions

$$h^+ = h^- = f \text{ on } \partial B_1.$$

Since we are primarily interested in local regularity, the boundary conditions are not of any importance as long as $L \neq \emptyset$.

The difficulty is, as pointed out before, that $F \neq C^2$. In order to work with something definite we will choose

$$F(x, z, p) = |p - a(x)|^2 \text{ for } a \in C^\alpha. \tag{1.4}$$

This choice is somewhat arbitrary. Similar constraints arise in micromagnetics (see [**1**]). In micromagnetics the natural constraint would be $|\nabla u - a|^2 \leq 1$, not that u is trapped between two solutions of that equation. In reality, as far as our method is concerned, we could have picked $F(x, r, p)$ to be any reasonable function, strictly and uniformly convex in p, increasing in r. However, the proof is quite involved as it is, so we deem it expedient to work with a simple fixed F.

Our main result is (in order to avoid certain technicalities we will only discuss the proof of a simplified version here):

MAIN THEOREM. *Let u be a minimizer of the Dirichlet energy,*

$$\int_{B_1} |\nabla u|^2,$$

in the set $\{u \in W^{1,2} \mid h^- \leq u \leq h^+,\ u = f \in C^\alpha(\partial B_1) \text{ on } \partial B_1\}$, where $h^\pm$ are viscosity solutions to

$$\begin{array}{ll} \pm|\nabla h^\pm - a|^2 = \pm 1 & \text{in } B_1, \\ h^\pm = f & \text{on } \partial B_1, \end{array} \tag{1.5}$$

with $[a]_{C^\alpha} \leq A \leq C_0$ for some C_0. Assume furthermore that $u(0) = h^+(0)$ or $u(0) = h^-(0)$. Then

$$\begin{array}{ll} \operatorname{osc}_{x \in B_r(0)} \big(u(x) - u(0) - \nabla u(0) \cdot x\big) \leq C\sqrt{A} r^{1+\frac{\alpha}{2}} & \text{if } r \leq A^{\frac{1}{2-\alpha}}, \\ \operatorname{osc}_{x \in B_r(0)} \big(u(x) - u(0) - \nabla u(0) \cdot x\big) \leq C r^2 & \text{if } r > C(\alpha) A^{\frac{1}{2-\alpha}}. \end{array} \tag{1.6}$$

Remark. In equation (1.5) we again use somewhat counterintuitive notation. Once again, we may not cancel the minus signs in the equations for h^-. They indicate that we are interested, in the viscosity sense, in the minimal solution, whereas we are interested in the maximal solution when we consider h^+.

That the gradient of u exists, and therefore that (1.6) makes sense, is a direct consequence of Lemma 3.5 and Proposition 3.6. Also using that proposition together with the main theorem will directly give that $u \in C^{1,\alpha/2}(B_{1/2})$. This regularity is optimal as the following example shows.

Example. If $a(x) = |x_2|^\alpha e_1$, then $u(x_1, x_2) = x_1 + b(x_2)$ is a solution to $|\nabla u - a(x)|^2 = 1$ for

$$b(x_2) = \int_0^{x_2} \sqrt{2|t|^\alpha - |t|^{2\alpha}} dt.$$

Here $a \in C^\alpha$ and $u \in C^{1,\alpha/2}$ but $u \notin C^{1,\beta}$ for any $\beta > \alpha/2$. We would like to thank Stefan Müller for providing us with this example.

We believe that similar techniques as explained here will also yield regularity for $F \in C^{1,\alpha}$ and $F \in C^0$, together with some structural assumptions on F. It is reasonable to conjecture that the minimizer $u \in C^{1,\frac{1+\alpha}{2}}$ when $F \in C^{1,\alpha}$ and $u \in C^1$ when $F \in C^0$. That would in particular recover the result that $u \in C^{1,1}$ when $F \in C^2$. At this point we have not tried to prove this since it would involve more complicated approximations of the Hamilton-Jacobi equations, and therefore very lengthy calculations.

In the next section, we will discuss some heuristics, or our strategy of proving the main theorem, as well as setting up a plan for the paper.

2. Heuristics

As noted in the introduction, our main difficulty will be to work with only C^α regular Hamilton-Jacobi equations. That precludes us from using methods used by previous authors. Instead we will use rescalings and blowup. In particular we would want to show that

$$\operatorname{osc}_{B_r(x^0)}\big(u(x) - p\cdot(x-x^0)\big) \le Cr^{1+\alpha/2} \tag{2.1}$$

for each $x^0 \in B_{1/2}$ and some $p = p_{x^0}$; from here on, we will always assume that $x^0 = 0$. If we can prove (2.1), then $C^{1,\alpha/2}$ regularity follows.

It is not that difficult to show that (2.1) would follow if we can show that there exists $p \in \mathbb{R}^n$ such that

$$\sup_{B_r(x^0)} \big(h^+(x) - h^+(x^0) - p\cdot(x-x^0)\big) \le C_0 r^{1+\alpha/2} \tag{2.2}$$

and

$$\inf_{B_r(x^0)} \big(h^-(x) - h^-(x^0) - p\cdot(x-x^0)\big) \ge -C_0 r^{1+\alpha/2}$$

for each $r > 0$ and $x^0 \in B_{1/2}$. In particular, wherever $u(x) = h^+(x)$, then heuristically u would be subharmonic at x and subharmonicity controls the second derivatives from below and (2.2) controls the second derivatives from above.

The problem is that when $F \neq C^2$, or with our choice of F in (1.4) when $a \neq C^2$, there are no good estimates such as (2.2) for solutions to Hamilton-Jacobi equations. When $F \in C^2$ the Bernstein technique implies (2.2) with $\alpha = 2$. These good estimates for C^2 equations suggest that we could prove regularity by freezing coefficients. We will go down this road in section 3. The problem is that as we consider smaller and smaller balls the control of the C^2-oscillations will deterioriate too fast and we cannot get optimal regularity by a naive freezing of coefficients. The method will give us $C^{1,\beta}$-estimates, $\beta = \alpha/(2+\alpha)$, but that is of lesser importance. This regularity will serve as the starting point for the iteration argument that constitutes the heart of the proof of our main theorem.

Our main argument will be a bootstrap argument based on blowup and scalings. It is therefore important to understand how u and $h^\pm$ scale. Since u minimizes a Dirichlet energy, its governing equation is the Laplacian; therefore the "right" scaling of u is

$$u \to \frac{u(rx)}{S_r},$$

where S_r is some normalizing constant. But the "right" scaling for $h^\pm$ is, heuristically and after a rotation of the coordinate system so that $\nabla h^\pm(0) + a(0) = e_n$,

$$h^\pm \to \frac{h^\pm(rx', r^{1-\beta}x_n)}{r^{1+\beta}}. \tag{2.3}$$

Since the scalings are different we have to make two separate arguments. In the first one we scale $h^\pm$ and consider u to be any $C^{1,\beta}$ function touching $h^\pm$ from below (or above). By correctly scaling $h^\pm$ we hope to get some better regularity of that function. Then we scale u and consider $h^\pm$ to be any function that is $C^{1,\beta}$ touching u from above (below).

The first problem we encounter is that from the freezing of coefficients we only get that

$$\sup_{B_r}\big(h^+(x) - h^+(0) - p\cdot x\big) \le Cr^{1+\beta}.$$

Since we scale nonhomogeneously in the x_n directions, this is not enough to show that the rescaled functions in (2.3) are bounded. In section 4 we will use the blow-up method to show that the rescaled functions (slightly modified) are actually bounded.

Interestingly enough if we set

$$h_j(x) = \frac{h^+(r_j x', r_j^{1-\beta} x_n)}{r_j^{1+\beta}}$$

and $h = \lim_{r_j \to 0} h_j$ for some sequence $r_j \to 0$, then heuristically, h will solve a parabolic Hamilton-Jacobi equation

$$|\nabla' h|^2 - 2\frac{\partial h}{\partial x_n} = 0.$$

We will use this observation, together with a freezing-of-coefficients argument, in section 5 to improve the regularity of $h^\pm$ in the x' directions. In particular, $h^\pm$ will in some sense have one-sided $C^{1,\beta+\epsilon}$-estimates in the x' directions. Here $\epsilon > 0$ as long as $2\beta < \alpha$.

We may then use the better regularity of $h^\pm$ in sections 4 and 5 to improve the regularity of u. This is done by a simple blow-up argument.

So far we have used $C^{1+\beta}$-estimates to deduce $C^{1,\beta+\epsilon}$-estimates of u. Setting $\beta_1 = \beta + \epsilon$ we may repeat the above argument with β_1 and gain another $\epsilon_1 > 0$ in the Hölder exponent. In particular, we may construct a sequence $\beta < \beta_1 < \beta_2 < \cdots < \beta_j \to \alpha/2$ such that $u \in C^{1,\beta_j}$. If we could control the norms $\|u\|_{C^{1,\beta_j}}$ uniformly, then the main theorem would follow.

It is not hard to show that if

$$[a]_{C^\alpha(B_1)} = \sup_{x,y \in B_1} \frac{|a(x) - a(y)|}{|x - y|^\alpha} \equiv A$$

is small enough, smaller than some universal constant ξ, then we will get uniform control of $\|u\|_{C^{1,\beta_j}}$. If $A \geq \xi$, then we may rescale the problem by a factor $\tau = (\xi/A)^{1/\alpha}$:

$$u_\tau(x) = \frac{u(\tau x)}{\tau},$$
$$a_\tau(x) = a(\tau x).$$

The new function u_τ will solve a similar problem with $[a_\tau]_{C^\alpha} = \xi$. We would get $C^{1,\alpha/2}$ estimates for the new function u_τ. Scaling back would give the main theorem.

Here there is another slight complication. That is, the oscillations of u will depend on the scale. On small scales ($r < A^{1/(2-\alpha)}$), the regularity of a will affect the regularity of u and on large scales when a is almost constant (compared to the scale), then the oscillations of u will be governed by the second derivative estimates of $h^\pm$ that we get from freezing the coefficients. This splitting of the regularity will make the proofs much longer and more difficult to follow. We will therefore make a simplifying assumption for the rest of this paper. The general case as formulated in the main theorem follows similarly, but only after much longer calculations.

Simplifying Assumption. We will assume that

$$A = [a]_{C^\alpha(B_1)} = \sup_{x,y \in B_1} \frac{|a(x) - a(y)|}{|x - y|^\alpha}$$

is a universal constant bounded away from 0 and ∞. That is, we will assume that the Hölder semi-norm of a has a specific value A depending only on n. This constant A will be determined at a later point. Even though this assumption is artificial we may recover the general case by a simple scaling as pointed out above.

With this simplifying assumption the main theorem can be stated as:

SIMPLIFIED MAIN THEOREM. *Let u be a minimizer of the Dirichlet energy,*

$$\int_{B_1} |\nabla u|^2,$$

in the set $\{u \in W^{1,2} \mid h^- \leq u \leq h^+,\ u = f \in C^\alpha(\partial B_1) \text{ on } \partial B_1\}$, where $h^\pm$ are viscosity solutions to

$$\begin{aligned} \pm|\nabla h^\pm - a|^2 &= \pm 1 \quad \text{in } B_1, \\ h^\pm &= f \qquad\ \ \text{on } \partial B_1, \end{aligned}$$

with $[a]_{C^\alpha} \leq A$, where A is a universal constant depending only on n. Assume furthermore that $u(0) = h^+(0)$. Then

$$\operatorname{osc}_{x\in B_r(0)}\big(u(x) - u(0) - \nabla u(0)\cdot x\big) \leq C(\alpha) r^{1+\frac{\alpha}{2}}. \tag{2.4}$$

3. Background Regularity

In this section we will gather some, more or less, known regularity results for viscosity solutions and obstacle problems.

3.1. Hamilton-Jacobi Equations. Let us first define what we mean by a viscosity solution.

DEFINITION 3.1. We say that $u \in C(\Omega)$ is a viscosity solution to

$$H(x, u, \nabla u) = f(x) \text{ in } \Omega$$

if for every $x^0 \in \Omega$ and $\phi \in C^1(B_r(x^0))$ such that $u(x^0) = \phi(x^0)$ and

(1) if $u(x) - \phi(x)$ has a local maximum at x^0, then $H(x^0, \phi(x^0), \nabla\phi(x^0)) \leq f(x^0)$, and
(2) if $u(x) - \phi(x)$ has a local minimum at x^0, then $H(x^0, \phi(x^0), \nabla\phi(x^0)) \geq f(x^0)$.

From this definition, general existence and uniqueness results follow. Our goal here is not to give a general exposition of the entire theory of viscosity solutions but only to gather some results that we will be needing later on. The first result is one-sided regularity for viscosity solutions for C^2 Hamilton-Jacobi equations. The proof of this lemma and all other results in this subsection (except the final lemma) can be found, somewhat scattered, throughout Lions' book [**7**].

LEMMA 3.2. *Let $u \in C(B_1 \times (-1, 0))$ be a viscosity solution to*

$$\frac{\partial u}{\partial t} + H(\nabla u) = m(x)$$

with $p \cdot D^2H(p) \cdot p \geq c|p|^2$ for some constant $c > 0$ and $m \in C^{1,1}$. Then

$$\frac{\partial^2 u}{\partial x_i^2} \leq C_0 \text{ in } B_{1/2} \times (-1/2, 0),$$

for all spatial directions x_i; here we interpret the derivatives in the distributional sense. C_0 depends only on $\|D^2 m\|_{L^\infty(B_R)}$, $\sup_{B_R}|\nabla u|$ and H.

A similar statement naturally holds for the time-independent case, that is, if $u \in C(B_1)$ is a viscosity solution to

$$H(\nabla u) = m(x)$$

with $p \cdot D^2 H(p) \cdot p \geq c|p|^2$ for some constant $c > 0$ and $m \in C^{1,1}$, then

$$\frac{\partial^2 u}{\partial x_i^2} \leq C_0 \text{ in } B_{1/2},$$

for all spatial directions x_i. C_0 depends only on $\|D^2 m\|_{L^\infty(B_R)}$, $\sup_{B_R} |\nabla u|$ and H.

In order to use these results in a freezing-of-coefficients argument we also need some good comparison principles.

LEMMA 3.3. *Let $u \in C^{0,1}(\overline{B}1)$ and $v \in C^{0,1}(\overline{B}1)$ be viscosity solutions in B_R to*

$$H(x, \nabla u) = n(x) \text{ and}$$
$$H(x, \nabla v) = m(x),$$

where $p \cdot D^2 H(x,p) \cdot p \geq c|p|^2$ for some constant $c > 0$, H is Lipschitz, $m > n \geq \lambda > 0$, $u, v \in C^{0,1}(B_1)$ and $\|u\|_{L^\infty(B_1)}, \|u\|_{L^\infty(B_1)} \leq K$. Then

$$u \leq v \leq u + C(\lambda, K) \max\big(\sup_{B_1}(m-n), \sup_{\partial B_1}(v-u)\big).$$

LEMMA 3.4. *Let $u \in C^{0,1}(\overline{B}_1 \times (-1,0))$ and $v \in C^{0,1}(\overline{B}_1 \times (-1,0))$ be viscosity solutions in $B_1 \times (-1,0)$ to*

$$\frac{\partial u}{\partial t} + H(\nabla u) = m(x) \text{ and}$$

$$\frac{\partial v}{\partial t} + H(\nabla v) = n(x),$$

where $p \cdot D^2 H(p) \cdot p \geq c|p|^2$ for some constant $c > 0$, H is Lipschitz, $n > m$, $u = v$ on ∂B_R and $u(0,-1) = v(0,-1)$. Then

$$u \leq v \leq u + CR \sup_{B_R \times (-R^2, 0)} (n - m).$$

We end this subsection by showing one-sided $C^{1,\beta}$-estimates for the Hamilton-Jacobi equations.

LEMMA 3.5. *Let h be a viscosity solution to $|\nabla h - a|^2 = 1$ in B_1, $a \in C^\alpha(\mathbb{R}^n; \mathbb{R}^n)$ and $[a(x)]_{C^\alpha} \leq A|x|^\alpha$. Then h^+ satisfies one-sided $C^{1,\alpha/(\alpha+2)}(B_{1/2})$-estimates: there is a vector p in the super differential of h such that*

$$\sup_{B_r(x^0)} (h^+(x) - p \cdot (x - x^0) - h^+(x^0)) \leq C r^{1+\alpha/(2+\alpha)}.$$

Proof: We will show the oscillation estimate for $x^0 = 0$. Let us fix a small r and ball $B_r(0)$. We would want to find a barrier function w, solving a Hamilton-Jacobi equation with constant coefficients, in a larger ball $B_\delta(0)$. Then we are going to use one-sided estimates for w and comparison principles to estimate h.

In particular, let

$$\begin{aligned} |\nabla w|^2 &= 1 + C\delta^\alpha && \text{in } B_\delta(0), \\ w &= h && \text{on } \partial B_\delta(0). \end{aligned}$$

Then we get from Lemma 3.3 that $h \le w \le h + C\delta^{1+\alpha}$. Using Lemma 3.2, for w, we may deduce that for some $p \in \mathbb{R}^n$,

$$h(x) \le w(x) \le w(0) + p \cdot x + \frac{C}{\delta}|x|^2 \le h(0) + p \cdot x + C\big(\frac{1}{\delta}|x|^2 + \delta^{1+\alpha}\big).$$

Rearranging terms and taking the supremum in B_r on both sides results in

$$\sup_{B_r(0)} \big(h(x) - h(0) - p \cdot x\big) \le C\big(\frac{1}{\delta}r^2 + \delta^{1+\alpha}\big). \tag{3.1}$$

The trick is to choose δ such that the left-hand side in (3.1) becomes as small as possible. Choosing $\delta = r^{\frac{2}{2+\alpha}}$ (then $\delta > r$ since $r < 1$) will give the best balance between terms. With this δ we get

$$\sup_{B_r(x^0)} \big(h^+(x) - p \cdot (x - x^0) - h^+(x^0)\big) \le C r^{1+\alpha/(2+\alpha)},$$

and the lemma follows. □

3.2. Obstacle Problems. In this section we review some results for the regularity of obstacle problems. In particular we will need the following result, of which we sketch the idea of the proof.

PROPOSITION 3.6. *Let $u \in C^{1,\beta}(B_R)$ be a minimizer of the Dirichlet energy*

$$\int_{B_R} |\nabla u|^2,$$

among all functions in the convex set $K = \{u \in W^{1,2} \mid h^- \le u \le h^+\}$.

Assume furthermore that for every $x^0 \in \Lambda = B_R \cap \overline{\{\Delta u \neq 0\}}$ (here $\{\Delta u \neq 0\}$ is interpreted in the distributional sense) we have the following two estimates: (i) If $u(x^0) = h^+(x^0)$, then

$$\sup_{B_r(x^0)} \big(u(x) - \nabla u(x^0) \cdot (x - x^0) - u(x^0)\big) \le C_0 r^{1+\gamma} \text{ for all } r > 0$$

and some $\gamma < 1$. (ii) If $u(x^0) = h^-(x^0)$, then

$$\inf_{B_r(x^0)} \big(u(x) - \nabla u(x^0) \cdot (x - x^0) - u(x^0)\big) \ge -C_0 r^{1+\gamma} \text{ for all } r > 0$$

and some $\gamma < 1$.

Then $u \in C^{1,\gamma}(B_{R/2})$ and $\|u\|_{C^{1,\gamma}} \le C(C_0, \|u\|_{L^\infty}(B_R))$.

Proof: We argue by contradiction and assume that there is a sequence u_j of minimizers as in the proposition and points x^j (which we may without loss of generality assume to be the origin, that is, $x^j = 0$) and $r_j > 0$ such that

$$\inf_{p \in \mathbb{R}^n} \operatorname{osc}_{B_1} \frac{u_j(r_j x) - r_j p \cdot x}{j r_j^{1+\gamma}} = 1. \tag{3.2}$$

We may also choose our r_j to be the largest r such that equation (3.2) holds. Next we define

$$u^j(x) = \frac{u_j(r_j x) - r_j p_j \cdot x - u_j(0)}{j r_j^{1+\gamma}},$$

where p_j is the vector that minimizes the expression in equation (3.2). Then $u^j \in C^{1,\beta}$ minimizes a similar problem. In particular $\Delta u = 0$ in some set $\Omega_j \subset B_{R/r_j}$ and for every point $x^0 \in \Lambda \equiv B_{R/r_j} \setminus \Omega_j$ we have

$$\sup_{B_r(x^0)} \left(u^j(x) - \nabla u^j(x^0)\cdot(x-x^0) - u^j(x^0)\right) \le \frac{C_0}{j} r^{1+\gamma} \tag{3.3}$$

or

$$\inf_{B_r(x^0)} \left(u^j(x) - \nabla u^j(x^0)\cdot(x-x^0) - u^j(x^0)\right) \ge -\frac{C_0}{j} r^{1+\gamma}. \tag{3.4}$$

Let us write $\Lambda = \Lambda^+ \cup \Lambda^-$ with (3.3) satisfied for points in Λ^+ and (3.4) satisfied in Λ^-.

Now we consider some subsequence $u^j \to u^0$, where u^0 is also a minimizer of the Dirichlet energy. Furthermore by equation (3.3) we have

$$u^0 \le \inf_{x^0 \in \Lambda^+} \left(u^0(x^0) + \nabla u^0(x^0)\cdot(x-x^0)\right)$$

and a similar estimate from below.

That is, u^0 solves the double obstacle problem with a concave upper obstacle and a convex lower obstacle. It follows that $\Delta u^0 = 0$ in $\mathbb{R}^n$. But by our choice of r_j we also have that

$$\inf_{p\in\mathbb{R}} \operatorname{osc}_{B_r}\left(u^0(x) - p\cdot x\right) \le 1 + r^{1+\gamma}.$$

It follows that u^0 is a linear function, but $u^0(0) = 0$ and

$$\inf_{p\in\mathbb{R}^n} \operatorname{osc}_{B_r}\left(u^0(x) - p\cdot x\right) = 1$$

by equation (3.2). This is a contradiction if $u^j \to u^0$ uniformly. We can actually prove uniform convergence. However, there is a slight complication that makes the proof somewhat technical, so we leave that part out. □

From Lemma 3.5 and Proposition 3.6 it follows that if u is as in the (simplified) main theorem, then $u \in C^{1,\frac{\alpha}{2+\alpha}}$.

4. Regularity in the Nondegenerate Directions

Having some basic regularity for h^+ and u we can now begin our main argument for optimal regularity. As pointed out in section 2 we need to be able to control the supremum of natural scalings

$$h_j^+(x) \to \frac{h^+(r_j x', r_j^{1-\beta} x_n)}{r_j^{1+\beta}}.$$

In the next proposition we will do that (with a slight modification, the function g in the proposition) at points where the graph of u touches the graph of h^+ from below. Since $u \in C^{1,\beta}$, that u touches h^+ from below implies the assumption (4.3) below.

The proposition is as follows.

PROPOSITION 4.1. *Let h^+ be a viscosity solution to $|\nabla h^+ - a|^2 = 1$ in B_R, $h^+(0) = 0$, and have one-sided $C^{1,\beta}$-estimates:*

$$\sup_{B_r(x^0)} \left(h^+(x) - h^+(x^0) - p\cdot(x-x^0)\right) \le CA^{\frac{1-\beta}{2}} r^{1+\beta} \tag{4.1}$$

for every p in the super-differential of h^+ at x^0. Furthermore we assume that

$$h^+(x', x_n) \le h^+(0, x_n) + p \cdot x' + CA^{\frac{1-\beta}{2}} |x'|^{1+\beta} \tag{4.2}$$

for each fixed x_n,

$$h^+(x) \ge -CA^{\frac{1-\beta}{2}} |x|^{1+\beta}. \tag{4.3}$$

Also assume that $[a]_{C^\alpha} \le A$, where A satisfies $A \le \tilde{C}$ for some $\tilde{C}$. We also assume that $a(x) = e_n + b(x)$, where $|b(x)| \le A|x|^\alpha$ and that $\alpha/(2+\alpha) \le \beta \le \alpha/2$.

Denote, for $r \le \inf(A^{1/2} R^{\frac{1}{1-\beta}}, R)$,

$$K(r) = \Big\{ \big(\frac{x'}{r}, \frac{A^{\frac{1-\beta}{2}} x_n}{r^{1-\beta}} \big) \in B_1 \Big\}.$$

Then,

$$\sup_{K(r)} |h^+(x) + g(x_n)| \le CA^{\frac{1-\beta}{2}} r^{1+\beta},$$

where $g(0) = g'(0) = 0$, $g'(t) = e_n \cdot b(0, \dots, 0, t)$.

Sketch of Proof: We will do this by blowup and contradiction. If the proposition is not true, then there exists a sequence of h_j^+,r_j, u^j, b^j and a^j such that

$$\sup_{K(r_j)} |h_j^+ + g_j| = j r_j^{1+\beta}.$$

Here g_j is defined as $g_j(0) = g'(0) = 0$ and $g_j'(t) = e_n \cdot b_j(0, \dots, 0, t)$. We would like to use this sequence to define a blowup sequence

$$h_j(x) = \frac{h_j^+(r_j x', r_j^{1-\beta} x_n) + g_j(r_j^{1-\beta} x_n)}{j r_j^{1+\beta}},$$

pass to the limit $j \to \infty$ and find a contradiction. The problem with this is that the above blowup does not scale as the Hamilton-Jacobi equation. Our first step is to modify the blowup somewhat to get a sequence that reflects the scaling of the equation. In particular we make the following claim.

Claim 1: *Under the above assumptions there exists a sequence $\sigma_j \to \infty$ such that if h_j is defined according to*

$$h_j(x) = \frac{h_j^+(\sqrt{\sigma_j} r_j x', r_j^{1-\beta} x_n) + g_j(r_j^{1-\beta} x_n)}{\sigma_j r_j^{1+\beta}}, \tag{4.4}$$

then $\sup_{B_1} |h_j| = 1$.

This is not difficult to prove; start by $\sigma_j = 1$ and continuously increase σ_j until we get the right supremum. For simplicity of notation we will assume $\sigma_j = j$. The scaling of h_j has two great advantages. First it is easy to see that the scaling is right. In particular h_j will solve the following Hamilton-Jacobi equation:

$$|\nabla' h_j|^2 - 2(h_j)_n = \frac{b}{2\sqrt{j} r_j^\beta} \cdot \nabla' h_j + 2(h_j)_n (b \cdot e_n - g') \tag{4.5}$$

$$+ \frac{|b - e_n g'|^2}{j r_j^{2\beta}} + \frac{e_n \cdot b - g'}{j r_j^{2\beta}} - j r_j^{2\beta} (h_j)_n^2 = T_1 \cdot \nabla' h_j + T_2 + T_3 + T_4 - T_5.$$

Secondly, the scaling in the x' directions is worse than $C^{1,\beta}$. By this we mean:

Claim 2: *For every x_n and p_j in the super-differential of h_j at (x', x_n) we have for each R, and any q_j in the super-differential of h_j^+ at $(0, r_j^{1-\beta})$,*

$$\sup_{B'_R}\left(h_j - h_j(0,x_n) - p_j\cdot x'\right) \le \sup_{B'_{\sqrt{j}r_jR}} \frac{h_j^+ - h_j^+(0,x_n) + q_j\cdot x'}{jr_j^{1+\beta}} \le C\frac{R^{1+\beta}}{j^{\frac{1-\beta}{2}}} \to 0.$$

If $x_n = 0$, then $\operatorname{osc}_{B'_R}(h_j(x',0) - h_j(0) - p_j\cdot x') \to 0$.

Proof of Claim 2: The first statement is proved in the claim; the second is proved similarly by reversing the inequalities and utilizing that $h_j^+(x',0) \ge u(x',0) \ge -CA_j^{\frac{1-\beta}{2}}|x'|^{1+\beta}$ in the second inequality.

We also need some control over the terms $T_1, \dots, T_5$. We can prove the following claim.

Claim 3: *We have that $T_i \to 0$ locally uniformly as $j \to \infty$, for $i = 1, \dots, 5$. T_i is defined in equation (4.5).*

The proof of this claim is naturally somewhat technical, but pretty straightforward.

In particular, Claim 3 implies that $h_j \to h_0$ locally uniformly, where

$$|\nabla' h_0|^2 - 2(h_0)_n = 0; \tag{4.6}$$

also $h_j \to 0$ on $\{x_n = 0\}$, by Claim 2. Using uniqueness for parabolic Hamilton-Jacobi equations we deduce that $h_0 = 0$ for $x_n \le 0$ and $h_0 \ge 0$ in $\{x_n \ge 0\}$. By Claim 2 we also know that h_0 is concave in the x' directions. This together with the bound from below implies that h_0 is constant for each $x_n \ge 0$. In particular if $h_0(x', x_n)$ is not constant for some $x_n = t > 0$, then $h_0(x',t) \le h_0(0,t) + p\cdot x'$ for some nonzero $p \in \mathbb{R}^{n-1}$. It follows that there is a point x_0' such that $h_0(x_0', t) < 0$. But from (4.6) it follows that $\partial_n h_0 \ge 0$; this is a contradiction. Therefore $h_0(x) = h_0(x_n)$. Using (4.6) again we see that $\partial_n h_0 = 0$ in $\{x_n \ge 0\}$, and thus $h_0 = 0$. This contradicts $\sup_{B_1}|h_0| = 1$ and the proposition follows. □

5. Regularity in the Orthogonal Directions

Proposition 4.1 provides very good control of the oscillations of $h^\pm$ in one direction. To get regularity in the directions orthogonal to that direction we need another argument provided in the next proposition.

PROPOSITION 5.1. *Let h^+ be a viscosity solution to $|\nabla h^+ - a|^2 = 1$ in B_1 and assume that h^+ has one-sided $C^{1,\beta}$-estimates from above:*

$$\sup_{B_r(x^0)}\left(h^+(x) - h^+(x^0) - p\cdot(x - x^0)\right) \le Cr^{1+\beta}. \tag{5.1}$$

Furthermore let

$$h^+(x) \ge -C|x|^{1+\beta} \tag{5.2}$$

and

$$h^+(x',0) \ge h^+(0) + p\cdot x' + C|x'|^{1+\beta}. \tag{5.3}$$

Also assume that $[a]_{C^\alpha} \le A$, where $A \le \tilde{C}$ for some fixed $\tilde{C}$. We also assume that $a(x) = e_n + b(x)$, where $|b(x)| \le A|x|^\alpha$.

Then for every x_n we have for some $p \in \mathbb{R}^n$ and every $r \le c_0$ (c_0 universal)

$$\sup_{B'_r(0,x_n)}\left(h^+(0) - h^+(0,x_n) - p\cdot x'\right) \le CA^{\frac{1}{2} - \frac{\alpha\beta}{2(2+\alpha-2\beta)}} r^{1+\frac{\alpha}{2+\alpha-2\beta}}.$$

Sketch of Proof: From Proposition 4.1 we will get that

$$h(x) = \frac{h^+(rx', A^{(1-\beta)/2}r^{1-\beta}x_n) - g(A^{(1-\beta)/2}r^{1-\beta}x_n)}{A^{(1-\beta)/2}r^{1+\beta}}$$

is bounded from above. Here we have scaled slightly differently since we need to take the value of the constant A into consideration. However since, by our simplifying assumption, A is a universal constant this changes nothing.

From the proof of Proposition 4.1 we also know that h solves (with slightly reordered terms as compared with equation (4.5))

$$\begin{aligned}(5.4)\qquad & \Big|\nabla' h - \frac{b'(0, A^{(1-\beta)/2}r^{1-\beta}x_n)}{A^{(1-\beta)/2}r^{\beta}}\Big|^2 - 2h_n \\ &= -A^{1-\beta}r^{2\beta}|h_n|^2 - 2\frac{b(rx', A^{(1-\beta)/2}r^{1-\beta}x_n) - b(0, A^{(1-\beta)/2}r^{1-\beta}x_n)}{A^{(1-\beta)/2}r^{\beta}} \cdot \nabla' h \\ &\quad - \frac{b(rx', A^{(1-\beta)/2}r^{1-\beta}x_n)\cdot e_n - g'}{A^{1-\beta}r^{2\beta}} - \big(b(rx', A^{(1-\beta)/2}r^{1-\beta}x_n)\cdot e_n - g'\big)h_n \\ &\quad - \frac{|b(rx', A^{(1-\beta)/2}r^{1-\beta}x_n) - g'e_n|^2}{A^{1-\beta}r^{2\beta}} \\ &= -I - II - III - IV - V.\end{aligned}$$

In order to use this Hamilton-Jacobi equation we once again need to estimate the nonconstant ingredients in the equation. This is quite straightforward using the assumptions in the proposition. In particular one can show that h is a subsolution to

$$\begin{aligned} |\nabla' w|^2 - 2w_n &= q + f(x_n) && \text{in } Q_1, \\ w &= h && \text{on } \partial Q_1 \setminus \{x_n = 0\}, \end{aligned}$$

where $Q_1 = \{x \mid |x'| \le 1, \ -1 < x_n < 0\}$ is the parabolic cube and $f(x_n)$ is some function only depending on x_n. The constant q is given by

$$q = C\sup\big(A^{1-\beta}r^{2\beta}, A^{\beta}r^{\alpha-2\beta}\big).$$

However since we have $C^{1,\beta}$ regularity with $\beta = \frac{\alpha}{2+\alpha}$ and that A is a universal constant we can assume that $q = CA^{\beta/2}r^{\alpha-2\beta}$. Arguing as in Lemma 3.5, but using parabolic comparison (Lemma 3.4) and parabolic one-sided estimates (Lemma 3.2) we can estimate h according to, for p in the super-differential of w at x_0,

$$\begin{aligned} h(x) \le w(x) &\le w(x_0) + p\cdot(x' - x_0') + C|x' - x_0'|^2 \\ &\le h(x_0) + p\cdot(x' - x_0') + C|x' - x_0'|^2 + Cq. \end{aligned}$$

Rescaling back to h^+ we see that this implies that (for some p in the super-differential of h^+)

$$h^+(x', x_n) \le h^+(0, x_n) + p\cdot x' + C\big(A^{(1-\beta)/2}r^{1+\beta}q + A^{(1-\beta)/2}r^{\beta-1}|x'|^2\big).$$

Using this estimate in an optimal way will give the proposition. In particular, slightly rearranging terms and taking the supremum in B'_δ on both sides will result in

$$(5.5)\quad \sup_{B'_\delta}\big(h^+(x', x_n) - h^+(0, x_n) - p\cdot x'\big) \le C\big(A^{(1-\beta)/2}r^{1+\beta}q + A^{(1-\beta)/2}r^{\beta-1}\delta^2\big).$$

We want to find the right balance between r and δ to optimize this estimate. Choosing $\delta = A^{\beta/2}r^{1+\alpha/2-\beta}$ and using this choice in equation (5.5) we get

$$\sup_{B'_\delta}\big(h^+(x', x_n) - h^+(0, x_n) - p\cdot x'\big) \le CA^{\frac{1}{2} - \frac{\alpha\beta}{2(2+\alpha-2\beta)}}\delta^{1+\frac{\alpha}{2+\alpha-2\beta}}.$$

This is the estimate we want. □

6. Proof of the Regularity Theorem

We are finally in the position to sketch a proof of the simplified form of the main theorem.

With no loss of generality we may assume that $|\nabla u(0)| = |u(0)| = 0$ and that $a(x) = e_n + b(x)$ with $|b(x)| \leq A|x|^\alpha$. If this is not true we may subtract $u(0) + \nabla u(0) \cdot x$ from u and f, add $\nabla u(0)$ to a and rotate the coordinate system to attain this situation.

From Lemma 3.5 and Lemma 3.6 we have $\mathrm{osc}_{B_r} u \leq C r^{1+\alpha/(2+\alpha)}$. Also from Lemma 3.5 we get $\sup_{B_r} h^+ \leq C r^{1+2/(2+\alpha)}$. That is enough to apply Proposition 5.1 with $\beta_0 = \alpha/(2+\alpha)$, at least for $r \leq c_0$.

In particular for every x_n and some $p(x_n) = p$ in the super-differential of h^+ at $(0, x_n)$ we have

$$\sup_{x' \in B'_r} \left(h^+(x', x_n) - h^+(0, x_n) - p \cdot x'\right) \leq C A^{\frac{1}{2} - \frac{\alpha\beta_0}{2(2+\alpha-2\beta_0)}} r^{1+\frac{\alpha}{1+\alpha-2\beta_0}} \tag{6.1}$$

if $r \leq c_0$. This estimate gives better regularity in the x' direction, but Proposition 4.1 gives good control in the x_n direction. A simple blow-up argument (similar to the proof of Lemma 3.6) will prove the following claim.

Claim: *Under the assumptions in the theorem there is a constant $\tilde{C}$ such that*

$$\mathrm{osc}_{B_r} u \leq \tilde{C} A^{\frac{1}{2} - \frac{\alpha\beta_0}{2(2+\alpha-2\beta_0)}} r^{1+\frac{\alpha}{2+\alpha-2\beta_0}} \tag{6.2}$$

for $r \leq c_0$. Here $\tilde{C}$ is independent of A.

So far we have shown that if $u \in C^{1,\beta}$, then $u \in C^{1,\frac{\alpha}{2+\alpha-2\beta_0}}$; moreover if A is small enough, then

$$\mathrm{osc}_{B_r} u \leq r^{1+\frac{\alpha}{2+\alpha-2\beta_0}}. \tag{6.3}$$

In order not to confuse our readers we want to point out that we are not claiming that by sending $A \to 0$ we will get that u is a constant. That is because the constant c_0 will depend on A. As pointed out in the paragraph before the simplifying assumption when $A \to 0$, then we have to consider different estimates for different scales.

Now if we set $\beta_1 = \frac{\alpha}{2+\alpha-2\beta_0}$ we can repeat the entire argument and deduce

$$\mathrm{osc}_{B_r} u \leq r^{1+\frac{\alpha}{2+\alpha-2\beta_1}}.$$

Inductively we will get a sequence $\beta_0 < \beta_1 < \beta_2 < \cdots < \beta_j \to \alpha/2$ such that $\beta_{j+1} = \frac{\alpha}{2+\alpha-2\beta_j}$ and

$$\mathrm{osc}_{B_r} u \leq r^{1+\frac{\alpha}{2+\alpha-2\beta_j}}$$

for all j. It is easy to see that $\lim_{j\to\infty} \beta_j = \alpha/2$. The theorem follows. □

References

[1] J. Andersson, B. Karmakar, and S. Müller. *Regularity in gradient constrained problems and micromagnetics.* Oberwolfach Report, 35, 2007.

[2] J. Andersson, H. Shahgholian, and G.S. Weiss. *Regularity below the C^2 threshold for a torsion problem; a Hamilton-Jacobi approach.* In Preparation.

[3] Haïm R. Brezis and Guido Stampacchia. *Sur la régularité de la solution d'inéquations elliptiques.* Bull. Soc. Math. France, **96** (1968), 153–180. MR0239302 (39:659)

[4] Luis A. Caffarelli and Avner Friedman. *The free boundary for elastic-plastic torsion problems.* Trans. Amer. Math. Soc., **252** (1979), 65–97. MR534111 (80i:35059)
[5] Hi Jun Choe and Yong-sun Shim. *Degenerate variational inequalities with gradient constraints.* Ann. Scuola Norm. Sup. Pisa Cl. Sci. (4), **22**(1995), no. 1, 25–53. MR1315349 (95k:49018)
[6] Robert Jensen. *Regularity for elastoplastic type variational inequalities.* Indiana Univ. Math. J., **32** (1983), no. 3, 407–423. MR697646 (84e:49011)
[7] Pierre-Louis Lions. *Generalized solutions of Hamilton-Jacobi equations*, Pitman, Boston, MA, 1982. MR0667669 (84a:49038)
[8] Tsuan Wu Ting. *Elastic-plasitc torsion of a square bar.* Trans. Amer. Math. Soc., **123** (1966), 369–401. MR0195316 (33:3518)
[9] ———, *Elastic-plastic torsion problem.* II. Arch. Rational Mech. Anal. **25** (1967), 342–366. MR0214325 (35:5176)
[10] ———, *Elastic-plastic torsion problem.* III. Arch. Rational Mech. Anal. **34** (1969), 228–244. MR0264889 (41:9479)

Department of Mathematics and Statistics, University of Jyväskylä, Finland
E-mail address: johnande@jyu.fi

Department of Mathematics, Royal Institute of Technology, 100 44 Stockholm, Sweden
E-mail address: henriksh@math.kth.se
URL: http://www.math.kth.se/~henriksh/

Graduate School of Mathematical Sciences, University of Tokyo, 3-8-1 Komaba, Meguro-ku, Tokyo-to, 153-8914 Japan,
E-mail address: gw@ms.u-tokyo.ac.jp
URL: http://www.ms.u-tokyo.ac.jp/~gw/

Amer. Math. Soc. Transl.
(2) Vol. **229**, 2010

Signorini-type Problem in $\mathbb{R}^N$ for a Class of Quadratic Functionals

Arina Arkhipova

Dedicated to Nina Nikolaevna Uraltseva.

ABSTRACT. We study regularity of a solution to the variational problem with an obstacle at the boundary of a domain for quadratic functionals in R^N, $N > 1$. The principal matrix of the integrand is supposed to have a splitting structure, and it depends on the solution. The obstacle is given as a noncompact subset of R^N with C^2-smooth boundary or C^2-smooth hypersurface in R^N. Partial regularity of any solution near the boundary is proved and the Hausdorff measure of the singular set is estimated. The result is proved by a local penalty method.

Introduction

About twenty years ago, I had the pleasure to study the Signorini problem with my teacher Nina Nikolaevna Uraltseva [**3**]–[**5**]. This paper was initiated by our joint works.

Let Ω be a bounded domain in $\mathbb{R}^n$, $n \geq 2$, with sufficiently smooth boundary $\partial\Omega$, and u: $\Omega \to \mathbb{R}^N$, $u = (u^1, \ldots, u^N)$, $N > 1$, $u_x = \{u^k_{x_\alpha}\}^{k \leq N}_{\alpha \leq n}$. We fix a subset $\mathcal{K}$ in R^N with smooth boundary $\partial\mathcal{K}$ and consider the following variational problem:

$$F[u] = \int_\Omega \Big((A(x,u)u_x, u_x) + a(x)|u|^2 + f(x)u \Big)\, dx \to \min_{W_\mathcal{K}}, \tag{1}$$

where

$$W_\mathcal{K} = \{u \in W_2^1(\Omega; R^N) : \ u(x) \in \mathcal{K} \text{ a.e. } on\ \partial\Omega\}. \tag{2}$$

The elliptic matrix $A = \{A^{\alpha\beta}_{kl}(x,u)\}^{\alpha,\beta \leq n}_{k,l \leq N}$ is supposed to have the following splitting structure:

$$A^{\alpha\beta}_{kl}(x,u) = a^{\alpha\beta}(x) b_{kl}(x,u), \ \ (x,u) \in \Omega \times \mathbb{R}^N, \tag{3}$$

the functions a, f are fixed in Ω, $a(x) \geq a_0 > 0$. We study the regularity problem for minimizers of (1)-(3). (See Remark 1 at the end of the Introduction on the existence problem.)

2000 *Mathematics Subject Classification.* Primary 35J55; Secondary 35B65.
The author was supported by grants RFBR 09-01-00729 and Sci. Schools RF 227.2008.1.

The partial regularity of *free* minimizers u of quadratic functionals (1) ($a = f = 0$) was studied by M. Giaquinta and E. Giusti. They proved that every local minimum $u \in W_2^1(\Omega; \mathbb{R}^N)$ is a Hölder continuous function on an open subset $\Omega_o \subset \Omega$, and the singular set $\Sigma = \Omega \setminus \Omega_0$, which is in general nonempty, has Hausdorff dimension strictly less than $n-2$ [**21**]. Later the authors considered functionals (1) under the splitting condition (3) and proved that a singular set Σ of a *bounded* local free minimum u has Hausdorff dimension not greater than $n-3$, and it consists at most of isolated points for $n = 3$ [**22**]. Boundary regularity of a *bounded* minimum u of (1) on the set $\{v \in W_2^1(\Omega; \mathbb{R}^N),\ v - g \in \overset{o}{W}{}_2^1(\Omega; \mathbb{R}^N)\}$, where g is a given smooth function, was studied by Jost and Meier [**26**]. They proved the Hölder continuity of u in a neighborhood of $\partial\Omega$.

Being of a particular type, the quadratic functionals under restriction (3) ($a = f = 0$) are of interest in the theory of harmonic mappings of Riemannian manifolds.

The *obstacle* problem for functionals (1),(3) was also studied:

$$F[u] \mapsto \min_{V_{\mathcal{K}}},$$

$$V_{\mathcal{K}} = \{u \in W_2^1(\Omega; \mathbb{R}^N),\ u \in \mathcal{K} \text{ a.e. } in\ \Omega,\ u - g \in \overset{o}{W}{}_2^1(\Omega; \mathbb{R}^N)\}, \tag{4}$$

where g is a given function.

Different kinds of restrictions $u(\Omega) \subset \mathcal{K}$, where $\mathcal{K}$ is a subset of $\mathbb{R}^N$, were considered by S. Hildebrandt, K.-O. Widman, M. Fuchs, F. Duzaar, M. Wiegner, and other mathematicians (see [**10**], [**11**], [**15**]–[**18**], [**23**]–[**25**], [**38**] and the references therein). In particular, it was proved that $\dim_H \Sigma \le n - 3$ (only isolated singularities can occur in the case $n = 3$) provided that $\mathcal{K}$ is a compact subset of $\mathbb{R}^N$ with the C^3-smooth boundary $\partial\mathcal{K}$ [**11**]. The regularity of minimizers in $\overline{\Omega}$ was proved under some special geometrical conditions on the set $\mathcal{K}$ [**25**], [**24**], [**38**], [**18**]. Partial regularity up to the boundary of minimizers u of problem (1), (2), (4) for noncompact subsets $\mathcal{K}$ in $\mathbb{R}^N$ with C^2-smooth boundary was stated in [**1**]. The restriction $u(\Omega) \subset S$, where S is a C^2-smooth (possibly noncompact) hypersurface in $\mathbb{R}^N$ was also considered in [**1**]. It was proved that $\dim_H \Sigma < n-2$ in these cases. We do not mention here other works devoted to the obstacle problem in Riemannian spaces and variational problems for more general types of integrands.

The Signorini problem for a single equation ($N = 1$) has been studied intensively since the beginning of the 1970s (see [**7**], [**8**], [**14**], [**27**], [**32**]–[**33**] and a references therein). $C^{1,\alpha}$-regularity of the solutions up to the boundary $\partial\Omega$ with some $\alpha > 0$ was established under different assumptions on the operators by L. Caffarelli [**8**], D. Kinderlehrer [**27**], and N. Uraltseva [**32**]. The optimal smoothness of a solution of the Signorini problem ($u \in C^{1,1/2}(\overline{\Omega})$) for the Dirichlet integral was stated by I. Anthanasopoulos and L. Caffarelli [**6**].

The Signorini problem for vector-valued functions ($N > 1$) has also been studied. Under different assumptions on *linear* elliptic operators and on the set $\mathcal{K}$ in (4), the W_2^2-smoothness of solutions was proved by G. Fichera [**13**], J. Nečas [**29**], and N. Uraltseva [**34**]. R. Schumman proved $C^{1,\alpha}$-regularity of a solution of the Signorini problem for the linear elasticity system [**31**]. The regularity problem for variational inequalities under the Signorini-type convex restrictions was studied by the author and N. Uraltseva for diagonal linear and strongly nonlinear operators [**3**]–[**5**].

In this paper, we study the regularity of solutions u of the problem (1)-(3) and prove that $u \in C^{1,\alpha}(\overline{\Omega} \setminus \Sigma; \mathbb{R}^N)$ with some $\alpha > 0$ and $H_{n-2}(\Sigma) = 0$, where Σ is the closed singular set of the solution. In the last section 8, we consider the problem

$$(5) \qquad F[u] \mapsto \min_{W_S}, \quad W_S = \{u \in W_2^1(\Omega; \mathbb{R}^N),\ u \in S \text{ a.e. } on\ \partial\Omega\},$$

where S is a C^2-smooth hypersurface and prove the same smoothness result. It appears that the same type of approach (a local penalty method) can be applied to studying both problem (1)-(3) and problem (5) for the functional (1), (3). We formulate restrictions on the sets $\mathcal{K}$ and surfaces S in terms of the distance function.

In the first section, we describe assumptions on the obstacle set $\mathcal{K}$, the matrix $A(x, u)$, and $\partial\Omega$. We formulate here the main result for problem (1)-(3) (Theorem 1). We explain our notation at the end of this section. The problem (1)-(3) is transformed to a local setting in section 2. Here we describe variational penalty problems in half of a ball and prove that the solutions u^ϵ of the penalty problems converge to the solution u of the model problem in the W_2^1-norm. A monotonicity-type inequality for u^ϵ is derived in section 3. In section 4, we prove local smoothness of u^ϵ (but not uniform estimates in ϵ) in the vicinity of a fixed point provided that the "normalized local energy" of u is small enough near this point (see condition (12)).

In section 5, we prove that condition (12) guarantees the estimate

$$(6) \qquad \sup_{B_r^+} |u_x^\epsilon(x)| \le c, \quad \epsilon \le \epsilon_\star,$$

where the constant c and $r < R$ do not depend on ϵ. To get this estimate, we derive certain scalar inequalities for the functions $H(x) = \sum_{\tau=1}^{n-1} b_{kl}(x,u) u^l_{x_\tau} u^k_{x_\tau}$ and $p(x) = b_{kl}(x,u) u^l_{x_n} u^k_{x_n}$, $u = u^\epsilon(x)$, and apply a rescaling procedure.

The Hölder continuity of the gradient of u is proved in section 6. Here we derive two kinds of weighted integral inequalities for the gradient of u. Using the Signorini boundary condition on the flat part of half of a ball, we perform an iterative process in these inequalities to estimate the C^α-norm of u_x. We prove the main results on the smoothness of u in section 7 (Theorems 1 and 2).

We consider problem (5) in section 8 and prove Theorem 4 on the partial regularity up to $\partial\Omega$ of any solution of this problem.

REMARK 1. To be sure that a nontrivial solution of (1)-(3) does exist, we add the term $a(x)|u|^2 + f(x)u$ to the integrand. To guarantee existence of a nontrivial solution, we could also assume that $a = f = 0$ but $u|_\gamma = \phi(x)$, $\gamma \Subset \partial\Omega$, $\phi(x) \neq$ *const.*, $\phi(x)|_{\partial\gamma} \in \mathcal{K}$. Here we are interested in the *regularity* of minimizers. The term $a|u|^2 + fu$ in the integrand has no influence on the result, and we can study the simplified version of the problem: $a = f = 0$ in (1).

Henceforth, we adopt the following notation:

$$B_R(x^0) = \{x \in \mathbb{R}^n :\ |x - x^0| < R\},\ S_R(x^0) = \{x \in \mathbb{R}^n :\ |x - x_0| = R\},$$

$$S_R^+(x^0) = S_R \cap \{x_n > x_n^0\},\ \Gamma_R(x^0) = B_R(x^0) \cap \{x_n = x_n^0\},\ \Omega_R(x^0) = \Omega \cap B_R(x^0);$$

$|A| = \operatorname{meas}_n A$ for a Lebesgue measurable set A in $\mathbb{R}^n$, $\omega_n = \operatorname{meas}_n B_1(0)$;

$$\fint_{\Omega_r(x^0)} g\,dx = \frac{1}{|\Omega_r|} \int_{\Omega_r(x^0)} g\,dx,\ \ \not\!\!\int_{\Omega_r} g\,dx = \frac{1}{r^{n-2}} \int_{\Omega_r} g\,dx,$$

$$\fint_{\Gamma_r} f\, d\Gamma = \frac{1}{r^{n-2}} \int_{\Gamma_r} f\, d\Gamma.$$

We write $u \in B(\Omega)$ instead of $u \in B(\Omega; \mathbb{R}^N)$ if it does not cause misunderstanding.

1. The main assumptions and results

Here we formulate the main assumptions.

$[\mathcal{A}_\mathcal{K}]$. Let $\mathcal{K}$ be a subset of $\mathbb{R}^N$ with C^2-smooth boundary $\partial\mathcal{K}$. There exist numbers δ_0 and $M > 0$ such that the C^2-smooth distance function $d(u) = \operatorname{dist}(u, \partial\mathcal{K})$ is defined in the neighborhood $U_{2\delta_0}(\mathcal{K})$, and

$$\sup_{u \in U_{2\delta_0}(\mathcal{K})} \|d''(u)\| \leq M. \tag{7}$$

$[\mathcal{A}_1]$. The matrix $a(x) = \{a^{\alpha\beta}(x)\}^{\alpha,\beta \leq n}$ is defined and C^1-smooth in $\overline{\Omega}$, $a^{\alpha\beta}(x) = a^{\beta\alpha}(x)$, and

$$(a(x)\xi, \xi) \geq \nu_1 |\xi|^2, \ \xi \in \mathbb{R}^N, \ x \in \overline{\Omega}, \ \nu_1 = const. > 0.$$

$[\mathcal{A}_2]$. The matrix $b(x,u) = \{b_{kl}(x,u)\}_{k,l \leq N}$ is defined and C^1-smooth on $\overline{\Omega} \times \mathbb{R}^N$,

$$\sup_{\Omega \times \mathbb{R}^N} \{|b(x,u)| + |b'_x(x,u)| + |b'_u(x,u)|\} \leq \mu, \ \ b_{kl}(x,u) = b_{lk}(x,u),$$
$$(b(x,u)\eta, \eta) \geq \nu_2 |\eta|^2, \ \eta \in \mathbb{R}^N, \ (x,u) \in \overline{\Omega} \times \mathbb{R}^N, \ \nu_2 = const. > 0.$$

$[\mathcal{A}_3]$. $a \in L_{q/2}(\Omega)$, $f \in L_q(\Omega)$, $q > n$, $a(x) \geq a_0 > 0$. Ω is a bounded domain in $\mathbb{R}^n$, $n \geq 2$, with C^2-smooth boundary $\partial\Omega$.

REMARK 2. Obviously, assumption $[\mathcal{A}_\mathcal{K}]$ holds provided that $\mathcal{K}$ is a compact set in $\mathbb{R}^N$ with C^2-smooth boundary $\partial\mathcal{K}$.

THEOREM 1. *Let the assumptions $[\mathcal{A}_\mathcal{K}]$, $[\mathcal{A}_1] - [\mathcal{A}_3]$ hold, and let $u(x)$ supply the minimum in the problem (1)-(3). Then 1) $u \in C^{1,\beta}(\overline{\Omega} \setminus \Sigma)$ with some $\beta > 0$, and $H_{n-2}(\Sigma) = 0$, where Σ is a closed singular set of u; 2) the function u is a weak solution of some Neumann problem for the Euler operator of $F[u]$ (more exactly, see Remark 7).*

2. Local penalty problem

Let u be a solution of the problem (1)-(3). To simplify the proofs, we assume that $a = f = 0$ in what follows. We fix numbers θ_0 and R_0 and consider the set

$$\Sigma_{\theta_0,R_0} = \bigcap_{r \leq R_0} \{x^0 \in \overline{\Omega} : \fint_{\Omega_r(x^0)} e[u](x)\, dx \geq \theta_0^2\}, \ e[u] = \frac{1}{2}(A(x,u)u_x, u_x). \tag{8}$$

We define

$$\Omega_0 = \overline{\Omega} \setminus \Sigma_{\theta_0,R_0}. \tag{9}$$

According to (9),

$$\fint_{\Omega_{R_1}(x^0)} e[u](x)\, dx < \theta_0^2 \tag{10}$$

for some $R_1 \leq R_0$ provided that $x^0 \in \Omega_0$.

THEOREM 2. *There exist numbers θ_0 and R_0 such that Ω_0 defined by (9) is an open with respect to $\overline{\Omega}$ set, and $u \in C^{1,\beta}(\Omega_0)$ with some $\beta > 0$. The numbers θ_0 and R_0 depend on the parameters from the assumptions $[\mathcal{A}_\mathcal{K}]$, $[\mathcal{A}_1] - [\mathcal{A}_3]$.*

Below, we prove Theorem 2 in detail for the case $x^0 \in \partial\Omega \cap \Omega_0$ and give only explanations for the other locations of x^0.

Now let $x^0 \in \partial\Omega \cap \Omega_0$. Then condition (10) holds for some $R_1 \leq R_0$. It is easy to see that one can "straighten" a part of $\partial\Omega$ in the vicinity of x^0 and transform the problem (1)-(3) to the local setting in half of a ball $B^+_{R_2}(0)$, $R_2 = c_\Gamma R_1$, where c_Γ depends on the C^1-characteristics of $\partial\Omega$. In the new variables, the splitting structure of the matrix and the smallness condition with a parameter $\theta = c\theta_0$ in $B^+_{R_2}(0)$ will be preserved (the constant c depends on the data only). In what follows, we keep the same notation of the variables and functions and study the local model variational problem

$$F_1[w, B^+_{R_2}] = \frac{1}{2}\int_{B^+_{R_2}(0)} (A(x,w)w_x, w_x)\,dx \to \min_{\hat{W}_{\mathcal{K}}},$$

$$w \in \hat{W}_{\mathcal{K}} = \{v \in W^1_2(B^+_{R_2}(0)),\ v|_{\Gamma_{R_2}} \in \mathcal{K},\ v - u|_{S^+_{R_2}(0)} = 0\}. \tag{11}$$

Note that the boundary condition in (11) is defined by the solution u of (1)-(3) in the new coordinates. The smallness condition

$$\fint_{B^+_{R_2}(0)} |u_x|^2\,dx < \theta^2 \tag{12}$$

holds for the function u.

We may also assume that

$$a^{nn}(x) \equiv 1,\ x \in B^+_{R_2},\ \ a^{n\tau}|_{\Gamma_{R_2}} = 0,\ \tau = 1, \dots, n-1 \tag{13}$$

(see, for example, [**35**], p. 95).

THEOREM 3. *There exist constants R_0 and $\theta_0 > 0$ such that assumption (12) for a solution u of the problem (11) with some $\theta \leq \theta_0$ and $R_2 \leq R_0$ guarantees that $u \in C^{1,\beta}(B^+_{\tau R_2}(0))$ with some parameters $\beta \in (0, \min\{1 - n/q, 1/2\})$, $\tau \in (0, 1/2)$, depending on the data fixed by assumptions* $[\mathcal{A}_{\mathcal{K}}], [\mathcal{A}_1] - [\mathcal{A}_3]$.

Now we describe penalty problems for (11).

Let the scalar function $\chi(s)$ be defined by the relations. $\chi(s) = s$ for $s \leq \delta_0^2$, $\chi(s) = -\frac{1}{6}(\frac{s^2}{\delta_0^2} - 8s + \delta_0^2)$, $s \in (\delta_0^2, 4\delta_0^2)$; $\chi(s) = \frac{5}{2}\delta_0^2$ for $s > 4\delta_0^2$. The function $\chi(s) \in C^{1,1}([0,\infty))$, $\chi'(s) \geq 0$, and

$$|\chi''(s)|s \leq \frac{4}{3\delta_0^2}\chi(s). \tag{14}$$

Here δ_0 is the parameter from the assumption $[\mathcal{A}_{\mathcal{K}}]$.

For any $\epsilon \in (0, 1]$, we consider the following variational problem:

$$F_1^\epsilon[v] = \frac{1}{2}\int_{B^+_{R_2}} [(A(x,v)v_x, v_x) + |v-u|^2]\,dx + \frac{1}{\epsilon}\int_{\Gamma_{R_2}} \chi(d^2(v))\,d\Gamma \mapsto \min_V, \tag{15}$$

where

$$V = \{v \in W^1_2(B^+_{R_2}(0)) :\ v - u|_{S^+_{R_2}} = 0\}.$$

There exists a solution $u^\epsilon \in V$ to problem (15) for any fixed $\epsilon \in (0, 1]$. Since $u \in \mathcal{K}$ on Γ_{R_2}, then $\chi(d^2(u)) = 0$ on Γ_{R_2} and

$$F_1^\epsilon[u^\epsilon] \leq F_1^\epsilon[u] = F_1[u] = \frac{1}{2}\int_{B^+_{R_2}} (A(x,u)u_x, u_x)\,dx. \tag{16}$$

This guarantees that

$$\text{(17)}\qquad \|u^\epsilon\|_{W_2^1(B^+_{R_2})} \le c_1\|u\|_{W_2^1(B^+_{R_2})},\quad \frac{1}{\epsilon}\int_{\Gamma_{R_2}} \chi(d^2(u^\epsilon))\,d\Gamma \le c_1\|u_x\|_{2,B^+_{R_2}}.$$

From (17) it follows that u^ϵ tends to u weakly in $W_2^1(B^+_{R_2})$, and the limit function $u^0 \in \hat{W}_{\mathcal{K}}$ (see (11)). Moreover,

$$F_1[u^\epsilon] \le F_1^\epsilon[u^\epsilon] \le F_1[u]$$

and

$$\text{(18)}\qquad F_1[u^0] \le \liminf_\epsilon F_1[u^\epsilon] \le \limsup_\epsilon F_1[u^\epsilon] \le F_1[u] \le_{(a)} F_1[u^0],$$

where the inequality (a) is valid because of the minimality of u on $\hat{W}_{\mathcal{K}}$. The chain of inequalities (18) attains an equality. Hence, there exists

$$\text{(19)}\qquad \lim_{\epsilon\to 0} F_1[u^\epsilon] = F_1[u^0] = F_1[u].$$

Now from (16) and (19) it follows that

$$\lim_\epsilon \int_{B^+_{R_2}} |u^\epsilon - u|^2\,dx = \int_{B^+_{R_2}} |u^0 - u|^2\,dx = 0$$

and

$$\text{(20)}\qquad \lim \frac{1}{\epsilon}\int_{\Gamma_{R_2}} \chi(d^2(u^\epsilon))\,d\Gamma = 0.$$

This means that $u^0 = u$ in $B^+_{R_2}$. Weak convergence in $W_2^1(B^+_{R_2})$ and (19) ensure the convergence $u^\epsilon \mapsto u$ in the norm of $W_2^1(B^+_{R_2})$. Thus, the following statement has been proved.

PROPOSITION 1. *The solutions u^ϵ of the penalty problems (15) tend to the solution u of (11) in the space $W_2^1(B^+_{R_2})$ for some sequence $\epsilon \mapsto 0$, and relations (20) are valid.*

We write

$$\text{(21)}\qquad \chi^\epsilon(d^2(\cdot)) = \frac{1}{\epsilon}\chi(d^2(\cdot))$$

in what follows.

3. Monotonicity inequality for the normalized local energy

We put here

$$\text{(22)}\qquad \Phi^\epsilon(r, x^0) = \fint_{\omega_r(x^0)} e[u^\epsilon](x)\,dx + \fint_{\gamma_r(x^0)} \chi^\epsilon(d^2(u^\epsilon))\,d\gamma,$$

where $\omega_r(x^0) = B^+_{R_2}(0) \cap B_r(x^0)$, $\gamma_r(x^0) = \Gamma_{R_2}(0) \cap B_r(x^0)$, $x^0 \in B^+_{R_2}(0) \cup \Gamma_{R_2}(0)$, $r < \operatorname{dist}(x^0, S^+_{R_2}(0))$, and $e[v] = \frac{1}{2}[(A(x,v)v_x, v_x) + |v-u|^2]$, χ_ϵ is fixed by (21). In this section we prove the following proposition.

PROPOSITION 2. *There exist $\tau_1 \in (0, 1/2]$ and $c_2 > 0$ such that*

$$\text{(23)}\qquad \Phi^\epsilon(\rho, x^0) \le c_2 \Phi^\epsilon(r, x^0),\quad x^0 \in \overline{B^+_{R_3}(0)},\quad \rho \le r \le R_3,$$

for any $\epsilon \in (0, 1]$, and $R_3 = \tau_1 R_2$.

PROOF. Let u^ϵ yield the minimum to problem (15). Then it is a critical point of F_1^ϵ both with respect to variations of the target and with respect to variations of the domain. The last assertion means that we can fix any smooth family of diffeomorphisms

$$\Psi_\tau(x) = x + \tau\xi(x) = x^{(\tau)},\ x \in B^+ = B^+_{R_2}(0),$$

such that $\Psi_\tau : B^+ \mapsto B^+$, $0 < \tau \ll 1$. Here $\xi = (\xi^1, \dots, \xi^n)$, $\xi \in C^{0,1}(\overline{B^+})$, $spt\,\xi \subset B^+ \cup \Gamma$, $\Gamma = \Gamma_{R_2}(0)$, $\xi^n \geq 0$ in B^+, $\xi^n|_\Gamma = 0$, and $\Psi_\tau(\Gamma) \subset \Gamma$. For a fixed $\epsilon > 0$, we put $u^{(\tau)}(x) = u^\epsilon(x^{(\tau)})$ and evaluate the expression $I_\epsilon(\tau) = \tau^{-1}(F_1^\epsilon[u^{(\tau)}] - F_1^\epsilon[u^\epsilon])$.

There exists a finite $\lim_{\tau\mapsto+0} I_\epsilon(\tau) = \frac{d}{d\,\tau}F_1^\epsilon[u^{(\tau)}]|_{\tau=+0}$, and

$$\frac{d}{d\,\tau}F_1^\epsilon[u^{(\tau)}]|_{\tau=+0} \geq 0 \tag{24}$$

due to the minimality of u^ϵ.

To save space, we omit further the term $\frac{|u^\epsilon - u|^2}{2}$ in the expression $e[u^\epsilon]$ (it does not influence the result) and write u instead of u^ϵ.

Then inequality (24) is realized as the following integral inequality:

$$\begin{aligned} &-\frac{1}{2}\int_{B^+} e[u]\operatorname{div}\xi\,dx + \int_{B^+} A^{\alpha\beta}_{kl} u^l_{x_\beta} u^k_{x_\gamma} \xi^\gamma_{x_\alpha}\,dx \\ &-\frac{1}{2}\int_{B^+} (A^{\alpha\beta}_{kl})'_{x_\gamma} u^l_{x_\beta} u^k_{x_\alpha} \xi^\gamma\,dx - \int_\Gamma \chi_\epsilon(d^2(u))\operatorname{div}'\xi\,d\gamma \geq 0, \end{aligned} \tag{25}$$

where $\xi \in C^{0,1}(B^+), \xi^n \geq 0$ in $B^+, \xi^n|_\Gamma = 0$, $\xi|_{S^+} = 0, \operatorname{div}'\xi = \Sigma_{\alpha=1}^{n-1}\xi^\alpha_{x_\alpha}$.

Now let $x^0 \in \Gamma_{1/2R_2}$ and $R \leq (1/2)R_2$. We fix a function ξ with $spt\,\xi \subset B^+_R(x^0) \cup \Gamma_R(x^0)$, and consider (25) in $B^+_R(x^0)$ but not in B^+.

We write in (25),

$$e[u] = e_0[u] + \frac{1}{2}\Delta a^{\alpha\beta} b_{kl}(x,u) u^l_{x_\beta} u^k_{x_\alpha},$$

$$A^{\alpha\beta}_{kl}(x,u) u^l_{x_\beta} u^k_{x_\gamma} = a^{\alpha\beta}(x^0) b_{kl}(x,u) u^l_{x_\beta} u^k_{x_\gamma} + \Delta a^{\alpha\beta} b_{kl}(x,u) u^l_{x_\beta} u^k_{x_\gamma}, \tag{26}$$

where

$$\Delta a = a(x) - a(x^0),\ |\Delta a| \leq c|x - x^0|. \tag{27}$$

We change the variables in (25) and put $z_j = \frac{\mathbf{C}_{jk}(x_k - (x^0)_k)}{\sqrt{\lambda_j}}$, where $\mathbf{C}$ is an $[n \times n]$-orthogonal matrix transforming the matrix $a(x^0)$ to the diagonal one: $D = \mathbf{C}a(x^0)\mathbf{C}'$, $D = \{\lambda_1, \dots, \lambda_n\}Id$, $\lambda_- = \min_{i\leq n}\lambda_i > 0$. The map $z = z(x)$ transforms $B^+_R(x^0)$ to half of the ellipsoid $P_R(0)$, and we derive from (25) the inequality for $q = (q^1, \dots, q^n)$, $q^j(z) = \frac{\mathbf{C}_{jk}\xi^k(x(z))}{\sqrt{\lambda_j}}$:

$$\begin{aligned} &-\frac{1}{2}\int_{P_R(0)} b_{kl} u^l_{z_\alpha} u^k_{z_\alpha} \operatorname{div} q(z)\,dz + \int_{P_R(0)} b_{kl} u^l_{z_i} u^k_{z_s} q^s_{z_i}\,dz \\ &+ \int_{P_R(0)} \widehat{\Delta a^{\delta\gamma}} b_{kl} u^l_{z_\gamma}(-\frac{1}{2}u^k_{z_\delta}\operatorname{div} q(z) + u^k_{z_s} q^s_{z_\delta})\,dz - \frac{1}{2}\int_{P_R(0)} M^{is,j}_{kl} u^l_{z_i} u^k_{z_s} q^j\,dz \\ &-\frac{1}{m}\int_{\gamma'_R} \chi_\epsilon \operatorname{div} q\,d\gamma + \frac{1}{m}\int_{\gamma'_R} \chi_\epsilon \mathbf{C}_{in}\sqrt{\lambda_i} q^i_{z_k} \frac{\mathbf{C}_{kn}}{\sqrt{\lambda_k}}\,d\gamma \geq 0, \end{aligned} \tag{28}$$

where $m = (\Sigma_{k\leq n}(\mathbf{C}_{kn})^2\lambda_k)^{1/2} \in [\sqrt{\lambda_-}, \sqrt{\lambda_+}]$, $\lambda_+ = \max_{j\leq n}\lambda_j$.

Some functions M in (28) are bounded in $P_R(0)$ and $|\widehat{\Delta a}(z)| \leq c|z|$ due to (27). Here we keep the notation u for u^ϵ in the new coordinates. Note that the "flat" part γ'_R of the surface $\partial P_R(0)$ is in the hyperplane

$$L = \{z \in \mathbb{R}^n : \Sigma_{j\leq n}\mathbf{C}_{jn}\sqrt{\lambda_j}z_j = 0\}. \tag{29}$$

Relation (28) is the base for deriving inequality (23). Following the idea by L.-C. Evans [**12**], we fix numbers r and $h > 0$ such that $r + h \leq r_0$, $r_0 \leq \frac{R}{\sqrt{\lambda_+}}$, and consider the cutoff function

$$\phi_h(s) = \begin{cases} 1, & s \leq r, \\ \frac{r+h-s}{h}, & s \in (r, r+h]. \end{cases} \tag{30}$$

We denote $\omega_\rho(0) = B_\rho(0) \cap P_R(0)$, $S'_\rho(0) = S_\rho(0) \cap P_R(0)$, $\gamma'_\rho(0) = B_\rho(0) \cap L$. We put $q(z) = z\,\phi_h(|z|)$, $z \in P_R(0)$, in (28), then turn h to 0, and arrive at the relation for almost all $r < r_0$:

$$-(n-2)\int_{\omega_r(0)} \hat{e}\,dz + r\int_{S'_r(0)} \hat{e}\,ds - (n-1)\int_{\gamma'_r(0)} \hat{\chi}_\epsilon\,d\gamma + r\int_{\partial\gamma'_r} \hat{\chi}_\epsilon\,d\gamma$$

$$\geq -c_3 r\int_{\omega_r(0)} |u_z|^2\,dz - c_3 r^2\int_{S'_r(0)} |u_z|^2\,ds + \frac{1}{2r}\int_{S'_r(0)} b_{kl}(u^l_z, z)(u^k_z, z)\,ds. \tag{31}$$

Here $\hat{e} = \frac{1}{2}b_{kl}u^l_{z_\alpha}u^k_{z_\alpha}$, $u(z) = u^\epsilon(x(z))$, $\hat{\chi}_\epsilon = \frac{1}{m}\chi_\epsilon(d^2(u^\epsilon(x(z))))$. The constant c_3 does not depend on ϵ and r.

Note that we have transformed $a(x^0)$ to the identical matrix to obtain the nonnegativity of the last integral in (31). It is the point where the splitting structure of the matrix a is important. Now we define

$$p^\epsilon(r) = ⨍_{\omega_r(0)} \hat{e}\,dz, \quad k^\epsilon(r) = ⨍_{\gamma'_r(0)} \hat{\chi}_\epsilon\,d\gamma,$$

and after dividing (31) by r^{n-1}, obtain the inequality

$$(p^\epsilon(r) + k^\epsilon(r))' \geq -c_4 p^\epsilon(r) - c_4 r[(p^\epsilon)' + \frac{n-2}{r^{n-1}}\int_{\omega_r} \hat{e}\,dz]$$

$$\geq -c_4(n-1)p^\epsilon - c_4 r(p^\epsilon)' \geq -c_5 p^\epsilon - c_5 r(p^\epsilon)', \; c_5 = c_4(n-1).$$

Hence,

$$(p^\epsilon)'(1 + c_5 r) + c_5 p^\epsilon + (k^\epsilon)' \geq 0, \text{ a.a. } r \leq r_0. \tag{32}$$

We put $\psi_\epsilon(r) = (1 + c_5 r)p^\epsilon(r)$, and obtain from (32) that $(\psi^\epsilon + k^\epsilon)' \geq 0$ for a.a. $r < r_0$.

This supplies the inequality

$$\psi^\epsilon(\rho) + k^\epsilon(\rho) \leq \psi^\epsilon(r) + k^\epsilon(r), \; \rho \leq r \leq r_0. \tag{33}$$

Taking into account the definitions of ψ^ϵ, k^ϵ, and $z = z(x)$, we arrive at (23) for the case $x^0 \in \Gamma_{\frac{R_2}{2}}(0)$.

It is evident that one can simplify the proof by considering $x^0 \in B_{\frac{R_2}{2}}(0)$ and $r \leq \operatorname{dist}(x^0, \Gamma_{R_2})$. The sewing procedure supplies (23) in the general case. Proposition 2 is proved. □

REMARK 3. Let assumption (12) hold with a fixed $\theta > 0$ for the solution $u(x)$ of problem (11). By Proposition 1, we can assert that

$$(34)\qquad \fint_{B^+_{R_2}} |u^\epsilon_x|^2\,dx + \fint_{\Gamma_{R_2}} \chi_\epsilon(d(u^2(u^\epsilon))\,d\Gamma < \theta^2,\ \epsilon \le \epsilon_*,$$

for some $\epsilon_* > 0$. Due to inequalities (23) and (34),

$$\Phi^\epsilon(\rho, x^0) \le c_2\Phi^\epsilon(R_3, x^0) \le c_2\tau_1^{2-n}\Phi^\epsilon(R_2, 0) \le c_6\theta^2, \quad x^0 \in \overline{B^+_{R_3}(0)},\ R_3 = \tau_1 R_2.$$

Hence,

$$(35)\qquad \sup_{x^0\in B^+_{R_3}, \rho\le R_3} (\fint_{\omega_\rho(x^0)} |u^\epsilon_x|^2\,dx + \fint_{\gamma_\rho(x^0)} \chi_\epsilon(d^2(u^\epsilon))\,d\gamma) \le c_*\theta^2,\ \epsilon \le \epsilon_*,\ R_3 = \tau_1 R_2,$$

with some $\tau_1 \in (0, 1/2]$. The constant c_* does not depend on ϵ and ϵ_*.

REMARK 4. From (35) and the Poincaré inequality it follows that

$$(36)\qquad [u^\epsilon]^2_{\mathcal{L}^{2,n}(B^+_{R_3})} \le c_7\theta^2,\ \epsilon \le \epsilon_*,$$

where $[\cdot]$ is the seminorm in the Campanato space $\mathcal{L}^{2,n}(\cdot)$. Due to the isomorphism of the spaces $\mathcal{L}^{p,n}(\cdot)$ for different $p \ge 1$, we have the estimate

$$[u^\epsilon]_{\mathcal{L}^{p,n}(B^+_{R_3})} \le c(p, n)[u^\epsilon]_{\mathcal{L}^{2,n}(B^+_{R_3})} \le c(p, n)c_7\theta^2.$$

As a consequence, the limit function $u \in \mathcal{L}^{2,n}(B^+_{R_3})$, $u \in L_m(B^+_{R_3})$, and

$$\|u^\epsilon\|^m_{m,B^+_{R_3}} + \|u\|^m_{m,B^+_{R_3}} \le c(m, n, R_3^{-1}, \|u\|_{2,B^+_{R_3}}),\ 1 < m < +\infty.$$

4. On the smoothness of u^ϵ near the origin

We remark that in general, the solutions u^ϵ to problem (15) may have singular sets Σ_ϵ in $\overline{B^+_{R_2}}$. At the same time, the following proposition holds.

PROPOSITION 3. *There exist $\theta > 0$, $R_2 > 0$ and $\tau_2 < \tau_1$, such that $u^\epsilon \in C^{1,\beta}(\overline{B^+_{R_4}}) \cap W^2_2(B^+_{R_4})$, $\beta \in (0, 1)$, $R_4 = \tau_2 R_2$, $\epsilon \le \epsilon_*$, provided that (35) is valid with the parameter θ and $R_3 = \tau_1 R_2$. Moreover,*

$$(37)\qquad \|u^\epsilon\|_{C^{1,\beta}(\overline{B^+_{R_4}})} + \|u^\epsilon_{xx}\|_{2,B^+_{R_4}} \le K(\epsilon^{-1}),$$

where $K(\epsilon^{-1})$ may go to infinity when $\epsilon \to 0$.

To prove the Hölder continuity of u^ϵ, we take into account estimate (35) and apply the direct method ([**19**], Ch.6). Here we consider the penalty term as an additional one to the quadratic functional $F_1[\cdot]$. First, we obtain an L_p-estimate of the gradient of u^ϵ with some $p > 2$, p does not depend on ϵ, and then freeze arguments of the matrix $A(x, u)$.

As a result, we arrive at the estimate

$$(38)\qquad \|u^\epsilon\|_{C^\beta(\overline{B^+_R})} \le K(R^{-1}, \epsilon^{-1}),\ R = \frac{R_3}{2},$$

with any $\beta \in (0, 1)$.

Now we can consider $u^\epsilon \in W^1_2(B^+_R) \cap C^\beta(\overline{B^+_R})$, $\epsilon \le \epsilon_*$, as a weak solution to the boundary-value problem

$$-(A^{\alpha\beta}_{kl}(x, u^\epsilon)(u^\epsilon)^l_{x_\beta})_{x_\alpha} + \frac{1}{2}(A^{\alpha\beta}_{ml})'_{u^k}(u^\epsilon)^l_{x_\beta}(u^\epsilon)^m_{x_\alpha} + ((u^\epsilon)^k - (u)^k) = 0,\ x \in B^+_R,$$

$$A^{n\beta}_{kl}(x,u^\epsilon)(u^\epsilon)^l_{x_\beta} = \frac{d\chi_\epsilon(d^2(u^\epsilon))}{du^k} \quad \text{on } \Gamma_R. \tag{39}$$

Taking into account that $a^{n\tau}=0$, $\tau\neq n$, on Γ_R, and $a_{nn}\equiv 1$, we can rewrite the boundary condition (39) in the form

$$b_{kl}(u^\epsilon)(u^\epsilon)^l_{x_n} = \chi'_\epsilon 2d\, d'_k(u^\epsilon) \text{ on } \Gamma_R,\ k\leq N. \tag{40}$$

The function u^ϵ satisfies the identity

$$\int_{B^+_R}[A^{\alpha\beta}_{kl}(x,u^\epsilon)(u^\epsilon)^l_{x_\beta}h^k_{x_\alpha} + \frac{1}{2}(A^{\alpha\beta}_{ml})'_{u^k}(u^\epsilon)^l_{x_\beta}(u^\epsilon)^m_{x_\alpha}h^k + ((u^\epsilon)^k-u^k)h^k]\,dx$$

$$+\int_{\gamma_R}\frac{d\chi_\epsilon}{du^k}h^k\,d\gamma = 0,\ h\in W^1_2(B^+_R)\cap L_\infty(B^+_R),\ h|_{S^+_R}=0. \tag{41}$$

The estimate of the $C^{1,\alpha}$-norm of weak Hölder continuous solutions of the Neumann problem (39) was derived in [**2**] a local estimate of the $C^{1,\alpha}$-norm to solutions of strongly nonlinear systems was obtained in [**20**]). Thus,

$$\|u^\epsilon_x\|_{C^\alpha(\overline{B^+_{R_4}})} \leq K(\epsilon^{-1},\alpha,R_4^{-1}), \tag{42}$$

where $\alpha\in(0,1)$, $R_4=\frac{R}{2}=\tau_2R_2$, $\tau_2<\tau_1$.

Now we may consider u^ϵ as a solution to the linear problem to assert that $\|u^\epsilon_{xx}\|_{2,B^+_{R_4}}\leq K(\epsilon^{-1})$. We emphasize that R_4 does not depend on $\epsilon>0$.

In what follows, we accept that $R_{i+2}=\tau_iR_2$, $\tau_{i+1}<\tau_i$, $\tau_1\leq 1/2$. All parameters τ_i and R_j do not depend on $\epsilon\leq\epsilon_*$.

5. Uniform in $\epsilon\leq\epsilon_*$ estimate of $|u^\epsilon_x|$

We introduce here a modification of the method applied earlier for the investigation of harmonic mappings ([**30**], [**9**]). The splitting structure (3) of the matrix $A(x,u)$ is essential; it allows us to apply some approaches appropriate for scalar equations. The main point here is to make an appropriate scaling of the problem. We prove the following statement for solutions u^ϵ to the problem (39).

PROPOSITION 4. *There exist θ and $R_5=\tau_3R_2$ such that the estimate*

$$\sup_{\overline{B^+_{R_5}}}|u^\epsilon_x|^2\leq c_8(\theta^{-1},\,R_5^{-1}),\qquad \epsilon\leq\epsilon_*, \tag{43}$$

is valid for solutions u^ϵ provided that condition (35) holds with the parameters θ and $R_3=\tau_1R_2$. The constant c_8 does not depend on $\epsilon\leq\epsilon_*$.

PROOF. By Proposition 3, $u^\epsilon\in C^{1,\alpha}(\overline{B^+_{R_4}})\cap W^2_2(B^+_{R_4})$, $\epsilon\leq\epsilon_*$. We consider the problem

$$\max_{0\leq\sigma\leq R}\{(R-\sigma)^2\max_{\overline{B^+_\sigma(0)}}e[u^\epsilon](x)\} = (R-\sigma_0)^2\max_{\overline{B^+_{\sigma_0}(0)}}e[u^\epsilon](x), \tag{44}$$

where

$$e[u^\epsilon](x)=\frac{1}{2}b_{kl}(x,u)(u^\epsilon)^l_{x_\alpha}(u^\epsilon)^k_{x_\alpha},\ R\leq R_4. \tag{45}$$

To save space, we omit the term with the function $u^\epsilon-u$ in (39) and (41) in what follows. It is not essential for our considerations to take into account the estimates for u^ϵ and u stated in Remark 4.

Let x^0 be a point of the maximum of $e[u^\epsilon](x)$ in $\overline{B^+_{\sigma_0}}$, $\sigma_0>0$.

We put

$$\max_{B^+_{\sigma_0}} e[u^\epsilon](x) = e[u^\epsilon](x^0) = e_0,\ \rho_0 = \frac{R-\sigma_0}{2},\ \omega_{\rho_0}(x^0) = B^+_R \cap B_{\rho_0}(x^0),$$

$$\gamma_{\rho_0}(x^0) = \Gamma_R(0) \cap B_{\rho_0}(x^0). \tag{46}$$

Let $y = \sqrt{e_0}(x - x^0)$ be the new coordinates, and let

$$v^\epsilon(y) = u^\epsilon(x^0 + \frac{y}{\sqrt{e_0}}), \qquad r_0 = \rho_0\sqrt{e_0}. \tag{47}$$

We put $y(\omega_{\rho_0}(x^0)) = \hat\omega_{r_0}(0),\ \ y(\gamma_{\rho_0}(x^0)) = \hat\gamma_{r_0}$. The function $v = v^\epsilon(y)$ satisfies the identity

$$\int_{\hat\omega_{r_0}(0)} [\hat a^{\alpha\beta}(y)\hat b_{kl}(y,v)v^l_{y_\beta}h^k_{y_\alpha} + \frac{1}{2}\hat a^{\alpha\beta}(\hat b_{ml})'_{v^k}v^l_{y_\beta}v^m_{y_\alpha}h^k + \frac{(v^k - v^k_0)}{e_0}h^k]\,dy$$

$$+ \int_{\hat\gamma_{r_0}} \frac{d\hat\chi_\epsilon}{dv^k}h^k\,d\hat\gamma = 0, \quad \hat\chi_\epsilon(d^2) = \frac{\chi_\epsilon(d^2)}{\sqrt{e_0}}, \tag{48}$$

$h \in C^1(\hat\omega_{r_0}),\ h|_{\partial'\hat\omega_{r_0}} = 0,\ \partial'\hat\omega_{r_0} = \partial\hat\omega_{r_0} \setminus \hat\gamma_{r_0}$. By the "hat" we have marked the functions $a^{\alpha\beta}$ and b_{kl} in the new coordinates; $v_0(y) = u(x^0 + \frac{y}{\sqrt{e_0}})$ is the limit function. The following relations are valid:

$$\sup_{\omega_{\rho_0(x^0)}} e[u^\epsilon] \le \sup_{B^+_{\rho_0+\sigma_0}(0)} e[u^\epsilon] \le 4e_0,$$

and, as a consequence,

$$\hat e[v^\epsilon](0) = 1, \ \ \sup_{\hat\omega_{r_0}(0)} \hat e[v^\epsilon](y) \le 4, \ \text{for}\ \hat e[v^\epsilon] = \frac{e[u^\epsilon]}{e_0}. \tag{49}$$

Below we study the case

$$\frac{1}{e_0} \le \theta^2. \tag{50}$$

.

In the other case, $\max_{B^+_{R/2}(0)} e[u^\epsilon] \le 4\sup_{B^+_{\sigma_0(0)}} e[u^\epsilon] = 4e_0 < \frac{4}{\theta^2}$, and estimate (43) follows with $R = R_4$.

Under the assumption (50), we prove below that

$$r_0 \le 2. \tag{51}$$

This gives us estimate (43). Indeed, we can address (44) with $\sigma = R/2$ and obtain the inequality

$$(R/2)^2 \max_{B^+_{R/2}(0)} e[u^\epsilon](x) \le e_0(2\rho_0)^2 = 4r_0^2 \le 16.$$

Estimate (43) follows from the last relation with $R = R_4$. Thus, to prove the proposition, we need to justify (51).

Suppose that

$$r_0 > 2, \tag{52}$$

and define the following scalar functions for $v = v^\epsilon$:

$$H[v] = \frac{1}{2}\Sigma_{\tau\le n-1}\hat b_{kl}(y,v)v^l_{y_\tau}v^k_{y_\tau}, \quad p[v] = \frac{1}{2}\hat b_{kl}(y,v)v^l_{y_n}v^k_{y_n}, \ \ \hat e[v] = H[v] + p[v].$$

We will prove that

$$H[v^\epsilon](0) \leq \lambda_1(\theta), \quad p[v^\epsilon](0) \leq \lambda_2(\theta) \tag{53}$$

hold with some functions $\lambda_i(\theta) \to 0$ when $\theta \to 0$, $i = 1, 2$, provided that (52) is valid. Relations (53) with $\theta \ll 1$ supply the contradiction with equality $\hat{e}[v^\epsilon](0) = 1$. Then (52) does not hold, and Proposition 4 will be proved.

Thus, to prove Proposition 4, we should justify relations (53). (As was said, we can omit the term with the function $(v - v_0)$ in identity (48) in what follows.)

First, we take (48) with the function $h = (v_{y_\tau}\phi)_{y_\tau}$, $v = v^\epsilon(y)$, $\tau \leq n-1$, $spt\,\phi \in \hat{\omega}_{r_0} \cup \hat{\gamma}_{r_0}$. After integrating by parts, we get the inequality

$$\int_{\hat{\omega}_2(0)} |v_{y'y}|^2 \phi\, dy + \int_{\hat{\omega}_2(0)} \hat{a}^{\alpha\beta} H_{y_\beta} \phi_{y_\alpha}\, dy + \int_{\hat{\gamma}_2(0)} [\frac{d\hat{\chi}_\epsilon}{dv^k}]'_{y_\tau} v^k_{y_\tau} \phi\, dy$$

$$\leq \int_{\hat{\omega}_2(0)} [G^\alpha(y)\phi_{y_\alpha} + g(y)\phi]\, dy, \;\; \phi \geq 0,\; spt\,\phi \subset \hat{\omega}_2 \cup \hat{\gamma}_2. \tag{54}$$

Here and below we denote $|v_{y'y}|^2 = \Sigma_{\tau \leq n-1, \alpha \leq n} |v_{y_\tau y_\alpha}|^2$. The functions G and g are bounded and satisfy the condition

$$|G(y)| + |g(y)| \leq c|v_y|^2, \;\; y \in \hat{\omega}_2(0). \tag{55}$$

Moreover,

$$\int_{\hat{\omega}_2(0)} \hat{e}[v]\, dy = 2^{n-2} \fint_{\omega_{\frac{2}{\sqrt{e_0}}}(x^0)} e[u^\epsilon]\, dx \leq c\theta^2,$$

$$\int_{\hat{\gamma}_2(0)} \hat{\chi}_\epsilon\, d\gamma = 2^{n-2} \fint_{\gamma_{\frac{2}{\sqrt{e_0}}}(x^0)} \chi_\epsilon\, d\gamma \leq c\theta^2, \tag{56}$$

because $2/\sqrt{e_0} < \rho_0$ by (52), and (35) holds. We also remark that the estimate

$$\sup_{\hat{\gamma}_2(0)} \hat{\chi}_\epsilon(d^2(v)) \leq c \tag{57}$$

is valid. Indeed, it follows from the boundary condition for v^ϵ that

$$\hat{\chi}'_\epsilon 2d = \hat{b}_{kl} v^l_{y_n} d'_k \leq c|v_y| \leq_{(49)} c, \;\; y \in \hat{\gamma}_2(0). \tag{58}$$

It follows from (58) that the tangential gradient $[\hat{\chi}_\epsilon]_{y'}$ is bounded on $\hat{\gamma}_2$. This fact and (56) guarantee estimate (57). We know also that $\hat{\chi}' \geq 0$ and $\hat{\chi}''_\epsilon(d^2)d^2 \leq_{(14)} c(\delta_0)\hat{\chi}_\epsilon(d^2)$. From what has been said it follows that the penalty term T_ϵ in (54) is estimated in the following way: $T_\epsilon \geq -c\int_{\hat{\gamma}_2(0)} |v_{y'}|^2\, d\gamma$, $|v_{y'}|^2 = \Sigma_{\tau \leq n-1} |v_{y_\tau}|^2$, and we transform this term to the volume integral.

Taking into account (49), (55), (57), and the estimate of T_ϵ, we derive from the inequality for the function H the estimate

$$\int_{\hat{\omega}_{3/2}(0)} |v_{y'y}|^2\, dy \leq c\theta^2, \;\; v = v^\epsilon,$$

and estimating $|v_{y_n y_n}|^2$ from the system, we arrive at the relation

$$\int_{\hat{\omega}_{3/2}(0)} |v_{yy}|^2\, dy \leq c\theta^2. \tag{59}$$

The second consequence from inequality (54) is the relation

$$\int_{\hat{\omega}_2} \hat{a}^{\alpha\beta} H_{y_\beta} \phi_{y_\alpha}\, dy \le \int_{\omega_2} [G^\alpha(y)\phi_{y_\alpha} + g(y)\phi]\, dy, \tag{60}$$

where the bounded functions G and g satisfy (55). We put $\phi = (H-k)_+\xi^2$ in (60), where $(H-k)_+ = \max\{H-k, 0\}$, and ξ is a cutoff function for $B_\rho(0)$, $\xi = 1$ in $B_{\rho(1-\sigma)}(0)$, $|\xi_y| \le c/\rho\sigma$, and $\rho \in (1/2, 1]$, $\rho(1-\sigma) \ge 1/2$.

Taking into account (49) and (55), we arrive at the inequality

$$\int_{A_{k,\rho}} H_y^2 \xi^2\, dy \le \frac{c}{(\sigma\rho)^2} \int_{A_{k,\rho}} (H-k)_+^2\, dy + c|A_{k,\rho}|,$$

$$A_{k,\rho} = \{y \in \hat{\omega}_\rho(0) : H(y) > k\}, \ k \ge \hat{k}. \tag{61}$$

Inequalities (61) with $\hat{k} = \theta^2$ and the first estimate (56) allow us to prove that there exists a number $k_0 = C_* \hat{k}$, $C_* = c\theta^{-\frac{2(n+1)}{(n+2)}}$, such that

$$\sup_{\hat{\omega}_{1/2}(0)} H[v^\epsilon](y) \le 2k_0 = c\theta^{2/(n+2)}. \tag{62}$$

The proof of (62) is a standard one (see, for example, [**28**], Ch.2, Theorem 5.3), and we omit simple calculations. The first relation (53) follows from (62).

Now we put in (48) $h = (v_{y_n}\phi)_{y_n}$, $v = v^\epsilon(y)$, where ϕ is a cutoff function for $B_2(0)$. After integrating by parts, we derive the inequality

$$\int_{\hat{\omega}_2} |(v_{y_n})_y|^2 \phi\, dy + \int_{\hat{\omega}_2} \hat{a}^{\alpha\beta} p_{y_\beta}\phi_{y_\alpha}\, dy + \int_{\hat{\gamma}_2} \hat{a}^{\tau\mu}\hat{b}_{kl} v^l_{y_\mu}(v^k_{y_n}\phi)_{y_\tau}\, d\gamma$$

$$\le \int_{\hat{\omega}_2} (\tilde{G}^\alpha(y)\phi_{y_\alpha} + \tilde{g}\phi)\, dy, \quad spt\phi \subset \hat{\omega}_2 \cup \hat{\gamma}_2, \ \phi \ge 0. \tag{63}$$

Here $\tau, \mu \le n-1$, and functions $\tilde{G}$ and $\tilde{g}$ satisfy relation (55). The boundary integral J_γ in (63) does exist. Indeed, we can differentiate the boundary condition for $v = v^\epsilon$ in the tangential directions. Then

$$\hat{b}_{kl} v^l_{y_n y_\tau} = [d\hat{\chi}_\epsilon/dv^k]'_{y_\tau} - [\hat{b}_{kl}]'_{v^m} v^m_{y_\tau} v^l_{y_n} \text{ on } \hat{\gamma}_2. \tag{64}$$

Equality (64) allows us to introduce J_γ in the following way:

$$J_\gamma = \int_{\hat{\gamma}_2} \hat{a}^{\tau\mu}[d\hat{\chi}_\epsilon/dv^l]'_{y_\tau} v^l_{y_\mu}\, d\gamma - \int_{\hat{\gamma}_2} [\hat{b}_{kl}]'_{v^m} v^m_{y_\tau} v^k_{y_n} v^l_{y_\mu}\phi\, d\gamma$$

$$+ \int_{\hat{\gamma}_2} \hat{a}^{\tau\mu}\hat{b}_{kl} v^l_{y_\mu} v^k_{y_n}\phi_{y_\tau}\, d\gamma = j_1 + j_2 + j_3.$$

Inequalities (14), (49), and (57) help us to derive the estimate

$$j_1 + j_2 \ge -c\int_{\hat{\gamma}_2} |v_{y'}|^2\phi\, d\gamma \ge -c\int_{\hat{\gamma}_2} \phi\, d\gamma. \tag{65}$$

From (63) and (65) it follows that

$$\int_{\hat{\omega}_2} \hat{a}^{\alpha\beta} p_{y_\beta}\phi_{y_\alpha}\, dy + \int_{\hat{\gamma}_2} \hat{a}^{\tau\mu}\hat{b}_{kl} v^l_{y_\mu} v^k_{y_n}\phi_{y_\tau}\, d\gamma \le c\int_{\hat{\gamma}_2} \phi\, d\gamma + \int_{\hat{\omega}_2} (\tilde{G}^\alpha\phi_{y_\alpha} + \tilde{g}\phi)\, dy. \tag{66}$$

We put $\phi = (p-k)_+\xi^2(y)$ in (66), where $k \geq \hat{k}$ and ξ are the same as in (61), and obtain the inequality

$$\int_{A_{k,\rho}} |p_y|^2\xi^2\,dy \leq c_1\{(\sigma\rho)^{-2}\int_{A_{k,\rho}} (p-y)_+^2\,dy + |A_{k,\rho}|\}$$

$$(67)\qquad +c_2\{(\sigma\rho)^{-2}\int_{l_{k,\rho}} (p-k)_+^2\,d\gamma + |l_{k,\rho}|\}, \qquad l_{k,\rho} = \{y \in \hat{\gamma}_\rho :\ p(y) > k\}.$$

To prove the second relation (53), we introduce two functions

$$(68)\qquad Q_k(\rho) = \int_{A_{k,\rho}} (p(y)-k)_+^2\,dy,\ \ I_k(\rho) = \int_{l_{k,\rho}} (p(y)-k)_+^2\,d\gamma$$

and remark that

$$Q_k(\rho) \leq c\int_{\hat{\omega}_1(0)} p\,dy \leq_{(56)} c\theta^2, \qquad I_k(\rho) \leq I_0(1) \leq c\int_{\hat{\gamma}_{3/2}(0)} p\eta\,d\gamma$$

$$(69)\qquad \leq c|\int_{\hat{\omega}_{3/2}(0)} (p\eta)_{y_n}|\,dy \leq c\int_{\hat{\omega}_{3/2}(0)} (|v_{yy}|^2 + |v_y|^2)\,dy \leq_{(56),(59)} c\theta^2,$$

where η is a cutoff function for $B_{3/2}(0)$, $\eta = 1$ in $B_1(0)$.

We now put $\hat{k} = \theta^2$, take into account (67), (69), and perform an iteration process for the functions $Z_k(\rho) = Q_k(\rho) + I_k(\rho)$, $\rho \in [1/2, 1]$, $k \in [k_0, 2k_0]$, $k_0 = c_*\hat{k}$, with an appropriate $c_* = c_*(\theta)$. The idea of the proof is just the same as in [**28**], Ch.2, Theorem 5.3 for $Z_k = Q_k$. As a result, we obtain that

$$(70)\qquad \sup_{\hat{\omega}_{1/2}(0)} p(y) \leq 2k_0,$$

provided that $c_* = c\theta^{(\delta-2)}$, with some constants $c \gg 1$ and $\delta = \delta(n) < 1$, all parameters not depending on ϵ. Thus, $k_0 = c\theta^\delta$ in (70), and the second relation (53) is proved. As was said, relations (53) supply a contradiction to assumption (52). Thus, estimate (43) is proved. □

REMARK 5. Let the assumptions of Proposition 4 hold. Then

$$(71)\qquad \sup_{\Gamma_{R_5}} \chi_\epsilon(d^2(u^\epsilon)) \leq c(\theta^{-1}, R_5^{-1}),\ \epsilon \leq \epsilon_*.$$

This estimate follows from (20), the boundary condition (40), and (43).

Inequality (71) guarantees that $\sup_{\Gamma_{R_5}} \chi(d^2(u^\epsilon)) \leq c\epsilon < \delta_0^2$, $\epsilon \leq \epsilon_*$, provided ϵ_* is small enough, δ_0 is the parameter from the assumption $[\mathcal{A}_\mathcal{K}]$ and the definition of $\chi(\cdot)$. Thus, $\chi(d^2(u^\epsilon)) = d^2(u^\epsilon)$, $\epsilon \leq \epsilon_*$, and the boundary condition (40) is realized as the equality

$$(72)\qquad b_{kl}(x, u^\epsilon)(u^\epsilon)^l_{x_n} = \frac{2d(u^\epsilon)}{\epsilon} d'_{u^k}(u^\epsilon) \text{ on } \Gamma_{R_5},\ k \leq N.$$

From (72) and (43) it follows that

$$(73)\qquad \sup_{\Gamma_{R_5}} d(u^\epsilon)/\epsilon \leq c,\ \ \epsilon \leq \epsilon_*.$$

Thus, $\frac{2d(u^\epsilon)}{\epsilon} \to \lambda(x)\mathbf{1}_{u\in\mathcal{K}}$ weakly in $L_m(\Gamma_{R_5})$, $m < \infty$, and weakly-$*$ in $L_\infty(\Gamma_{R_5})$. Here $\lambda(x)$ is some nonnegative scalar bounded on Γ_{R_5} function, and $\mathbf{1}_{u\in\partial\mathcal{K}}$ is the characteristic function of the set $\{x \in \Gamma :\ u(x) \in \partial\mathcal{K}\}$. Moreover, $d'(u^\epsilon) \to \nu(u)$

uniformly on Γ_{R_5}, $\epsilon \to 0$, where $\nu(u)$ is a unit normal vector field on $\partial\mathcal{K}$. Taking into account (71) and (43), it is not difficult to obtain from (41) the estimate

$$\|u^\epsilon_{xx}\|_{2,B^+_{R_6}} \le c,\ \epsilon \le \epsilon_*,\ R_6 = \frac{1}{2}R_5. \tag{74}$$

The derived estimates allow us to go to the limit in identity (41) and to obtain the following relation for the limit function u:

$$\int_{B^+_{R_6}} [a^{\alpha\beta}(x)b_{kl}(x,u)u^l_{x_\beta}h^k_{x_\alpha} + \frac{1}{2}a^{\alpha\beta}(b_{ml})'_{u^k}u^m_{x_\beta}u^l_{x_\alpha}h^k]\,dx$$

$$+ \int_{\Gamma_{R_6}} \lambda(x)\mathbf{1}_{u\in\partial\mathcal{K}}\nu(u)h\,d\Gamma = 0,\ \ h \in C^1(\overline{B^+_{R_6}}),\ spt\,h \subset B^+_{R_6} \cup \Gamma_{R_6}. \tag{75}$$

Thus, $u \in W^2_2(B^+_{R_6})$ satisfies the Euler system of equations $Lu = 0$ for the functional $F_1[u]$ in $B^+_{R_6}$ under the following boundary condition:

$$b(x,u)u_{x_n} = \lambda(x)\mathbf{1}_{u\in\partial\mathcal{K}}\,\nu(u)\ \text{ on } \Gamma_{R_6}. \tag{76}$$

We recall that relations (75), (76) are valid provided that the smallness assumption (12) holds in $B^+_{R_2}$.

6. Hölder continuity of the gradient of the limit function

PROPOSITION 5. *There exist $\theta > 0$ and $R_2 > 0$ such that the condition (35) with these parameters guarantees that the solution u of problem (11) is a $C^{1,\beta}$-smooth function in $\overline{B^+_{\tau R_2}}$, and*

$$\|u_x\|_{C^\beta(\overline{B^+_{\tau R_2}})} \le c, \tag{77}$$

with some $\beta \in (0,1/2)$ and $\tau \in (0,1/2)$. The constant c depends on β and the parameters from the assumptions of Theorem 3.

PROOF. The function $v = u^\epsilon_{x_\tau}$, $\tau \le n-1$, satisfies the identity

$$\int_{B^+_R} a^{\alpha\beta}b_{kl}v^l_{x_\beta}\eta^k_{x_\alpha} + 2\int_{\Gamma_R} \frac{(d',v)(d',\eta)}{\epsilon}\,d\Gamma + 2\int_{\Gamma_R} \frac{d}{\epsilon}d''_{km}v^m\eta^k\,d\Gamma$$

$$= \int_{B^+_R} (\Phi_\alpha\eta_{x_\alpha} + M(x)\eta)\,dx,\ \ \eta \in W^1_2(B^+_R),\ \eta|_{S^+_R} = 0. \tag{78}$$

Here $R \le R_6 = \tau_4 R_2$, $\tau_4 \in (0,1/2)$, $\epsilon \le \epsilon_*$, and some functions Φ_α and M are bounded in B^+_R.

Now we fix $M = \sup_{\overline{B^+_{R_6}}} |u^\epsilon(x)|$, and denote by $S_M = \partial\mathcal{K} \cap B_M(0)$ the compact part of $\partial\mathcal{K}$. There is $\delta_1 > 0$ such that the smooth projection $Pr_{S_M}w = w_p$ is defined for w in the tubular neighborhood $V_{\delta_1}(S_M)$. There are two possibilities for the limit function u:

1) $\operatorname{dist}(u(0), S_M) \le \frac{\delta_1}{2}$ and

2) $\operatorname{dist}(u(0), S_M) > \frac{\delta_1}{2}$. In the second case, $\operatorname{dist}(u^\epsilon(x), S_M) > 0$ for $\epsilon \le \epsilon_*$ and $x \in \overline{B^+_{R_6}(0)}$ provided that ϵ_* and R_2 are fixed small enough. In this case, it follows from (72) that $u^\epsilon_{x_n} = 0$ on $\Gamma_{R_6}(0)$. We can consider each component of the vector function u^ϵ as a solution of the scalar elliptic equation satisfying the simplest Neumann condition on Γ_{R_6}. Taking into account estimate (43), we can estimate

$\|u^\epsilon_x\|_{C^\beta(\overline{B^+_{R_7}(0)})} \le c$, $\epsilon \le \epsilon_*$, $R_7 = \frac{1}{2}R_6$, with any $\beta \in (0,1)$. Inequality (77) follows in this case.

Now we consider the more interesting first situation. Obviously, the projection $u^\epsilon_p(x)$ is well defined in this case for $x \in \overline{B_{R_6}}$ and $\epsilon \le \epsilon_*$ provided that ϵ_* and R_2 are fixed in an appropriate way.

We introduce a smooth moving coordinate system

$$(\lambda_1(w), \dots, \lambda_{N-1}(w), \nu(w)),\ w \in S_M,$$

where each $\lambda_j(w)$ belongs to the tangential plane $T_w(\partial\mathcal{K})$, and $\nu(w)$ is a normal vector to $T_w(\mathcal{K})$ (outer to the set $\mathcal{K}$). Note that

$$\sup_{w \in V_{\delta_1}(S_M)} (|\nabla_w \lambda(w)| + |\nabla_w \nu(w)|) \le c, \tag{79}$$

due to the assumption $[\mathcal{A}_\mathcal{K}]$. It follows from (43) and (79) that

$$\sup_{x \in \overline{B^+_R}} (\Sigma_{j \le N-1} |(\lambda_j(u^\epsilon_p(x)))'_x| + |(\nu(u^\epsilon_p))'_x|) \le c,\ \epsilon \le \epsilon_*,\ 2R \le R_6. \tag{80}$$

Now let $G^{\hat{x}}_\rho(x)$ be a solution of the problem

$$-(a^{\alpha\beta}(G^{\hat{x}}_\rho(x))_{x_\beta})_{x_\alpha} = \frac{\mathbf{1}_{\omega_\rho(\hat{x})}(x)}{2|\omega\rho|},\ x \in B^+_R,$$

$$(G^{\hat{x}}_\rho)_{x_n}|_{\Gamma_R} = 0,\ G^{\hat{x}}_\rho|_{S^+_R} = 0,\ \omega_\rho(\hat{x}) = B^+_R \cap B_\rho(\hat{x}),\ \rho \le R/8. \tag{81}$$

Now we consider the case $\hat{x} \in \Gamma_{R/4}(0)$. We can extend functions $a^{\alpha\beta}(x)$ in $B^-_R(0)$ in the following way: $\tilde{a}^{\alpha\beta}(x', x_n) = a^{\alpha\beta}(x', -x_n)$, $\alpha, \beta \le n-1$, $\alpha = \beta = n$, and $\tilde{a}^{\alpha\beta}(x', x_n) = -a^{\alpha\beta}(x', -x_n)$, $\alpha \le n-1, \beta = n$. Then the even continuation of $G^{\hat{x}}_\rho$ in $B^-_R(0)$ is a regularization of the Green function of the Dirichlet problem in the ball B_R for the operator $Lw = -(\tilde{a}^{\alpha\beta}(x) w_{x_\beta})_{x_\alpha}$. The properties of such functions are well studied ([**25**], [**36**], [**37**]).

Now we fix constants $l_1, \dots, l_{N-1}$ and assume that the estimate

$$\sup_{j \le N-1} |l_j| \le \sup_{x \in B^+_{R_6}} |(u^\epsilon)_x| \le_{(43)} c,\ \epsilon \le \epsilon_*, \tag{82}$$

holds. We put $\tilde{v} = v - \Sigma_{j \le N-1} l_j \lambda_j(u^\epsilon_p(x))$ and consider (78) with $\eta^k = \tilde{v}^k G^{\hat{x}}_\rho \xi^2$, $k \le N$, where ξ is a cutoff function for $B_{R/2}(\hat{x})$, $\xi = 1$ in $B_{R/4}(\hat{x})$). Then the first surface integral in (78) is nonnegative because $(d'(u^\epsilon), \lambda_j(u^\epsilon_p)) = 0$, and the second one we denote by J_Γ and estimate it by (73) and (82):

$$|J_\Gamma| \le c \int_{\Gamma_{R/2}(\hat{x})} |\tilde{v}|^2 G^{\hat{x}}_\rho \xi^2\, d\Gamma + cR. \tag{83}$$

Transforming the integral in (83) to the volume one, and using properties of $G^{\hat{x}}_\rho$, we deduce the estimate

$$|J_\Gamma| \le \frac{\nu}{4} \int_{B^+_{R/2}} |\tilde{v}_x|^2 G^{\hat{x}}_\rho \xi^2\, dx + cR,\ \nu = \nu_1\nu_2. \tag{84}$$

Now we put $H(x) = \frac{1}{2} b_{kl}(x, u^\epsilon) \tilde{v}^l \tilde{v}^k$, $\hat{\omega}_R = B^+_{R/2}(\hat{x})$. Taking into account (80) and (84), we derive the inequality

$$\frac{\nu}{2} \int_{\hat{\omega}_R} |\tilde{v}_x|^2 G^{\hat{x}}_\rho \xi^2\, dx + \int_{\hat{\omega}_R} a^{\alpha\beta} H_{x_\beta} (G^{\hat{x}}_\rho)_{x_\alpha} \xi^2\, dx$$

$$\leq \frac{c}{R^n}\int_{T_R} |\tilde v|^2\,dx + cR, \;\; T_R = B^+_{R/2}(\hat x)\setminus B^+_{R/4}(\hat x).$$

After going to the limit as $\epsilon \to 0$ in the last inequality, we arrive at the following relation for the limit function $\tilde v^0 = u_{x_\tau} - \Sigma_{j\leq N-1} l_j\lambda_j(u_p)$:

$$\frac{\nu}{2}\int_{\hat\omega_R} |\tilde v^0_x|^2 G^{\hat x}_\rho \xi^2\,dx + \int_{\hat\omega_R} a^{\alpha\beta} H^o_{x_\beta}(G^{\hat x}_\rho)_{x_\alpha}\xi^2\,dx$$

$$\leq \frac{c}{R^n}\int_{T_R} |\tilde v^0|^2\,dx + cR, \tag{85}$$

where $H^o(x) = \frac{1}{2}b_{kl}(x,u)(\tilde v^0)^l(\tilde v^0)^k$.

We estimate the integral with the function H^o in the last inequality exploiting the integral identity for the function $G^{\hat x}_\rho$ (see the details in [**25**]):

$$\int_{B^+_R} a^{\alpha\beta}(G^{\hat x}_\rho)_{x_\beta}\psi_{x_\alpha}\,dx = \frac{1}{2}\fint_{\hat\omega_\rho(\hat x)} \psi\,dx, \;\; \psi\in W^1_2(B^+_R), \;\; \psi|_{S^+_R} = 0.$$

First, we recall that $a_{\alpha\beta} = a_{\beta\alpha}$, put $\psi = H^o\xi^2$ in this identity and obtain that

$$J = \int_{\hat\omega_R(\hat x)} a^{\alpha\beta}(G^{\hat x}_\rho)_{x_\alpha} H^o_{x_\beta}\xi^2\,dx = -\int_{\hat\omega_R} a^{\alpha\beta}(G^{\hat x}_\rho)_{x_\alpha} H^o 2\xi\xi_{x_\beta}\,dx$$

$$+\frac{1}{2}\fint_{\hat\omega_\rho} H^o\xi^2\,dx \geq -c\int_{T_R} |(G^{\hat x}_\rho)_x| H^o\xi|\xi_x|\,dx = -cP.$$

We estimate the integral P by the Cauchy inequality:

$$P \leq \int_{T_R} \frac{|(G_\rho)_x|^2}{G^{\hat x}_\rho}|\tilde v^0|^2\,dx + \frac{c}{R^2}\int_{T_R} G^{\hat x}_\rho|\tilde v^0|^2\,dx$$

$$\leq cR^{n-2}\int_{T_R} |(G_\rho)_x|^2|\tilde v^0|^2\,dx + \frac{c}{R^n}\int_{T_R}|\tilde v^0|^2\,dx = (a)+(b).$$

To estimate the integral (a), we address the integral identity for $G^{\hat x}_\rho$ once more, and fix $\psi = G^{\hat x}_\rho|\tilde v^0|^2\eta^2(|x-\hat x|)$, where the cutoff function $\eta(t) = 1$, $t\in[R/4,R/2]$, $\eta(t) = 0$ when $t\leq R/8$ and $t\geq R$.

We have the equality

$$\int_{D_R} a^{\alpha\beta}(G_\rho)_{x_\beta}(G_\rho|\tilde v^0|^2\eta^2)_{x\alpha}\,dx = 0, \;\; D_R = B^+_R(\hat x)\setminus B^+_{R/8}(\hat x),$$

and, as a consequence, the estimate

$$\int_{D_R} |(G_\rho)_x|^2|\tilde v^0|^2\eta^2\,dx \leq c\int_{D_R}(G_\rho)^2\eta^2|(\tilde v^0)_x|^2\,dx + c\int_{D_R}(G_\rho)^2|\eta_x|^2|\tilde v^0|^2\,dx.$$

The last inequality allows us to estimate the integral P:

$$P \leq c\int_{D_R} G^{\hat x}_\rho|(\tilde v^0)_x)|^2\,dx + \frac{c}{R^n}\int_{D_R}|\tilde v^0|^2\,dx.$$

Now we go back to the inequality for the integral J and derive the following estimate for $\tilde v^0$ from (85):

$$\int_{B^+_{R/8}(\hat x)} G^{\hat x}_\rho|\tilde v^0_x|^2\,dx \leq c\int_{D_R(\hat x)} G^{\hat x}_\rho|\tilde v^0_x|^2\,dx + \frac{c}{R^n}\int_{D_R}|\tilde v^0|^2\,dx + cR. \tag{86}$$

Here

$$(87)\quad \int_{D_R} |\tilde{v}^0|^2\,dx = \int_{D_R} (u_{x_\tau}, \nu(u_p))^2\,dx + \Sigma_{j\le N-1} \int_{D_R} ((u_{x_\tau}, \lambda_j(u_p)) - l_j)^2\,dx.$$

Now we fix $l_j = \fint_{D_R} (u_{x_\tau}, \lambda_j(u_p))\,dx$ and estimate the last integral in (87) by the Poincaré inequality. In a result, we have

(88)

$$\int_{B^+_{R/8}(\hat{x})} G^{\hat{x}}_\rho |u_{x'x}|^2\,dx \le \frac{c}{R^{n-2}} \int_{D_R} |u_{x'x}|^2\,dx + cR + \frac{c}{R^n} \Sigma_{\tau\le n-1} \int_{D_R} (u_{x_\tau}, \nu(u_p))^2\,dx.$$

Functions $G^{\hat{x}}_\rho \to G^{\hat{x}}$ a.e. in $B^+_{R/8}$, when $\rho \to 0$, and we obtain by the Fatou Lemma that

$$(89)\quad \int_{B^+_{R/8}} G^{\hat{x}} |u_{x\prime x}|^2\,dx \le c \int_{D_R} G^{\hat{x}} |u_{x'x}|^2\,dx + cR + \frac{c}{R^n} \int_{D_R} (u_{x_\tau}, \nu(u_p))^2\,dx.$$

Now we assert that the following inequality is valid for the limit function u:

$$(90)\quad \int_{B^+_{R/8}(\hat{x})} G^{\hat{x}} |u_{x_n x}|^2\,dx \le c \int_{D_R} G^{\hat{x}} |u_{x_n x}|^2\,dx + cR + \frac{c}{R^n} \int_{D_R} |u_{x_n}|^2\,dx,$$

where $|u_{x_n x}| = |(u_{x_n})_x|$. To prove it, we put $\eta = (u^\epsilon_{x_n} G^{\hat{x}}_\rho \xi^2)_{x_n}$ in (41) with the same functions $G^{\hat{x}}_\rho$ and ξ as earlier, and we obtain the following relation for $Z(x) = \frac{1}{2} b_{kl}(x, u^\epsilon)(u^\epsilon)^l_{x_n} (u^\epsilon)^k_{x_n}$:

$$\int_{\hat{\omega}_R} G^{\hat{x}}_\rho |u^\epsilon_{x_n x}|^2 \xi^2\,dx + \int_{\hat{\omega}_R} a^{\alpha\beta} Z_{x_\beta} (G^{\hat{x}}_\rho)_{x_\alpha} \xi^2\,dx$$

$$(91)\qquad + \int_{\Gamma_R} a^{\tau\mu} b_{kl}(u^\epsilon)^l_{x_\mu} ((u^\epsilon)^k_{x_n} G^{\hat{x}}_\rho \xi^2)_{x\tau}\,d\Gamma \le \frac{c}{R^n} \int_{T_R} |u^\epsilon_{x_n}|^2\,dx + cR.$$

To estimate the surface integral L^ϵ_Γ in (91), we differentiate the boundary condition (72) in the direction $\mu \le n-1$ and express $(u^\epsilon)^k_{x_n x_\mu}$ on Γ_R. As a result,

$$(92)\qquad L^\epsilon_\Gamma \ge_{(73)} -cR + \int_{\Gamma_R} a^{\tau\mu} b_{kl}(u^\epsilon)^l_{x_\tau} (u^\epsilon)^k_{x_n} [G^{\hat{x}}_\rho \xi^2]'_{x_\mu}\,d\Gamma.$$

The last integral j_ϵ in (92) is too strong to obtain a reasonable integral estimate for $|(u^\epsilon_{x_n})_x|$, but we note that

$$\lim_{\epsilon\to 0} j_\epsilon = \int_{\Gamma_{R/2}(\hat{x})} a^{\tau\mu}(u_{x_\tau}, d')\lambda(x) \mathbf{1}_{\{u\in\partial\mathcal{K}\}} [G^{\hat{x}}_\rho \xi^2]_{x_\mu}\,d\Gamma = 0$$

because $d(u) \cdot \mathbf{1}_{\{u\in\partial\mathcal{K}\}} = 0$ and, as a consequence, $(d', u_{x_\tau}) = 0$ when $u \in \partial\mathcal{K}$. What has been said above allows us to go to the limit in (91) as $\epsilon \to 0$ and to obtain inequality (90) for the limit function u. (The term with the function $Z^o(x) = \frac{1}{2} b_{kl}(x, u) u^l_{x_n} u^k_{x_n}$ can be estimated in the same way as the similar integral with H^o was estimated in (85).)

The limit function u satisfies inequalities (89) and (90). We put $t_R(\hat{x}) = \Gamma_R(\hat{x}) \setminus \Gamma_{R/8}(\hat{x})$ and

$$t^{(1)}_R = \{x \in t_R(\hat{x}) :\ u(x) \in \partial\mathcal{K}\}, \quad t^{(2)}_R = \{x \in t_R(\hat{x}) :\ u \in \operatorname{int}\mathcal{K}\}.$$

For a fixed $R \le \frac{1}{2}R_6$ two cases are possible: a) $|t_R^{(1)}|_{n-1} \ge \frac{1}{2}|t_R|$, and b) $|t_R^{(2)}| \ge \frac{1}{2}|t_R|$. In case a), $d(u(x)) = 0$ on $t_R^{(1)}$ and, as a consequence, $(d', u_{x_\tau}) = 0,\ \tau \le n-1$, on the "fat" set $t_R^{(1)}$. By the Poincaré inequality,

$$\int_{D_R(\hat{x})} (u_{x'}, \nu(u_p))^2\, dx \le cR^2 \int_{D_R} |[(u_{x'}, \nu(u_p))]'_x|^2\, dx.$$

Then it follows from (89)and (43) that

$$\int_{B^+_{R/8}(\hat{x})} G^{\hat{x}} |u_{x'x}|^2\, dx \le c \int_{D_R} G^{\hat{x}} |u_{x'x}|^2\, dx + cR. \tag{93}$$

In case b), the boundary condition guarantees that $u^m_{x_n} = 0,\ m \le N,\ \ x \in t_R^{(2)}$, and by the Poincaré inequality,

$$\int_{D_R(\hat{x})} |u_{x_n}|^2\, dx \le cR^2 \int_{D_R} |u_{x_n x}|^2\, dx.$$

From (90) and the last inequality we obtain the relation

$$\int_{B^+_{R/8}(\hat{x})} G^{\hat{x}} |u_{x_n x}|^2\, dx \le c \int_{D_R(\hat{x})} G^{\hat{x}} |u_{x_n x}|^2\, dx + cR. \tag{94}$$

Now we apply the "hole-filling procedure" to derive from (93) and (94) the inequality

$$\int_{B^+_{R/8}} G^{\hat{x}} w\, dx \le q \int_{B^+_R} G^{\hat{x}} w\, dx + cR, \tag{95}$$

with some parameter $q \in (0,1)$. In (95) $w = |u_{x'x}|^2$ or $w = |u_{x_n x}|^2$ for each fixed $R \le \frac{1}{2}R_7$. We now denote

$$\psi(r, w) = \int_{B^+_r(\hat{x})} G^{\hat{x}}(x) w(x)\, dx,$$

and write (95) in the form:

$$\psi(R/8, w) \le q\psi(R, w) + cR, \tag{96}$$

where w is one of the functions defined above.

To perform an iterative process, we fix a radius $\rho \in (0, R)$ and consider the sequence $R_j = \frac{R}{8^j},\ j = 0, \dots, M+1$, where the number M is fixed by the condition $R_{M+1} \le \rho < R_M$. Further we fix such a function w that inequality (95) holds for at least $[M/2]$ radii $R_j,\ j \le M$. We denote such a subsequence of R_j as $\hat{R}_s = R_{j_s},\ s \le m,\ [M/2] \le m \le M$, and make iterations for the chosen w and $\hat{R}_s,\ s \le m$.

As a result, we obtain either

$$\int_{B^+_\rho(\hat{x})} G^{\hat{x}} |u_{x'x}|^2\, dx \le c(\rho/R)^\alpha \Big[\int_{B^+_R} G^{\hat{x}} |u_{x'x}|^2\, dx + 1\Big] \tag{96}$$

or

$$\int_{B^+_\rho(\hat{x})} G^{\hat{x}} |u_{x_n x}|^2\, dx \le c(\rho/R)^\alpha \Big[\int_{B^+_R(\hat{x})} G^{\hat{x}} |u_{x_n x}|^2\, dx + 1\Big], \tag{97}$$

where $R \le \frac{1}{2}R_6,\ \hat{x} \in \Gamma_{R/4}(0),\ \rho \le R$, and some $\alpha \in (0,1)$.

It is not difficult to check that the expressions in the square brackets of (96) and (97) are bounded, and these inequalities and the system for $u(x)$ supply the estimate

$$\sup_{\hat{x}\in\Gamma_{R_6/2}(0)} \int_{B_\rho^+} G^{\hat{x}}(x)|u_{x_n x}|^2\,dx \le c(R_6^{-1})\rho^\alpha, \quad \rho \le R, \tag{98}$$

in any case.

A similar estimate is derived for $\hat{x} \in B^+_{R/2}(0)$, $R \le \min\{\frac{1}{2}R_6, \operatorname{dist}(\hat{x}, \Gamma_R(0))\}$.

It follows from the "sewing procedure" that estimate (98) holds for all $\hat{x} \in B^+_{R_7}$, $R_7 = R_6/4$. The last estimate in $\overline{B^+_{R_7}}$ allows us to assert that $u_{x_n} \in C^\beta(\overline{B^+_{R_7}(0)})$, and $\beta = \alpha/2$.

Now we can consider $u^m(x), m \le N$ as a solution of the scalar problem

$$-(a^{\alpha\beta}(x)u^m_{x_\beta})_{x_\alpha} = g^m(x) \text{ a.e. in } B^+_{R_7},$$
$$u^m_{x_n}|_{\Gamma_{R_7}} \in C^\beta(\Gamma_{R_7}), \quad g^m \in L_\infty(B^+_{R_7}). \tag{99}$$

The theory of linear boundary value problems gives that $u^m \in C^{1,\beta}(\overline{B^+_{R_8}})$, $R_8 = R_7/2 = \tau R_2$, with some $\tau < 1/2$. Proposition 5 is proved. □

REMARK 6. Remark 3 and Proposition 5 imply that condition (12) with small enough R_2 and θ guarantees estimate (77) for the solution u of the model problem (11), and Theorem 3 is proved.

7. Proofs of Theorem 1 and Theorem 2

To prove Theorem 2, we go back to the initial setting (1)-(3). Taking into account Remark 6, we assert that the following proposition is valid:

PROPOSITION 6. *Let u be a solution to problem (1)-(3). There exist R_0 and θ_0 in the definition (8) of Σ_{θ_0,R_0} such that the smallness condition (10) for u with some radius $R_1 = R_1(x^0)$, $x^0 \in \partial\Omega \cap \Omega_0$, ensures $C^{1,\beta}$-smoothness of u in the vicinity of x^0 and the estimate $\|u\|_{C^{1,\beta}(\overline{\Omega_{\tau R_1}(x^0)})} \le c$ with some exponent $\beta < \min\{1/2, 1-n/q\}$ and some $\tau \in (0, 1/2)$. The parameters β and τ do not depend on the fixed point x^0.*

The smoothness of u inside Ω for $x^0 \in \Omega$ under smallness condition (10) is well known [**21**]. We can sew the boundary and internal estimates to assert that any point $x^0 \in \Omega_0 = \overline{\Omega} \setminus \Sigma_{\theta_0,R_0}$ is a regular one for the solution u under consideration. It should be noted that all restrictions on θ in (12) and, as a consequence, on θ_0 in (10) (to guarantee smoothness of u in the vicinity of a fixed point) depend on the data of problem (1)-(3) but not on the fixed point.

Obviously, the set Ω_0 is open with respect to $\overline{\Omega}$, and the singular set $\Sigma = \Sigma_{\theta_0,R_0}$ is closed. Theorem 2 is proved.

The estimate of the Hausdorff measure of Σ is the standard one. It follows from the definition (8) that $H_{n-2}(\Sigma) = 0$. The first part of Theorem 1 is proved.

We conclude the proof of Theorem 1 with the following remark.

REMARK 7. Up to now, we studied problem (1)-(3) in the vicinity of a *regular* point. Now let u be a solution of (11) in B^+_R but with the smallness assumption (12) removed. We denote by $\Sigma_R = \Sigma \cap \overline{B^+_R}$ the closed singular set of u in

$\overline{B_R^+}$, $H_{n-2}(\Sigma_R) = 0$. Further we apply a modification of the proof of Theorem 3.1 [**9**].

For an arbitrary $\eta > 0$ we fix a finite family of balls $B_{R_j}(x^j)$, $x^j \in \Sigma_R$, $R_j < \eta$, $\Sigma_R \subset T = \bigcup_{j\le M} B_{R_j}(x^j)$. Obviously, $|\bigcup_{j\le M} B_{R_j}| \le \omega_n \eta^2 \Sigma_{j\le M} R_j^{n-2} \to 0$, $\eta \to 0$. It follows from Remark 5 that u satisfies the Euler system $Lu = 0$ on $B_R^+ \setminus T \cap B_R^+$. Boundary condition (76) guarantees that $(\frac{\partial u}{\partial n_A}, \nu(u)) \le 0$ on $\Gamma_R^{(\eta)} = \Gamma_R \setminus (T \cap \Gamma_R)$, and $\frac{\partial u}{\partial n_A} = 0$ when $u(x) \in \operatorname{int} K$, $x \in \Gamma_R^{(\eta)}$; here $\frac{\partial u}{\partial n_A} = -b_{kl} u^l_{x_n}$. We fix some function $\psi_\eta \in C^{0,1}(\overline{B_R^+})$ with the properties: $\psi_\eta \ge 0$, $\psi_\eta = 0$ on $T \cap B_R^+$, $\psi_\eta = 1$ on $\overline{B_R^+} \setminus B_R^+ \cap (\bigcup_{j\le M} B_{2R_j}(x^j))$ (see the details in [**9**]). Then it follows from (75) that

$$L_{B_R^+}(u;\phi\psi_\eta) \equiv \int_{B_R^+} [A^{\alpha\beta}_{kl} u^l_{x_\beta} (\phi^k \psi_\eta)_{x_\alpha} + \frac{1}{2} (A^{\alpha\beta}_{ml})'_{u^k} u^l_{x_\beta} u^m_{x_\alpha} (\phi^k \psi_\eta)]\, dx$$

$$= \int_{\Gamma_R} (\frac{\partial u}{\partial n_a}, \phi) \psi_\eta \, d\Gamma, \ \phi \in C^1(B_R^+; \mathbb{R}^N),\ \phi_{S_R^+} = 0. \tag{100}$$

Taking into account that $H_{n-2}(\Sigma_R) = 0$, we prove that there exists the finite $\lim_{\eta\to 0} L_{B_R^+}(u;\phi\psi_\eta) = L_{B_R^+}(u;\phi)$. Then there exists the finite limit of the boundary integral in (100) as $\epsilon \to 0$. Moreover, this integral in (100) with $\phi = \Phi(x)\nu(u(x))$ (Φ is a scalar function, $\Phi \ge 0, \Phi|_{\Gamma_R} \in C_0^1(\Gamma_R)$) is nonpositive and $-\int_{\Gamma_R} \Phi(x) (\frac{\partial u}{\partial n_A}, \nu(u)) \mathbf{1}_{u\in\partial\mathcal{K}} \psi_\eta \, d\Gamma \ge 0$. Using the Riesz representation theorem, after going to the limit in (100), we arrive at the relation

$$L_{B_R^+}(u;\phi) + \int_{\Gamma_R} (\phi, \nu(u)) \mathbf{1}_{u\in\partial\mathcal{K}}\, d\mu = 0,\ \phi \in C^1(B_R^+; \mathbb{R}^N), \tag{101}$$

where μ is some Radon measure on $\Gamma_R \subset \mathbb{R}^{N-1}$. Since $\partial\Omega$ is a compact smooth surface, we can go back to the initial setting of the problem in Ω (preserving the notation of the variables and the functions) and assert that u satisfies the identity

$$L_\Omega(u;\phi) + \int_{\partial\Omega} (\phi, \nu(u)) \mathbf{1}_{u\in\partial\mathcal{K}}\, d\mu_\Gamma = 0, \quad \phi \in C^1(\overline{\Omega}; \mathbb{R}^N), \tag{102}$$

where $\nu(u)$ is a C^1-smooth unit vector field on $\partial\mathcal{K}$, orthogonal to the tangent plane to $\partial\mathcal{K}$ at the point u, and μ_Γ is some surface measure on $\partial\Omega$.

8. Regularity of minimizers with a hypersurface-obstacle at the boundary $\partial\Omega$

In this section, we analyze a variational problem with an obstacle $u(\partial\Omega) \subset S$, where S is a smooth given hypersurface in R^N.

We define the set

$$W_S = \{u \in W_2^1(\Omega; R^N),\ u(x) \in S \text{ a.e. on } \partial\Omega\},$$

and assume that the C^2-smooth surface S satisfies the following assumption:

$[\mathcal{A}_S]$ There exist numbers δ_0 and $M > 0$ such that the distance function $d(u) = \operatorname{dist}(u, S)$ is defined in the tubular neighborhood $U_{2\delta_0}(S)$ of the hypersurface S, $U_{2\delta_0}(S) = U^{(1)}_{2\delta_0}(S) \cup S \cup U^{(2)}_{2\delta_0}(S)$, $d \in C^2(\overline{U^{(i)}_{2\delta_0}(S)})$, $i = 1, 2$, and

$$\sup_{\overline{U^{(i)}_{2\delta_0}(S)}} |d''(u)| \le M,\ i = 1, 2. \tag{103}$$

REMARK 8. For a noncompact surface S, it follows from $[\mathcal{A}_S]$ that S does not become glued at infinity, and the principal curvatures of S are uniformly bounded for $u \in S$.

We consider the variational problem (5) and state the following result.

THEOREM 4. *Let the assumptions* $[\mathcal{A}_S]$, $[\mathcal{A}_1] - [\mathcal{A}_3]$ *be valid, and let* u *supply the minimum in problem (1), (3), (5). Then the assertions of Theorem* 1 *hold but the characteristic function is absent in (101) and (102).*

To prove Theorem 4, we consider the same model setting in half of a ball, and we study the behavior of solutions u^ϵ for the same penalty problem as the one defined in Section 2. Certainly, the penalty function $\chi(d^2)$ now depends on $d = d(u, S)$. We can repeat all arguments of sections 1-4 and now explain only the difference in the derivation of estimate (77). In this case, $u|_{\Gamma_R} \in S$, where $u(x) = \lim_{\epsilon\to 0} u^\epsilon(x)$. This means that $d(u(x)) = 0$ on Γ_R and, as a consequence, $(d', u_{x_\tau}) = 0$, $\tau \le n-1$, on Γ_R. Thus, we can estimate the last integral in (89) by the Poincaré inequality and arrive at an inequality similar to (89) where the last integral vanishes. This allows us to prove that

$$\psi(R/8) \le q\psi(R) + cR,\ q \in (0,1),\ \psi(r) = \int_{B_r^+(\hat{x})} G^{\hat{x}} |u_{x'x}|^2\, dx$$

immediately.

After iterations in this inequality, we obtain that

$$\psi(\rho) \le c(R^{-1}, \beta)\rho^{2\beta},\ \rho \le R,\ 2\beta \in (0,1). \tag{104}$$

It follows from (104) and the system for u that

$$\int_{B_\rho^+(\hat{x})} |u_{x_n x}|^2 G^{\hat{x}}\, dx \le c\rho^{2\beta},\ \hat{x} \in \Gamma_{R/2}(0).$$

Then we can repeat the final part of the proof of Proposition 5. All other steps of the proofs of Theorem 4 and Theorem 1 coincide.

References

[1] Arkhipova A. *Variational problem with an obstacle in R^N for a class of quadratic functionals.* Zapiski Nauchn. Semin. POMI, St.Petersburg, **362** (2008), 1–34.

[2] ———, *On the regularity of solutions of the boundary-value problems for quasilinear elliptic systems with quadratic nonlinearities.* in Russian: Problemy Mat. Analiza, **15**, St.Petersburg State Univ., 1995, 3–32. MR1334137 (96j:35067)

[3] Arkhipova A. and Uraltseva N. *The regularity of solutions of diagonal elliptic systems under convex boundary constraints.* J. of Soviet Math., **40**, no. 5 (1988), 591–599. MR869237 (87m:35107)

[4] ———, *The regularity of solutions of variational inequalities with convex boundary constraints for nonlinear operators with diagonal main part.* Vestnik Leningr. Univ., **15** (1987), 13–19. MR928154 (89f:35088)

[5] ———, *The best possible smoothness for solutions of variational inequalities with convex boundary constraints.* J. of Soviet Math., **49**, no. 5 (1990), 1121–1128. MR918937 (88m:49001)

[6] Athanasopoulos I. and Caffarelli L. *Optimal regularity of lower dimensional obstacle problems.* Zapiski Nauchn. Semin. POMI, **310** (2004), 49–66. MR2120184 (2006i:35053)

[7] Beirao da Veiga H. and Conti F. *Equazioni ellittiche non lineari con ostacoli sottili.* Annali Scuola Norm. Super. Pisa, **26**, no. 2 (1972).

[8] Caffarelli L. *Further regularity for the Signorini problem.* Commun. Partial Diff. Equations, **4** (1979), 1067–1075. MR542512 (80i:35058)

[9] Chen Y. and Struwe M. *Existence and partial regularity results for the heat flow of harmonic maps.* Math. Z., **201** (1989), 83–103. MR990191 (90i:58031)
[10] Duzaar F. *Variational inequalities and harmonic maps.* J. Reine Angew. Math., **374** (1987), 39–60. MR876220 (88m:58038)
[11] Duzaar F. and Fuchs M. *Optimal regularity theorem for variational problems with obstacles.* Manuscripta Math., **56** (1986), 209–234. MR850371 (87k:49015)
[12] Evans L.-C. *Partial regularity for stationary harmonic maps into spheres.* Arch. Rat. Mech. Anal., **116** (1991), 101–113. MR1143435 (93m:58026)
[13] Fichera G. *Existence Theorems in Elasticity Theory,* Mir, Moscow, 1974.
[14] Frehse J. *On Signorini's problem and variational problems with thin obstacles.* Ann. Scuola Norm. Super. Pisa, **4** (1977), 343–362. MR0509085 (58:22987)
[15] Fuchs M. *A regularity theorem for energy minimizing maps of Riemannian manifolds.* Commun. Partial Diff. Equations, **12**, no. 11 (1987). MR888462 (88g:58041)
[16] Fuchs M. *Some remarks on the boundary regularity for minima of variational problems with obstacles.* Manuscripta Math., **54** (1985), 107–119. MR808683 (87b:49050)
[17] Fuchs M. and Fusko N. *Partial regularity results for vector-valued functions which minimize certain functionals having nonquadratic growth under smooth side conditions.* J. Reine Angew. Math., **3** (1988). MR953677 (89h:49005)
[18] Fuchs M. and Wiegner M. *The regularity of minima of variational problems with graph obstacles.* Arch. Math., **53** (1988), 75-81. MR1005172 (90f:49004)
[19] Giaquinta M. *Multiple Integrals in the Calculus of Variations and Nonlinear Elliptic Systems.* Annals of Math. Studies **105**, Princeton Univ. Press, Princeton, NJ, 1983. MR717034 (86b:49003)
[20] Giaquinta M. and Giusti E. *Non-linear elliptic systems with quadratic growth.* Math. Z., **201** (1978), 323–349. MR0481490 (58:1606)
[21] ———, *On the regularity of the minima of variational integrals.* Acta Math. **148** (1982), 31–46. MR666107 (84b:58034)
[22] ———, *The singular set of the minima of certain quadratic functionals.* Ann. Scuola Super. Pisa, Cl.Sci. (4), **11** (1984), 45–55. MR752579 (86a:49086)
[23] Hildebrandt S. *Harmonic mappings of Riemannian manifolds.* Lecture Notes in Math., volume 1161, Springer, Heidelberg, 1984. MR821968 (87j:58030)
[24] Hildebrandt S., Kaul H., and Widman K.-O. *An existence theorem for harmonic mappings of Riemannian manifolds.* Acta Math., **138** (1977), 1–16. MR0433502 (55:6478)
[25] Hildebrandt S. and Widman K.-O. *Variational inequalities for vector-valued functions* J. Reine Angew. Math., **309** (1979), 191–220. MR542048 (81a:35023)
[26] Jost J. and Meier M. *Boundary regularity for minima of certain quadratic functionals.* Math. Ann. **262** (1983), 549–561. MR696525 (84i:35051)
[27] Kinderlehrer D. *The smoothness of the solution of the boundary obstacle problem.* J. Math. Pure Appl., **60** (1981), 193–212. MR620584 (84j:49011)
[28] Ladyzhenskaya O. A, Uraltseva N. N. *Linear and Quasilinear Equations of Elliptic Type,* second edition, Moscow, Nauka, 1973; English transl. of the first edition, Academic Press, New-York–London, 1968 MR0509265 (58:23009)
[29] Nečas J. *On regularity of solutions to nonlinear variational inequalities for second-order elliptic systems.* Rend. Mat., **8**, (1975), 481–498. MR0382827 (52:3709)
[30] Schoen R. and Uhlenbeck K. *A regularity theory for harmonic maps.* J. Diff. Geom., **17** (1982), 307–335. MR664498 (84b:58037a)
[31] Schumman R. *On the regularity of a contact boundary value problem.* Z. Anal. Anwend. 9, **5** (1990), 455–465. MR1119544 (93e:73040)
[32] Uraltseva N. *A problem with a one-side condition on the boundary for a quasilinear elliptic equation.* Problemy Mat. Analiza, St. Petersburg State Univ., no. 6, 1987, 151–174. (Russian)
[33] Uraltseva N. *On the regularity of solutions of variational inequalities.* Uspehi Mat. Nauk, **42** (1987), 151–174. MR933999 (90c:35033)
[34] ———, *Strong solutions of the generalized Signorini problem.* Sibirsk. Mat. Zh., **19** (1978), 1204–1212. MR508511 (80i:35166)
[35] ———, *Estimation on the boundary of the domain of derivatives of solutions of variational inequalities.* Probl. Mat. Anal., no. 10, 1986, 92–105. (Russian) MR860572 (87k:35106)
[36] Widman K.-O. *Inequalities for the Green functions of second order elliptic operators.* Univ. Linköping, Inst. für Math. 1972.

[37] ———, *Inequalities for the Green function and boundary continuity of the gradient of solutions of elliptic differential equations.* Math. Scandinav., **21** (1967), 17–37. MR0239264 (39:621)

[38] Wiegner M. *On minima of variational problems with some nonconvex constraints.* Manuscripta Math., **57** (1987), 149–168. MR871628 (88j:49002)

MATH. AND MECH. DEPARTMENT, ST.-PETERSBURG STATE UNIVERSITY, RUSSIA
E-mail address: ARINA@AA1101.SPB.EDU

Amer. Math. Soc. Transl.
(2) Vol. **229**, 2010

A 2D-variant of a Theorem of Uraltseva and Urdaletova for Higher-Order Variational Problems

Michael Bildhauer and Martin Fuchs

Dedicated to Prof. N. Uraltseva on the occasion of her anniversary

ABSTRACT. If Ω is a domain in $\mathbb{R}^2$ and if $u : \Omega \to \mathbb{R}$ locally minimizes the energy

$$\int_\Omega \left[h_1\left(|(\nabla^2 u)_I|\right) + h_2\left(|(\nabla^2 u)_{II}|\right)\right] dx\,,$$

where $(\nabla^2 u)_I, (\nabla^2 u)_{II}$ denotes a decomposition of the Hessian matrix $\nabla^2 u$, then we prove the higher integrability and even the continuity of $\nabla^2 u$ under rather general assumptions imposed on the N-functions h_1, h_2.

1. Introduction

In their paper [**UU**], Uraltseva and Urdaletova established the local Lipschitz regularity of bounded generalized solutions of certain degenerate, nonuniformly elliptic equations. In particular, their result applies to bounded local minimizers of the variational integral

$$\int_\Omega \sum_{i=1}^n \left|\frac{\partial u}{\partial x_i}\right|^{m_i} dx\,, \tag{1.1}$$

Ω denoting a domain in $\mathbb{R}^n, n \geq 2$, provided the exponents m_i satisfy $m_i \geq 2$ together with

$$\max\{m_1, \dots, m_n\} < 2m_i, \quad i = 1, \dots, n\,. \tag{1.2}$$

The reader should note that Giaquinta's counterexample (see [**Gi2**]) involves a functional of the form (1.1) with $m_1 = \cdots = m_{n-1} = 2$, $m_n = 4$, which means that in the limit case of (1.2), unbounded local minimizers can exist (at least if n is large enough), whereas (1.2) together with the boundedness assumption leads to a higher degree of regularity (e.g. local boundedness of the gradient) for this restricted class of local minimizers u of (1.1). However, having a global minimization problem in mind, the hypothesis $u \in L^\infty(\Omega)$ is not so unnatural and usually follows from the

2000 *Mathematics Subject Classification.* Primary 35J35, 49N60, 74G40.
Key words and phrases. Variational problems of higher order, general growth conditions, two-dimensional problems, regularity, splitting functionals.

maximum principle. If the boundedness of u is not required a priori, then Fusco and Sbordone [**FS**] and Marcellini [**Ma1**], [**Ma2**] showed how to get regularity of u under stronger assumptions than (1.2): suppose that the range of anisotropy is limited through an inequality of the form

$$\max\{m_1,\dots,m_n\} < c(n)m_i, \quad i=1,\dots,n, \tag{1.3}$$

for a suitable constant $c(n) > 1$ with $c(n) \to 1$ as $n \to \infty$. Then we still have that $|\nabla u| \in L^\infty_{\mathrm{loc}}(\Omega)$. In our recent paper [**BFZ**] we returned to the point of view of Uraltseva and Urdatelova and proved that their hypothesis $u \in L^\infty(\Omega)$ (or even $u \in L^\infty_{\mathrm{loc}}(\Omega)$) is a very strong assumption in the sense that it already implies higher regularity of u without further restrictions of the form (1.2) or (1.3). The purpose of the present note is the investigation of the regularity problem for the higher-order variant of (1.1) at least in a special case. To be precise let us consider the variational integral

$$I[u,\Omega] = \int_\Omega H(\nabla^2 u)\,dx\,, \tag{1.4}$$

where $\nabla^2 u = (\partial_\alpha\partial_\beta u)_{1\le\alpha,\beta\le n}$ is the Hessian matrix of the function $u:\Omega\to\mathbb{R}$. We here already note that with similar arguments we can replace $\nabla^2 u$ by $\nabla^k u$ for some $k \ge 2$, and that it is also possible to include vectorial functions $u:\Omega\to\mathbb{R}^M$, $M\ge 2$. As a matter of fact a discussion of bounded local I-minimizers now seems to be artificial, since no maximum principle is available for the higher-order case, but as will be outlined below, it is possible to obtain regularity results without extra assumptions on u at least in the 2D-case. So let $n=2$ and suppose that $\nabla^2 u$ is represented by the vector $(\partial_1\partial_1 u, \partial_1\partial_2 u, \partial_2\partial_2 u) =: \xi$. With $(\nabla^2 u)_I$ and $(\nabla^2 u)_{II}$ we denote two arbitrary vectors in $\mathbb{R}^3$ formed of r respectively s entries of ξ filled up by 0, if necessary, where $r,s \in \{0,1,2,3\}$ and where in case $r=0$ or $s=0$ we just have the zero vector in $\mathbb{R}^3$. The only requirement is the following: the set generated by the totality of all entries of $(\nabla^2 u)_I$ and $(\nabla^2 u)_{II}$ contains all entries of ξ; for example:

$$\begin{array}{lll}
(\nabla^2 u)_I = (\partial_1\partial_1 u, 0, 0), & (\nabla^2 u)_{II} = (0, \partial_1\partial_2 u, \partial_2\partial_2 u) & \text{or}\\
(\nabla^2 u)_I = \nabla^2 u, & (\nabla^2 u)_{II} = (0, \partial_1\partial_2 u, 0) & \text{or}\\
(\nabla^2 u)_I = (\partial_1\partial_1 u, \partial_1\partial_2 u, 0), & (\nabla^2 u)_{II} = (0, \partial_1\partial_2 u, \partial_2\partial_2 u) & \text{or}\\
(\nabla^2 u)_I = \nabla^2 u, & (\nabla^2 u)_{II} = (0,0,0), & \text{etc.}
\end{array}$$

Returning to (1.4) we assume that the energy density H is of the form

$$H(\nabla^2 u) = h_1\left(|(\nabla^2 u)_I|\right) + h_2\left(|(\nabla^2 u)_{II}|\right), \tag{1.5}$$

where for instance

$$h_i(t) = (\mu_i + t^2)^{\frac{m_i}{2}}, \quad i = 1,2, \tag{1.6}$$

with $\mu_i \ge 0$ and exponents $m_i \ge 2$. A natural class for local I-minimizers then is the Sobolev space $W^2_{2,\mathrm{loc}}(\Omega)$ (see, e.g., [**Ad**] for a definition of these classes), and in [**BF2**] we proved:

THEOREM 1.1. *Let $u \in W^2_{2,\mathrm{loc}}(\Omega)$ denote a local minimizer of the functional I from (1.4) with H defined according to (1.5). Suppose further that (1.6) holds together with*

$$\max\{m_1, m_2\} < 2\min\{m_1, m_2\}. \tag{1.7}$$

a) *If the nondegenerate case* μ_1, $\mu_2 > 0$ *is considered, then we have* $u \in C^{2,\alpha}(\Omega)$ *for any* $\alpha < 1$.
b) *In the degenerate case we still have* $u \in C^{1,\beta}(\Omega)$ *for any* $\beta < 1$.

REMARK 1.1. *For* $n = 2$, *condition* (1.2) *introduced by Uraltseva and Urdaletova is equivalent to* (1.7).

Here we are going to improve the results of Theorem 1.1 by showing

THEOREM 1.2. *The statements of Theorem* 1.1 *hold for any choices of exponents* m_1, $m_2 \geq 2$ *without requiring the bound* (1.7).

Theorem 1.2 will be a by-product of a more general result dealing with integrands H of splitting-type as described in (1.5) but replacing (1.6) by a larger class of functions h_1 and h_2. To be precise, let

$$H(E) := h_1\left(|(E)_I|\right) + h_2\left(|(E)_{II}|\right) \tag{1.8}$$

for symmetric (2×2)-matrices E with an obvious meaning of $(E)_{I,II}$. Suppose further that the functions h_1, $h_2 : [0, \infty) \to [0, \infty)$ are of class C^2 such that for $h = h_1$ and $h = h_2$ it holds

(A1) $\begin{cases} h \text{ is strictly increasing and convex together with } h''(0) > 0 \text{ and} \\ \lim_{t \downarrow 0} \frac{h(t)}{t} = 0; \end{cases}$

(A2) there exists a constant $\overline{k} > 0$ such that $h(2t) \leq \overline{k} h(t)$ for all $t \geq 0$;

(A3) $\begin{cases} \text{for an exponent } \omega \geq 0 \text{ and a constant } a \geq 0 \text{ it holds} \\ \frac{h'(t)}{t} \leq h''(t) \leq a(1+t^2)^{\frac{\omega}{2}} \frac{h'(t)}{t} \quad \text{for all } t \geq 0. \end{cases}$

Let us draw some conclusions from (A1)–(A3):

i) (A1) implies that $h(0) = 0 = h'(0)$ and $h'(t) > 0$ for $t > 0$. From (A3) it follows that $t \mapsto h'(t)/t$ is increasing; moreover we get $h(t) \geq h''(0)t^2/2$. In particular, h is an N-function (see [**Ad**]) of at least quadratic growth.
ii) The ($\Delta 2$)-property stated in (A2) implies that
$$h(t) \leq c(t^{\overline{m}} + 1)$$
for some exponent $\overline{m} \geq 2$; hence by the convexity of h,
$$h'(t) \leq c(t^{\overline{m}-1} + 1),$$
where here and in the following "c" denotes a constant whose value may vary from line to line.
iii) Combining (A2) with the convexity of h we see that
$$\overline{k}^{-1} h'(t)t \leq h(t) \leq t h'(t), \quad t \geq 0. \tag{1.9}$$
iv) For symmetric (2×2)-matrices Y, Z we have

$$\begin{aligned} &\min\left\{\frac{h_1'(|(Z)_I|)}{|(Z)_I|}, h_1''\left(|(Z)_I|\right)\right\} |(Y)_I|^2 \\ &\quad + \min\left\{\frac{h_2'(|(Z)_{II}|)}{|(Z)_{II}|}, h_2''\left(|(Z)_{II}|\right)\right\} |(Y)_{II}|^2 \\ &\leq \; D^2H(Z)(Y,Y) \leq \max\{\ldots\} \left|(Y)_I\right|^2 + \max\{\ldots\} \left|(Y)_{II}\right|^2, \end{aligned}$$

so that by (A3),

$$\begin{aligned}(1.10)\qquad &\frac{h_1'(|(Z)_I|)}{|(Z)_I|}|(Y)_I|^2+\frac{h_2'(|(Z)_{II}|)}{|(Z)_{II}|}|(Y)_{II}|^2\\ \leq\ & D^2H(Z)(Y,Y)\\ \leq\ & a(1+|(Z)_I|^2)^{\frac{\omega}{2}}\frac{h_1'(|(Z)_I|)}{|(Z)_I|}|(Y)_I|^2\\ &+a(1+|(Z)_{II}|^2)^{\frac{\omega}{2}}\frac{h_2'(|(Z)_{II}|)}{|(Z)_{II}|}|(Y)_{II}|^2,\end{aligned}$$

and for a suitable exponent $\overline{q}>2$, it follows that

$$(1.11)\qquad c|Y|^2\leq D^2H(Z)(Y,Y)\leq C(1+|Z|^2)^{\frac{\overline{q}-2}{2}}|Y|^2,$$

the first inequality being a consequence of i).

Now our main result is

THEOREM 1.3. *Let H satisfy (1.8) with functions h_1, h_2 for which (A1)–(A3) hold. Suppose further that $u\in W^2_{2,\mathrm{loc}}(\Omega)$ locally minimizes the functional I defined in (1.4). Then we have:*

a) *$\nabla^2 u$ belongs to the class $L^p_{\mathrm{loc}}(\Omega;\mathbb{R}^{2\times 2})$ for any finite p; in particular, $u\in C^{1,\alpha}(\Omega)$ for any $0<\alpha<1$.*

b) *If $\omega<2$ in (A3), then we get $u\in C^{2,\alpha}(\Omega)$ for all $\alpha<1$.*

REMARK 1.2. *Of course Theorem 1.3 applies to the special choice of the functions h_i as stated in (1.6) $\mu_i>0$, i.e., we obtain the $C^{2,\alpha}$-regularity result of Theorem 1.2 in the nondegenerate situation. To be precise, one has to replace h_i by $h_i-h_i(0)$, but this does not affect the arguments. The proof of Theorem 1.3 a) for the case that (1.6) holds with $\mu_i=0$ is left to the reader.*

REMARK 1.3. *Variational integrals of the form (1.4) defined on two-dimensional domains have some relation to the theory of elastic plates. For a discussion of this issue we refer to the paper* [**Fu**]*, in which Theorem 1.3 a) is established for the isotropic energy $\int_\Omega h(|\nabla^2 u|)\,dx$ with h satisfying (A1)–(A3). Of course the results of this paper do not apply to the situation at hand, since now our energy density is of the splitting form (1.8) showing a completely different ellipticity behaviour in comparison to the isotropic case studied in* [**Fu**].

2. Higher integrability of $\nabla^2 u$

Here we are going to prove Theorem 1.3 a). So let u denote a local I-minimizer and fix domains $\Omega_1\Subset\Omega_2\Subset\Omega$. Proceeding as in [**BF1**], [**BF2**], we denote by $\overline{u}_m$ the mollification of u with radius $1/m$, $m\in\mathbb{N}$; in particular, we have

$$\|\overline{u}_m-u\|_{W^2_2(\Omega_2)}\to 0$$

as $m\to\infty$. Moreover it follows that (compare (2.1) in [**BF1**])

$$I[\overline{u}_m,\Omega_2]\to I[u,\Omega_2]\,.$$

Recalling that on account of (1.11) the hypothesis (1.1) of [**BF1**] now holds for $p=2$, $q=\overline{q}$ we may define

$$\rho_m:=\|\overline{u}_m-u\|_{W^2_2(\Omega_2)}\left[\int_{\Omega_2}(1+|\nabla^2\overline{u}_m|^2)^{\frac{\overline{q}}{2}}\,dx\right]^{-1}$$

as well as the perturbed energy

$$I_m[w,\Omega_2] := \rho_m \int_{\Omega_2} (1+|\nabla^2 w|^2)^{\frac{\overline{q}}{2}}\, dx + I[w,\Omega_2]$$

with density

$$H_m := \rho_m (1+|\cdot|^2)^{\frac{\overline{q}}{2}} + H\,.$$

Finally we consider the unique solution u_m of

$$I_m[\cdot,\Omega_2] \to \min \text{ in } \overline{u}_m + \overset{\circ}{W}{}^2_{\overline{q}}(\Omega_2)$$

for which it was shown in [**BF1**] that

$$u_m \rightharpoonup u \text{ in } W^2_2(\Omega_2)\,, \quad I_m[u_m,\Omega_2] \to I[u,\Omega_2] \tag{2.1}$$

as $m \to \infty$. Moreover we proved in [**BF1**] (compare the inequality stated after (2.13)) the validity of

$$\begin{aligned}(2.2)\quad &\int_{\Omega_2} \eta^6 D^2 H_m(\nabla^2 u_m)(\partial_\alpha \nabla^2 u_m, \partial_\alpha \nabla^2 u_m)\, dx \\ &\leq -\int_{\Omega_2} D^2 H_m(\nabla^2 u_m)(\partial_\alpha \nabla^2 u_m, \nabla^2\eta^6 \partial_\alpha u_m + 2\nabla\eta^6 \odot \nabla\partial_\alpha u_m)\, dx\,,\end{aligned}$$

where $\eta \in C_0^\infty(\Omega_2)$ is arbitrary and where we use the summation convention w.r.t. greek indices repeated twice. In (2.2), "$\odot$" denotes the symmetric product of vectors, and we can justify (2.2) by an application of the difference quotient technique to the Euler equation satisfied by u_m. We note that the Caccioppoli-type inequality (2.2) also occurs in [**BF2**] (compare inequality (4.1)), being established along the same lines as in [**BF1**], but here we are going to exploit (2.2) in a completely different manner, leading to the improvement of the result from [**BF2**], which we mentioned before.

For notational simplicity we will drop the index m; i.e., we write u, H, I in place of u_m, H_m, I_m, but the reader should keep in mind that we actually work with an approximation. However, since we will prove estimates involving Ω_1 on the left-hand sides and with quantities such as $I_m[u_m,\Omega_2]$ on the right-hand sides, uniform bounds on Ω_1 will be a consequence of (2.1).

Now after these preparations we integrate by parts on the r.h.s. of (2.2) in order to get

$$\begin{aligned}(2.3)\quad &\int_{\Omega_2} \eta^6 D^2 H(\nabla^2 u)(\partial_\alpha \nabla^2 u, \partial_\alpha \nabla^2 u)\, dx \\ &\leq \int_{\Omega_2} DH(\nabla^2 u) : \partial_\alpha \left[\nabla^2\eta^6 \partial_\alpha u + 2\nabla\eta^6 \odot \nabla\partial_\alpha u\right] dx\,.\end{aligned}$$

Note that this integration by parts is justified since the "critical term" occurring in $DH(\nabla^2 u) : \partial_\alpha[\ldots]$ is of the form $|DH(\nabla^2 u)||\nabla^3 u|$. But since the r.h.s. of (2.2) is finite (for each m), we deduce from (1.11) that u $(= u_m)$ is of class $W^3_{2,\mathrm{loc}}(\Omega_2)$ and since $n = 2$, it follows that $|\nabla^2 u|$ is in $L^t_{\mathrm{loc}}(\Omega_2)$ for any finite t. Recalling that $|DH|$ is bounded in terms of a suitable power, the local integrability of $|D^2 H(\nabla^2 u)||\nabla^3 u|$

follows (but at this stage it is not necessarily uniform in m). Let us discuss the r.h.s. of (2.3): from (1.8) we get

$$\begin{aligned}
|\text{ r.h.s. of (2.3)}| &\leq c\int_{\Omega_2}\{h_1'\left(|(\nabla^2 u)_I|\right)+h_2'\left(|(\nabla^2 u)_{II}|\right)\}\\
&\quad\cdot\left[|\nabla^3\eta^6||\nabla u|+|\nabla^2\eta^6||\nabla^2 u|+|\nabla\eta^6||\nabla^3 u|\right]\,dx\\
&= c\Bigg(\int_{\Omega_2}\{\ldots\}|\nabla\eta^6||\nabla^3 u|\,dx\\
&\quad+\int_{\Omega_2}\{\ldots\}|\nabla^2\eta^6||\nabla^2 u|\,dx+\int_{\Omega_2}\{\ldots\}|\nabla^3\eta^6||\nabla u|\,dx\Bigg)\\
&=: c(T_1+T_2+T_3)\,.
\end{aligned}$$

To the terms T_i we apply Young's inequality:

$$T_1\leq\varepsilon\int_{\Omega_2}\eta^6|\nabla^3 u|^2\,dx+c(\varepsilon)\int_{\Omega_2}|\nabla\eta|^2\{\ldots\}^2\,dx\,,$$

where ε is arbitrary and where w.l.o.g. we assume $0\leq\eta\leq 1$. On account of (1.11), the first item on the r.h.s. of the above inequality can be absorbed in the l.h.s. of (2.3), provided ε is small enough. Here we emphasize that the value of ε can be chosen to be independent of the approximation parameter m. For T_2 we have

$$T_2\leq\int_{\Omega_2}|\nabla^2 u|^2\,dx+\int_{\Omega_2}|\nabla^2\eta^6|^2\{\ldots\}^2\,dx$$

and by (2.1) the first integral on the r.h.s. is bounded independently of m. For T_3 we argue in a similar way by observing that $u_m-\overline{u}_m$ is in the space $\overset{\circ}{W}{}^2_2(\Omega_2)$; hence we can apply Poincaré's inequality to get a uniform $L^2(\Omega_2)$-bound for $|\nabla u_m|$ in terms of the original energy. If we finally fix concentric discs $B_r(z)\subset B_R(z)\Subset\Omega_2$ and let $\eta=1$ on $B_r(z)$, $\operatorname{spt}\eta\subset B_R(z)$, $|\nabla^l\eta|\leq c(R-r)^{-l}$, $l=1$, 2, then we obtain from the above estimates,

$$\begin{aligned}
(2.4)\qquad &\int_{B_r(z)} D^2H(\nabla^2 u)(\partial_\alpha\nabla^2 u,\partial_\alpha\nabla^2 u)\,dx\\
&\leq c\Bigg(I[u,\Omega_2]\\
&\quad+(R-r)^{-\beta}\int_{B_R(z)}\{h_1'\left(|(\nabla^2 u)_I|\right)+h_2'\left(|(\nabla^2 u)_{II}|\right)\}^2\,dx\Bigg)\,.
\end{aligned}$$

Here β is a suitable positive exponent and c denotes a positive constant, both being independent of m. Let us have a closer look at the integrals involving $h_{1,2}'$ on the r.h.s. of (2.4): we have

$$\begin{aligned}
&\int_{B_R(z)} h_1'\left(|(\nabla^2 u)_I|\right)^2\,dx\\
&= \int_{B_R(z)\cap[|(\nabla^2 u)_I|\leq L]}\ldots dx+\int_{B_R(z)\cap[|(\nabla^2 u)_I|\geq L]}\ldots dx\\
&\leq h_1'(L)^2 2\pi R^2+cL^{-2}\int_{B_R(z)\cap[|(\nabla^2 u)_I|\geq L]} h_1^2\left(|(\nabla^2 u)_I|\right)\,dx\,,
\end{aligned}$$

where we have used inequality (1.9) for h_1 and where $L > 0$ is arbitrary. Obviously the same estimate is valid for h_2, and if we choose

$$L = \lambda^{-1}(R-r)^{-\frac{\beta}{2}}$$

for a parameter $\lambda > 0$ and recall the estimate of h_i' in terms of the power $\overline{m}-1$, then we get from (2.4) and the above inequalities

$$(2.5)\int_{B_r(z)} D^2H(\nabla^2 u)(\partial_\alpha \nabla^2 u, \partial_\alpha \nabla^2 u)\, dx \leq c\left[c(\lambda)(R-r)^{-\overline{\beta}} + \lambda^2 \int_{B_R(z)} \{h_1^2(|(\nabla^2 u)_I|)] + h_2^2(|(\nabla^2 u)_{II}|)\}\, dx\right]$$

for a new positive exponent $\overline{\beta}$ and a constant c involving the energy of u on Ω_2.

Suppose now that we have fixed a disc $B_R(z) \Subset \Omega_2$. If $\rho \in (0,R)$, we then let $r := \frac{1}{2}(\rho + R)$ and choose $\eta \in C_0^\infty(B_r(z))$ such that $0 \leq \eta \leq 1$, $\eta \equiv 1$ on $B_\rho(z)$, $|\nabla \eta| \leq c/(r-\rho)(= 2c/(R-\rho))$. Sobolev's inequality yields

$$\begin{aligned}
&\int_{B_\rho(z)} \left[h_1(|(\nabla^2 u)_I|)^2 + h_2(|(\nabla^2 u)_{II}|)^2\right] dx \\
&\leq \int_{B_r(z)} \left[(\eta h_1(|(\nabla^2 u)_I|))^2 + (\eta h_2(|(\nabla^2 u)_{II}|))^2\right] dx \\
&\leq c\left[\int_{B_r(z)} |\nabla \eta| \{h_1(|(\nabla^2 u)_I|) + h_2(|(\nabla^2 u)_{II}|)\}\, dx\right. \\
&\qquad \left. + \int_{B_r(z)} h_1'(|(\nabla^2 u)_I|)\, |\nabla(\nabla^2 u)_I|\, dx + \int_{B_r(z)} h_2'(|(\nabla^2 u)_{II}|)\, |\nabla(\nabla^2 u)_{II}|\, dx\right]^2 \\
&\leq c(R-\rho)^{-2}\left(\int_{B_R(z)} H(\nabla^2 u)\, dx\right)^2 \\
&\qquad + c\left[\int_{B_r(z)} h_1'(|(\nabla^2 u)_I|)|\nabla(\nabla^2 u)_I|\, dx\right. \\
&\qquad \left. + \int_{B_r(z)} h_2'(|(\nabla^2 u)_{II}|)|\nabla(\nabla^2 u)_{II}|\, dx\right]^2.
\end{aligned}$$

To the quantity $[\dots]^2$ we apply Hölder's ineqality in combination with (1.9):

$$\begin{aligned}
[\dots]^2 &\leq \int_{B_r(z)} \frac{h_1'(|(\nabla^2 u)_I|)}{|(\nabla^2 u)_I|}|\nabla(\nabla^2 u)_I|^2\, dx \int_{B_r(z)} h_1'(|(\nabla^2 u)_I|)|(\nabla^2 u)_I|\, dx \\
&\quad + \int_{B_r(z)} \frac{h_2'(|(\nabla^2 u)_{II}|)}{|(\nabla^2 u)_{II}|}|\nabla(\nabla^2 u)_{II}|^2\, dx \int_{B_r(z)} h_2'(|(\nabla^2 u)_{II}|)|(\nabla^2 u)_{II}|\, dx \\
&\leq c\int_{\Omega_2} H(\nabla^2 u)\, dx \left\{\int_{B_r(z)} \frac{h_1'(|(\nabla^2 u)_I|)}{|(\nabla^2 u)_I|}|\nabla(\nabla^2 u)_I|^2\, dx\right. \\
&\qquad \left. + \int_{B_r(z)} \frac{h_2'(|(\nabla^2 u)_{II}|)}{|(\nabla^2 u)_{II}|}|\nabla(\nabla^2 u)_{II}|^2\, dx\right\}.
\end{aligned}$$

If we use the first inequality from (1.10) with the choices $Z = \nabla^2 u$ and $Y = \partial_1 \nabla^2 u, \partial_2 \nabla^2 u$ and add the results, then we obtain

$$\{\ldots\} \le \int_{B_r(z)} D^2 H(\nabla^2 u)(\partial_\alpha \nabla^2 u, \partial_\alpha \nabla^2 u)\, dx;$$

hence

$$\begin{aligned}\int_{B_\rho(z)} & \left[h_1\left(|(\nabla^2 u)_I|\right)^2 + h_2\left(|(\nabla^2 u)_{II}|\right)^2\right] dx \\ &\le c\left\{(R-\rho)^{-2} + \int_{B_r(z)} D^2 H(\nabla^2 u)(\partial_\alpha \nabla^2 u, \partial_\alpha \nabla^2 u)\, dx\right\},\end{aligned}$$

and by applying (2.5) we find that (w.l.o.g. $\overline{\beta} \ge 2$)

$$\begin{aligned}(2.6)\int_{B_\rho(z)} & \left[h_1\left(|(\nabla^2 u)_I|\right)^2 + h_2\left(|(\nabla^2 u)_{II}|\right)^2\right] dx \\ &\le c\left[c(\lambda)(R-\rho)^{-\overline{\beta}} + \lambda^2 \int_{B_R(z)} \{h_1(|(\nabla^2 u)_I|)^2 + h_2(|(\nabla^2 u)_{II}|)^2\}\, dx\right]\end{aligned}$$

is valid for all discs $B_\rho(z) \subset B_R(z) \Subset \Omega_2$ and any $\lambda > 0$. Choosing $\lambda = 1/\sqrt{2c}$, a well-known lemma (see [**Gi1**], Lemma 3.1, p. 161) applies to (2.6) with the result that

$$h_1(|(\nabla^2 u)_I|)^2 + h_2(|(\nabla^2 u)_{II}|)^2 \in L^1_{\mathrm{loc}}(\Omega_2)$$

is true uniformly w.r.t. the hidden parameter m. But then (2.5) shows the same for $D^2 H(\nabla u)(\partial_\alpha \nabla^2 u, \partial_\alpha \nabla^2 u)$, and (1.11) implies that $u \in W^3_{2,\mathrm{loc}}(\Omega_2)$, which proves part a) of Theorem 1.3 by quoting Sobolev's embedding theorem one more time. □

3. Hölder continuity of the second derivatives

Assume now that the hypothesis of Theorem 1.3 b) holds. Keeping the notation from Section 2 and referring again to [**BF1**], [**BF2**], we first observe that u (more precisely u_m) can be replaced on the r.h.s. of (2.2) by $u - P$, where P is a polynomial of degree ≤ 2. Letting

$$\Phi^2 := D^2 H(\nabla^2 u)(\partial_\alpha \nabla^2 u, \partial_\alpha \nabla^2 u), \quad \sigma := DH(\nabla^2 u)$$

and choosing η such that $\eta = 1$ on $B_r(z_0)$, $\operatorname{spt}\eta \subset B_{2r}(z_0)$, $0 \le \eta \le 1$, $|\nabla^l \eta| \le c/r^l$, $l = 1, 2$, for a disc $B_{2r}(z_0) \Subset \Omega_1$, we obtain from (2.2),

$$\begin{aligned}(3.1) \quad \int_{B_{2r}(z_0)} & \eta^6 \Phi^2\, dx \\ &\le c \int_{B_{2r}(z_0)} |\nabla \sigma| \left[|\nabla^2 \eta^6||\nabla u - \nabla P| + |\nabla \eta^6||\nabla^2 u - \nabla^2 P|\right] dx\,.\end{aligned}$$

The Cauchy-Schwarz inequality applied to the bilinear form $D^2 H(\nabla^2 u)$ implies that

$$\begin{aligned}|\nabla \sigma|^2 &= D^2 H(\nabla^2 u)(\partial_\alpha \nabla^2 u, \partial_\alpha \sigma) \\ &\le D^2 H(\nabla^2 u)(\partial_\alpha \nabla^2 u, \partial_\alpha \nabla^2 u)^{1/2} D^2 H(\nabla^2 u)(\partial_\alpha \sigma, \partial_\alpha \sigma)^{1/2} \\ &\le \Phi |D^2 H(\nabla^2 u)|^{1/2} |\nabla \sigma|\,,\end{aligned}$$

i.e. $|\nabla\sigma| \le \Phi|D^2H(\nabla^2 u)|^{1/2}$, and the second inequality from (1.10) gives

$$\begin{aligned} |D^2H(\nabla^2 u)|^{1/2} &\le c\left[(1+|(\nabla^2 u)_I|^2)^{\frac{\omega}{4}}\sqrt{\frac{h_1'(|(\nabla^2 u)_I|)}{|(\nabla^2 u)_I|}}\right. \\ &\quad \left. + (1+|(\nabla^2 u)_{II}|^2)^{\frac{\omega}{4}}\sqrt{\frac{h_2'(|(\nabla^2 u)_{II}|)}{|(\nabla^2 u)_{II}|}}\right] \\ &=: c\left[\widetilde{\Psi}_1 + \widetilde{\Psi}_2\right]. \end{aligned}$$

Inserting these estimates into (3.1) we obtain, by letting $\widetilde{\Psi} := \left(\widetilde{\Psi}_1^2 + \widetilde{\Psi}_2^2\right)^{1/2}$,

$$\begin{aligned} \int_{B_r(z_0)} \Phi^2\,dx &\le c\left[\frac{1}{r}\int_{B_{2r}(z_0)} |\nabla^2 u - \nabla^2 P|\Phi\widetilde{\Psi}\,dx\right. \\ &\quad \left. + \frac{1}{r^2}\int_{B_{2r}(z_0)} |\nabla u - \nabla P|\Phi\widetilde{\Psi}\,dx\right]. \end{aligned} \tag{3.2}$$

Letting $\gamma = 4/3$ we can now follow the calculations in [**BF1**] leading from (2.18) to (2.21) in this reference, which means that after an appropriate choice of P and proper applications of the Sobolev-Poincaré and the Poincaré inequality on the r.h.s. of (3.2) we obtain from (3.2) the basic estimate

$$\left[\fint_{B_r(z_0)} \Phi^2\,dx\right]^{\frac{1}{2}} \le c\left[\fint_{B_{2r}(z_0)} (\Phi\widetilde{\Psi})^\gamma\,dx\right]^{\frac{1}{\gamma}}. \tag{3.3}$$

We recall one more time that (3.3) is actually valid for the approximations u_m, i.e. we have $\Phi = \Phi_m$, etc., but the constant c appearing in (3.3) is independent of m. Let us also note that during the derivation of (3.3) one needs the information that

$$|\nabla^3 u| \le c\Phi \le c\Phi\Psi,$$

which follows from (1.11). In order to continue as outlined after (2.21) in [**BF1**] we only have to check that

$$\exp(\beta\widetilde{\Psi}^2) \in L^1(\Omega_1) \tag{3.4}$$

holds for any $\beta > 0$, since $\Phi \in L^2(\Omega_1)$ (uniformly w.r.t. the index m) has already been shown in Section 2. Let us introduce the auxiliary functions

$$\Psi_1 := \int_0^{|(\nabla^2 u)_I|}\sqrt{\frac{h_1'(t)}{t}}\,dt, \quad \Psi_2 := \int_0^{|(\nabla^2 u)_{II}|}\sqrt{\frac{h_2'(t)}{t}}\,dt,$$

for which we have by the first inequality in (1.10),

$$|\nabla\Psi_1|^2 + |\nabla\Psi_2|^2 \le c\Phi^2;$$

moreover (1.9) implies that

$$\Psi_1^2 + \Psi_2^2 \le cH(\nabla^2 u),$$

so that Ψ_1, Ψ_2 belong to $W^1_2(\Omega_1)$ and therefore $\Psi := (\Psi_1^2 + \Psi_2^2)^{1/2}$ is in the same space (uniform in m). Thus we can apply Trudinger's inequality (see [**GT**], Theorem 7.15) and find $\beta_0 > 0$ such that for discs $B_\rho \subset \Omega_1$ we have

$$\int_{B_\rho} \exp(\beta_0 \Psi^2)\, dx \le c(\rho)\,. \tag{3.5}$$

On the set $[|(\nabla^2 u)_I| \ge 1]$ it follows that (recall (1.9))

$$\widetilde{\Psi}_1 \le c|(\nabla^2 u)_I|^{\frac{\omega}{2}-1} h_1(|(\nabla^2 u)_I|)^{\frac{1}{2}}\,,$$

whereas

$$\Psi_1 \ge \int_{|(\nabla^2 u)_I|/2}^{|(\nabla^2 u)_I|} \sqrt{\frac{h_1'(t)}{t}}\, dt \ge c h_1(|(\nabla^2 u)_I|)^{\frac{1}{2}};$$

thus

$$\widetilde{\Psi}_1 \le c|(\nabla^2 u)_I|^{\frac{\omega}{2}-1} \Psi_1 \quad \text{on} \quad [|(\nabla^2 u)_I| \ge 1]\,.$$

From ii) after (A3) we obtain

$$\Psi_1 \le c h_1(|(\nabla^2 u)_I|^{\frac{1}{2}} \le c|(\nabla^2 u)_I|^{\frac{\overline{m}}{2}}\,,$$

and for $\delta > 0$ it follows that

$$\widetilde{\Psi}_1 \le c\Psi_1^{1-\delta} |(\nabla^2 u)_I|^{\frac{\omega}{2}-1+\delta\frac{\overline{m}}{2}}$$

on $[|(\nabla^2 u)_I| \ge 1]$. Since we assume $\omega < 2$ we can fix $\delta > 0$ such that

$$\frac{\omega}{2} - 1 + \delta \frac{\overline{m}}{2} < 0$$

and Young's inequality gives for any $\mu > 0$,

$$\widetilde{\Psi}_1^2 \le \mu \Psi_1^2 + c(\mu)$$

on the relevant set.

On $[|(\nabla^2 u)_I| \le 1]$ this inequality is immediate, and clearly the same arguments apply to $\Psi_2, \widetilde{\Psi}_2$; hence we have a.e.

$$\widetilde{\Psi}^2 \le \mu \Psi^2 + c(\mu)\,, \tag{3.6}$$

and (3.4) follows from (3.5) and (3.6) with $\mu := \beta_0/\beta$. Now the proof of Theorem 1.3 b) can be completed exactly with the same arguments as applied in [**BF1**], p. 361. □

References

[Ad] Adams, R. A., *Sobolev spaces.* Academic Press, New York-San Francisco-London, 1975. MR0450957 (56:9247)

[BF1] Bildhauer, M. and Fuchs, M., *Higher-order variational problems on two-dimensional domains.* Ann. Acad. Sci. Fenn. Math. **31** (2006), 349–362. MR2248820 (2007d:49004)

[BF2] ———, *On the regularity of local minimizers of decomposable variational integrals on domains in* $\mathbb{R}^2$. Comment. Math. Univ. Carolin. **48** (2007), 321–341. MR2338100 (2008m:49187)

[BFZ] Bildhauer, M., Fuchs, M., and Zhong, X., *A regularity theory for scalar local minimizers of splitting-type variational integrals.* Ann. Scuola Norm. Sup. Pisa, Serie V, Vol. VI, Fasc. 3 (2007), 385–404. MR2370266 (2008j:49092)

[Fu] Fuchs, M., *Minimization of energies related to the plate problem.* Math. Meth. Appl. Sciences. **32** (2009), 773–782.

[FS] Fusco, N. and Sbordone, C., *Some remarks on the regularity of minima of anisotropic integrals.* Commun. Partial Diff. Equations **18** (1993), 153–167. MR1211728 (94e:49013)

[Gi1] Giaquinta, M., *Multiple integrals in the calculus of variations and nonlinear elliptic systems.* Ann. Math. Studies, vol. 105, Princeton Univ. Press, Princeton, NJ, 1983. MR717034 (86b:49003)

[Gi2] Giaquinta, M., *Growth conditions and regularity, a counterexample.* Manuscripta Math. **59** (1987), 245–248. MR905200 (88h:49034)

[GT] Gilbarg, D. and Trudinger, N. S., *Elliptic partial differential equations of second order.* Second ed., Springer-Verlag, Berlin-Heidelberg-New York, 1983. MR0737190 (86c:35035)

[Ma1] Marcellini, P., *Regularity of minimizers of integrals of the calculus of variations with non-standard growth conditions.* Arch. Rat. Mech. Anal. **105** (1989), 267–284. MR969900 (90a:49017)

[Ma2] Marcellini, P., *Everywhere regularity for a class of elliptic systems without growth conditions.* Ann. Scuola Norm. Sup. Pisa **23** (1996), 1–25. MR1401415 (97h:35048)

[UU] Ural'tseva, N.N. and Urdaletova, A.B., *The boundedness of the gradients of generalized solutions of degenerate quasilinear nonuniformly elliptic equations.* Vestn. Leningr. Univ. 1983, Mat. Mekh. Astron. no. 4 (1983), 50–56; English transl., Vestn. Leningr. Univ. Math **16** (1984), 263–270. MR725829 (85g:35048)

SAARLAND UNIVERSITY, DEPARTMENT OF MATHEMATICS, P.O. BOX 15 11 50, D–66041 SAARBRÜCKEN, GERMANY

E-mail address: bibi@math.uni-sb.de

E-mail address: fuchs@math.uni-sb.de

Amer. Math. Soc. Transl.
(2) Vol. **229**, 2010

The Linear Boltzmann Equation with Space Periodic Electric Field

Mihai Bostan, Irene M. Gamba, and Thierry Goudon

This paper is dedicated to Nina Uraltceva for her 75th birthday.

ABSTRACT. We investigate the well posedness of the stationary linear Boltzmann equation with space periodic electric field. We discuss the different behaviors that occur depending on whether the average electric field vanishes or not. The existence follows by perturbation techniques and stability arguments under uniform a priori estimates. The uniqueness of the weak solution holds for space periodic electric fields with nonvanishing average, under the constraint of the given current. The main ingredients of the proof rely on the dissipation properties of the linear collision operator and the derivation of refined estimates.

Introduction

This paper is concerned with the free space linear Boltzmann equation

$$v(p)\partial_x f + F(x)\partial_p f - Q(f), \qquad (x,p) \in \mathbb{R}^2. \tag{0.1}$$

The unknown $f = f(x,p)$ represents the number density of a population of charged particles, with $x \in \mathbb{R}$ the space variable and $p \in \mathbb{R}$ the momentum variable. The velocity $p \mapsto v(p)$ is defined by

$$v(p) = \frac{p}{m}\left(1 + \frac{p^2}{m^2c_0^2}\right)^{-1/2}, \tag{0.2}$$

where m is the mass of the particles and c_0 is the speed of light in a vacuum. The kinetic energy associated to $v(p)$ is then given by

$$\mathcal{E}(p) = mc_0^2\left(\left(1 + \frac{p^2}{m^2c_0^2}\right)^{1/2} - 1\right)$$

so that $\mathcal{E}'(p) = v(p), p \in \mathbb{R}$. The collision operator Q, which is an integral operator with respect to the variable p, is defined as follows:

$$Q(g)(p) = M_\theta(p)\langle g\rangle_s(p) - \sigma(p)g(p), \qquad p \in \mathbb{R}$$

2000 *Mathematics Subject Classification.* Primary 82D10, 78A35, 35Q99.
Key words and phrases. Stationary transport equations, plasma physics models.

with
(0.3)

$$\langle g\rangle_s(p) = \int_{\mathbb{R}} s(p,p')g(p')\,\mathrm{d}p' \qquad \text{and} \qquad \sigma(p) = \int_{\mathbb{R}} s(p,p')M_\theta(p')\,\mathrm{d}p' = \langle M_\theta\rangle_s(p)$$

and

$$M_\theta(p) = \exp\left(-\frac{\mathcal{E}(p)}{\theta}\right)\left(\int_{\mathbb{R}} \exp\left(-\frac{\mathcal{E}(p')}{\theta}\right)\,\mathrm{d}p'\right)^{-1}. \tag{0.4}$$

The force field is given by $F = qE$, where q is the charge of the particles and E is the electric field the particles are subject to.

The definition (0.2) means that we are dealing with relativistic particles, and it will be crucial for the analysis to observe that

$$-c_0 < v(p) < c_0. \tag{0.5}$$

We also remark that

$$mc_0^2 + \mathcal{E}(p) \geq c_0|p| \geq v(p)p \geq 0. \tag{0.6}$$

Throughout the paper the scattering cross section $s(p,p')$ is required to satisfy

$$s(p,p') = s(p',p)\,, \qquad 0 < s_0 \leq s(p,p') \leq s_1 < +\infty, \qquad p,p' \in \mathbb{R}. \tag{0.7}$$

Consequently, for any integrable function g we have

$$0 < \langle g\rangle s_0 \leq \langle g\rangle_s(p) \leq \langle g\rangle s_1 < +\infty, \qquad p \in \mathbb{R},$$

and in particular the collision frequency $\sigma(\cdot)$ satisfies

$$0 < s_0 \leq \sigma(p) \leq s_1 < +\infty, \qquad p \in \mathbb{R}.$$

The equation (0.1) models charge transport phenomena, with applications in semiconductor theory or plasma physics. We refer to [**4**] for further details on the model as well as for the basis of its mathematical analysis. The boundary value problem for (0.1) has been studied in [**6**]. The analysis of [**6**] uses comparison principles based on the observation that

$$\mathcal{M}_{\theta,\phi}(x,p) = \exp\left(-\frac{\mathcal{E}(p) + q\phi(x)}{\theta}\right), \qquad \phi' = -E$$

is a particular solution for (0.1), since this function makes vanish both the transport operator $v(p)\partial_x + F(x)\partial_p$ and the collision operator Q. Hence, existence results for the boundary value problem associated to (0.1) can be established dealing with incoming data comparable with $\mathcal{M}_{\theta,\phi}$. Existence theory was recently extended to general integrable data in [**3**].

The aim of this article is to analyze the free space problem (0.1). As said before, the function $\mathcal{M}_{\theta,\phi}$, and obviously all the multiple of $\mathcal{M}_{\theta,\phi}$, are solutions of (0.1). However we shall see that the equation admits other solutions. This can already be seen in the specific case where the electric field and the scattering function are constant: if $E(x) = E \neq 0$ and $s(p,p') = \frac{1}{\tau}$, τ being the relaxation time, we can find particle densities f depending only on the momentum by solving analytically the ordinary differential equation

$$\frac{1}{\tau}f(p) + qE\frac{df}{dp} = \frac{1}{\tau}M_\theta(p), \qquad \int_{\mathbb{R}} f(p)\,\mathrm{d}p = 1. \tag{0.8}$$

This example already appears in [**5**] in the nonrelativistic case (i.e., $v(p) = p/m$), and the properties of the solution can be used to investigate the associated boundary value and Milne problems; see [**5**], [**2**]. This remark raises the question of selecting the physically relevant solution of (0.1) by some appropriate criterion. Still in the specific case of a constant nonvanishing electric field, it seems reasonable to select the solution given by (0.8) which remains bounded on $\mathbb{R}^2$, while the distribution $\mathcal{M}_{\theta,\phi}$ becomes unbounded (as $x \to -\infty$ if $qE > 0$ and as $x \to +\infty$ if $qE < 0$). In some sense, the lack of boundary conditions has to be compensated by imposing a uniform behaviour of the solution with respect to the space variable, at least when dealing with bounded electric fields. We shall call such bounded solutions "permanent regimes" (with respect to the space variable). Observe also that since (0.1) is linear we can only expect uniqueness up to a multiplicative constant. Eventually this constant can be determined by imposing the current $j = q\int_{\mathbb{R}} v(p) f \, \mathrm{d}p$ since this quantity does not depend on the space variable: we have

$$\frac{\mathrm{d}}{\mathrm{d}x}\int_{\mathbb{R}} v(p) f \, \mathrm{d}p = 0, \qquad x \in \mathbb{R},$$

due to the conservative character of the collision operator. Hence, a legitimate uniqueness result for permanent regimes is

Uniqueness. *Let f, g be two permanent solutions for (0.1) having the same current*

$$q\int_{\mathbb{R}} v(p) f \, \mathrm{d}p = q\int_{\mathbb{R}} v(p) g \, \mathrm{d}p.$$

Then the solutions f, g coincide.

Having defined relevant criteria for uniqueness, we are nevertheless still left with the task of establishing the existence of such permanent solutions, in the general case of space-varying electric fields. In this work we restrict ourselves to the situation where the electric field is space periodic and we seek space periodic solutions. This does not answer the general question we address, but to our knowledge it is the first work in this direction. Besides the physical relevance of this case, the study is interesting from the mathematical point of view. Moreover we expect that similar results can be established for more general frameworks, such as the almost periodic one, by adapting the same techniques. Our main result is the following.

We assume that $F(x) - qE(x)$ is a given L-periodic bounded force field. We consider the periodic domain $\mathbb{T} = \mathbb{R}/(L\mathbb{Z})$.

THEOREM 0.1. *Assume that $E \in L^\infty(\mathbb{R})$ is a bounded L-periodic electric field.*

a) *If $\int_{\mathbb{T}} E(x) \, \mathrm{d}x = 0$, then all the periodic solutions of the linear Boltzmann equation (0.1) are of the form $k\mathcal{M}_{\theta,\phi}$ with $k \in \mathbb{R}$.*

b) *If $\int_{\mathbb{T}} E(x) \, \mathrm{d}x \neq 0$, then there exists a periodic weak solution f of the linear Boltzmann equation (0.1) such that $f(x,p) \geq 0$ and $\frac{1}{L}\int_{\mathbb{T}}\int_{\mathbb{R}} f(x,p) \, \mathrm{d}p \, \mathrm{d}x = 1$. Moreover, this solution satisfies*

$$\begin{cases} (1+\mathcal{E}(p))f \in L^1([0,L]\times\mathbb{R}) = L^1(\mathbb{T}\times\mathbb{R}), \\ f \in L^\infty([0,L]\times\mathbb{R}) = L^\infty(\mathbb{T}\times\mathbb{R}), \\ \langle f\rangle := \int_{\mathbb{R}} f \, \mathrm{d}p \in L^\infty(\mathbb{T}). \end{cases}$$

c) *The current associated to the solution exhibited in b) does not vanish and we have*

$$q\int_{\mathbb{R}} v(p)\, f(x,p) \, \mathrm{d}p \int_{\mathbb{T}} E(x) \, \mathrm{d}x > 0.$$

d) *Assume* $\int_{\mathbb{T}} E(x)\,\mathrm{d}x \neq 0$. *For any* $j \in \mathbb{R}$ *there is a unique periodic weak solution* f *of the linear Boltzmann equation* (0.1) *satisfying* $q\int_{\mathbb{R}} v(p) f\,\mathrm{d}p = j$. *The solution has constant sign, given by* $\operatorname{sgn}(f) = \operatorname{sgn}\left(j/\int_{\mathbb{T}} E(x)\,\mathrm{d}x\right)$.

Our paper is organized as follows. In Section 1 we set up a few definitions that will be necessary for the existence-uniqueness theory. Section 2 is devoted to the uniqueness result. The proof is based on new sharp dissipative properties for the linear collision operator. In the last section we discuss the existence of a periodic weak solution: we analyze a penalized periodic problem, we establish a priori estimates and we conclude the section with stability arguments.

1. Weak solutions

Let us introduce the corresponding notion of a weak solution for (0.1).

DEFINITION 1.1. Assume that F belongs to $L^\infty(\mathbb{T})$. We say that $f \in L^1(\mathbb{T}\times\mathbb{R})$ is a periodic weak solution for (0.1) iff

$$-\int_{\mathbb{T}}\int_{\mathbb{R}} f(x,p)(v(p)\partial_x\varphi + F(x)\partial_p\varphi)\,\mathrm{d}p\,\mathrm{d}x = \int_{\mathbb{T}}\int_{\mathbb{R}} Q(f)\varphi(x,p)\,\mathrm{d}p\,\mathrm{d}x \tag{1.1}$$

for any function $\varphi \in C_c^1(\mathbb{T}\times\mathbb{R})$.

It is easily seen that the formulation (1.1) holds true for any test function $\varphi \in C_b^1(\mathbb{T}\times\mathbb{R})$ (i.e., the set of bounded C^1 functions with bounded partial derivatives). Since f belongs to $L^1(\mathbb{T}\times\mathbb{R})$ and the relativistic velocity is bounded, see (0.5), the function $v(p)f \in L^1(\mathbb{T}\times\mathbb{R})$ and therefore the current $j(x) = q\int_{\mathbb{R}} v(p) f\,\mathrm{d}p$ is well defined for a.a. $x \in \mathbb{T}$. In particular, taking $\varphi = \varphi(x) \in C^1(\mathbb{T})$ in (1.1) yields

$$\int_{\mathbb{T}} \varphi'(x) j(x)\,\mathrm{d}x = 0,$$

implying that the current is preserved along $x \in \mathbb{T}$.

PROPOSITION 1.2. *Let* f *be a periodic weak solution of* (0.1). *Then the current* $j = q\int_{\mathbb{R}} v(p) f\,\mathrm{d}p$ *is constant.*

2. Uniqueness of the periodic weak solution

Consider $f, g \in L^1(\mathbb{T}\times\mathbb{R})$ to be two periodic weak solutions for (0.1). By linearity we have

$$v(p)\partial_x(f-g) + F(x)\partial_p(f-g) = Q(f-g), \qquad (x,p) \in \mathbb{T}\times\mathbb{R} \tag{2.1}$$

and by standard computations one gets in $\mathcal{D}'(\mathbb{T}\times\mathbb{R})$,

$$v(p)\partial_x|f-g| + F(x)\partial_p|f-g| - \operatorname{sgn}(f-g)Q(f-g) = 0. \tag{2.2}$$

After integration with respect to momentum we have

$$\frac{\mathrm{d}}{\mathrm{d}x}\int_{\mathbb{R}} v(p)|f-g|\,\mathrm{d}p - \int_{\mathbb{R}} \operatorname{sgn}(f-g)Q(f-g)\,\mathrm{d}p = 0, \qquad \text{in} \qquad \mathcal{D}'(\mathbb{T}). \tag{2.3}$$

Following the idea in [**1**] we can write
(2.4)

$$
\begin{aligned}
&-\int_{\mathbb{R}} \operatorname{sgn}(f-g) Q(f-g)\,\mathrm{d}p \\
&= \int_{\mathbb{R}} \operatorname{sgn}(f-g) \int_{\mathbb{R}} s(p,p')\Big\{M_\theta(p')(f-g)(x,p) - M_\theta(p)(f-g)(x,p')\Big\}\,\mathrm{d}p'\,\mathrm{d}p \\
&= \frac{1}{2}\int_{\mathbb{R}}\int_{\mathbb{R}} s(p,p')\Big\{M_\theta(p')(f-g)(x,p) - M_\theta(p)(f-g)(x,p')\Big\} \\
&\qquad \times\Big\{\operatorname{sgn}(M_\theta(p')(f-g)(x,p)) - \operatorname{sgn}(M_\theta(p)(f-g)(x,p'))\Big\}\,\mathrm{d}p'\,\mathrm{d}p \\
&\geq 0
\end{aligned}
$$

with equality iff $\operatorname{sgn}(f-g)$ is constant with respect to p. Now integrating (2.3) with respect to x and using the periodicity of f and g implies that

$$
-\int_{\mathbb{T}}\int_{\mathbb{R}} \operatorname{sgn}(f-g)Q(f-g)\,\mathrm{d}p\,\mathrm{d}x = 0.
$$

Therefore for a.a. $x \in \mathbb{T}$ we have $-\int_{\mathbb{R}} \operatorname{sgn}(f-g)Q(f-g)\,\mathrm{d}p = 0$, and thus $\operatorname{sgn}(f-g)$ depends only on x. Eventually (2.1) can be written as

$$
v(p)\partial_x|f-g| + F(x)\partial_p|f-g| = Q(|f-g|), \qquad (x,p) \in \mathbb{T}\times\mathbb{R},
$$

implying that

$$
\frac{\mathrm{d}}{\mathrm{d}x}\int_{\mathbb{R}} v(p)|f-g|\,\mathrm{d}p = 0, \qquad x \in \mathbb{T},
$$

but this is not enough in order to guarantee the uniqueness of the periodic weak solution. Actually we will see that, in the particular case of electric fields satisfying $\int_{\mathbb{T}} E(x)\,\mathrm{d}x = 0$, the above arguments allow us to determine all the periodic solutions. When the average of E does not vanish, a sharper estimate will be necessary.

2.1. Vanishing electric field average. Let us assume that $\int_{\mathbb{T}} E\,\mathrm{d}x = 0$ holds. In such a case, the potential $\phi(x) = -\int_0^x E(y)\,\mathrm{d}y$ is also L-periodic. Since for any fixed $c \in \mathbb{R}$ the function $c\mathcal{M}_{\theta,\phi}(x,p)$ solves (0.1) we can replace (2.1) by

$$
v(p)\partial_x(f-g-c\mathcal{M}_{\theta,\phi}) + F(x)\partial_p(f-g-c\mathcal{M}_{\theta,\phi}) = Q(f-g-c\mathcal{M}_{\theta,\phi}). \tag{2.5}
$$

Following the same steps as before we find for any $c \in \mathbb{R}$,

$$
-\int_{\mathbb{R}} \operatorname{sgn}(f-g-c\mathcal{M}_{\theta,\phi})Q(f-g-c\mathcal{M}_{\theta,\phi})\,\mathrm{d}p = 0, \qquad \text{a.e. } x \in \mathbb{T}. \tag{2.6}
$$

Notice that the periodicity of the potential is crucial when writing

$$
\int_{\mathbb{T}}\frac{\mathrm{d}}{\mathrm{d}x}\int_{\mathbb{R}} v(p)|f-g-c\mathcal{M}_{\theta,\phi}|\,\mathrm{d}p\,\mathrm{d}x = 0.
$$

Hence (2.6) implies that, for a.a. $x \in \mathbb{T}$ and any $c \in \mathbb{R}$, the function $p \mapsto (f-g-c\mathcal{M}_{\theta,\phi})(x,p)$ has a constant sign. In particular it holds true for $c = c(x)$ given by $\int_{\mathbb{R}}(f-g)(x,p)\,\mathrm{d}p = c(x)\int_{\mathbb{R}}\mathcal{M}_{\theta,\phi}(x,p)\,\mathrm{d}p$. Since the function $p \mapsto (f-g-c\mathcal{M}_{\theta,\phi})(x,p)$ has a constant sign and zero integral with respect to $p \in \mathbb{R}$ we have

$$
f(x,p) - g(x,p) = c(x)\mathcal{M}_{\theta,\phi}(x,p) = \langle f-g\rangle(x)M_\theta(p), \qquad (x,p)\in\mathbb{T}\times\mathbb{R}. \tag{2.7}
$$

Inserting now (2.7) in (2.1) we deduce that

$$
f(x,p) - g(x,p) = k\mathcal{M}_{\theta,\phi}(x,p), \qquad (x,p)\in\mathbb{T}\times\mathbb{R}
$$

for some real constant k. This conclusion holds for every two periodic solutions, and for $g=0$ it tells us that all the periodic solutions of (0.1) when $\int_{\mathbb{T}} E \, dx = 0$ are $k\mathcal{M}_{\theta,\phi}$, $k \in \mathbb{R}$. Clearly, these solutions remain bounded. Observe also that the current of these solutions vanishes since $\int_{\mathbb{R}} v(p)\mathcal{M}_{\theta,\phi} \, dp = 0$. This already proves part a) of Theorem 0.1.

2.2. Nonvanishing electric field average. Let us analyze the case of electric fields with nonvanishing average: from now on we assume that

$$E \text{ is } L\text{−periodic with } \int_0^L E(x) \, dx \neq 0.$$

Now $\mathcal{M}_{\theta,\phi}$ is not periodic with respect to the space variable. Let us show that there is at most one periodic solution with a given current.

PROPOSITION 2.1. *Assume that $E \in L^\infty(\mathbb{T})$ such that $\int_{\mathbb{T}} E \, dx \neq 0$ and let $f, g \in L^1(\mathbb{T} \times \mathbb{R})$ be two periodic weak solutions for (0.1) with the same current*

$$q \int_{\mathbb{R}} v(p) f \, dp = q \int_{\mathbb{R}} v(p) g \, dp.$$

Then we have $f = g$.

The proof exploits new dissipation properties of the linear collision operator Q. We have seen that the inequality (2.4) is not strong enough for our purposes. Actually a better minoration for the dissipation term $-\int_{\mathbb{R}} \operatorname{sgn}(f-g)Q(f-g) \, dp$ is available, at least in the relativistic case.

LEMMA 2.2. *Let $h = h(p)$ be a function of $L^1(\mathbb{R})$ with vanishing current $\int_{\mathbb{R}} v(p)h(p) \, dp = 0$. Then we have the following inequality:*

$$-\int_{\mathbb{R}} \operatorname{sgn}(h(p))Q(h)(p) \, dp \geq \frac{s_0}{c_0} \left| \int_{\mathbb{R}} v(p)|h(p)| \, dp \right|. \tag{2.8}$$

PROOF. We consider the sets

$$A_+ = \{p \in \mathbb{R} \, : \, h(p) \geq 0\}, \qquad A_- = \{p \in \mathbb{R} \, : \, h(p) < 0\}.$$

Since $\int_{\mathbb{R}} v(p)h(p) \, dp = 0$, we have

$$\int_{\mathbb{R}} v(p)|h(p)|\mathbf{1}_{A_+}(p) \, dp = \int_{\mathbb{R}} v(p)|h(p)|\mathbf{1}_{A_-}(p) \, dp = \frac{1}{2}\int_{\mathbb{R}} v(p)|h(p)| \, dp.$$

Observe that

$$\begin{aligned} &-\int_{\mathbb{R}} \operatorname{sgn}(h(p))Q(h)(p) \, dp \\ &= \int_{\mathbb{R}} \operatorname{sgn}(h(p)) \int_{\mathbb{R}} s(p,p')\Big(M_\theta(p')h(p) - M_\theta(p)h(p')\Big) \, dp' \, dp \\ &= \int_{\mathbb{R}}\int_{\mathbb{R}} s(p,p')h(p)M_\theta(p')\Big(\operatorname{sgn}(h(p)) - \operatorname{sgn}(h(p'))\Big) \, dp' \, dp. \end{aligned}$$

But for any $(p,p') \in \mathbb{R}^2$ we have

$$h(p)M_\theta(p')(\operatorname{sgn}(h(p)) - \operatorname{sgn}(h(p'))) \geq 0, \qquad s(p,p') \geq s_0 > 0,$$

and thus, by taking into account (0.5), we can write

$$h(p)M_\theta(p')(\operatorname{sgn}(h(p)) - \operatorname{sgn}(h(p'))) \geq \pm\frac{v(p)}{c_0} h(p)M_\theta(p')(\operatorname{sgn}(h(p)) - \operatorname{sgn}(h(p'))).$$

Combining these computations yields

$$
\begin{aligned}
-\int_{\mathbb{R}} \operatorname{sgn}(h)Q(h)\,\mathrm{d}p &\geq \frac{\pm s_0}{c_0}\int_{\mathbb{R}}\int_{\mathbb{R}} \mathrm{d}p\mathrm{d}p v(p)h(p)M_\theta(p')(\operatorname{sgn}(h(p)) - \operatorname{sgn}(h(p')))\,\mathrm{d}p' \\
&\geq \frac{\pm 2s_0}{c_0}\int_{\mathbb{R}} v(p)|h(p)|\mathbf{1}_{A_+}(p)\,\mathrm{d}p \int_{\mathbb{R}} M_\theta(p')\mathbf{1}_{A_-}(p')\,\mathrm{d}p' \\
&+ \frac{\pm 2s_0}{c_0}\int_{\mathbb{R}} v(p)|h(p)|\mathbf{1}_{A_-}(p)\,\mathrm{d}p \int_{\mathbb{R}} M_\theta(p')\mathbf{1}_{A_+}(p')\,\mathrm{d}p' \\
&\geq \frac{\pm s_0}{c_0}\int_{\mathbb{R}} v(p)|h(p)|\,\mathrm{d}p \int_{\mathbb{R}} M_\theta(p')(\mathbf{1}_{A_-} + \mathbf{1}_{A_+})(p')\,\mathrm{d}p' \\
&\geq \frac{\pm s_0}{c_0}\int_{\mathbb{R}} v(p)|h(p)|\,\mathrm{d}p.
\end{aligned}
$$

□

REMARK 2.3. When the speed of light c_0 becomes large, the inequality (2.8) degenerates to (2.4). In particular, in the nonrelativistic case, the conclusion of Lemma 2.2 reduces to the well-known inequality (2.4), which is not enough for the uniqueness of the periodic weak solution.

Proof of Proposition 2.1. Consider the function $h = f - g - c\mathcal{M}_{\theta,\phi}$ with $c \in \mathbb{R}$. This function belongs to $L^1([a,b] \times \mathbb{R})$ for any $a < b$. It has vanishing current $\int_{\mathbb{R}} v(p)h(x,p)\,\mathrm{d}p = 0, x \in \mathbb{R}$ and satisfies in $\mathcal{D}'(\mathbb{R}^2)$,

$$v(p)\partial_x h + F(x)\partial_p h = Q(h).$$

We obtain

$$v(p)\partial_x|h| + F(x)\partial_p|h| - \operatorname{sgn}(h)Q(h) = 0. \tag{2.9}$$

Integrating with respect to $p \in \mathbb{R}$ and combining with Lemma 2.2 yield

$$
\begin{aligned}
&\frac{\mathrm{d}}{\mathrm{d}x}\int_{\mathbb{R}} v(p)|h|\,\mathrm{d}p + \frac{s_0}{c_0}\left|\int_{\mathbb{R}} v(p)|h|\,\mathrm{d}p\right| \\
&\qquad \leq \frac{\mathrm{d}}{\mathrm{d}x}\int_{\mathbb{R}} v(p)|h|\,\mathrm{d}p - \int_{\mathbb{R}} \operatorname{sgn}(h)Q(h)\,\mathrm{d}p = 0.
\end{aligned}
\tag{2.10}
$$

Let us set

$$u(x) = \int_{\mathbb{R}} v(p)|h(x,p)|\,\mathrm{d}p.$$

This function is not periodic but satisfies the bounds

$$\sup_{n\in\mathbb{Z}} |u(x+nL)| < +\infty, \qquad \text{a.e. } x \in \mathbb{R}. \tag{2.11}$$

Indeed we have for any $n \in \mathbb{Z}$,

$$
\begin{aligned}
|u(x+nL)| &= \left| \int_{\mathbb{R}} v(p)\Big(|h(x+nL,p)| - |c\mathcal{M}_{\theta,\phi}(x+nL,p)|\Big)\,\mathrm{d}p \right| \\
&\le \int_{\mathbb{R}} |v(p)| \,\Big|\, |h(x+nL,p)| - |c\mathcal{M}_{\theta,\phi}(x+nL,p)| \,\Big|\,\mathrm{d}p \\
&\le c_0 \int_{\mathbb{R}} \Big| h(x+nL,p) + c\mathcal{M}_{\theta,\phi}(x+nL,p)\Big|\,\mathrm{d}p \\
&\le c_0 \int_{\mathbb{R}} \Big| f(x+nL,p) - g(x+nL,p)\Big|\,\mathrm{d}p \\
&\le c_0 \int_{\mathbb{R}} |f(x,p)|\,\mathrm{d}p + c_0 \int_{\mathbb{R}} |g(x,p)|\,\mathrm{d}p.
\end{aligned}
$$

We shall see now that $u(x)$ actually vanishes. By (2.10) we know that $u'(x) + \frac{\pm s_0}{c_0} u(x) \le 0$, $x \in \mathbb{R}$, which implies that

$$
\frac{\mathrm{d}}{\mathrm{d}x}\left\{ u(x)\exp\left(\frac{\pm s_0 x}{c_0}\right)\right\} \le 0, \qquad x \in \mathbb{R}. \tag{2.12}
$$

Consider first the case with the sign $+$. Let us integrate (2.12) between $x - nL$ and x with $n \in \mathbb{N}$. We deduce that

$$
u(x) \le u(x-nL)\exp\left(\frac{-nLs_0}{c_0}\right).
$$

We let $n \to +\infty$: by using (2.11) we get $u(x) \le 0$ for a.a. $x \in \mathbb{R}$. Similarly, for the case with the sign $-$, we integrate over $[x, x+nL]$ with $n \in \mathbb{N}$ and and we let $n \to +\infty$. We get $u(x) \ge 0$ for a.a. $x \in \mathbb{R}$. Therefore we have $u = 0$ and coming back to (2.10) we deduce that

$$
\int_{\mathbb{R}} \operatorname{sgn}(h) Q(h)\,\mathrm{d}p = 0, \qquad \text{a.e. } x \in \mathbb{R}. \tag{2.13}
$$

At this stage let us point out that one cannot obtain (2.13) as in the case of periodic potentials, by integrating (2.9) over $\mathbb{T} \times \mathbb{R}$. Indeed, in this case, h is not periodic and thus

$$
\int_{\mathbb{T}} \frac{\mathrm{d}}{\mathrm{d}x} \int_{\mathbb{R}} v(p)|h(x,p)|\,\mathrm{d}p\,\mathrm{d}x \ne 0.
$$

Therefore Lemma 2.2 is crucial when establishing (2.13) for nonperiodic potentials.

From now on we follow the same steps as for periodic potentials. We deduce that there is a constant $k \in \mathbb{R}$ such that

$$
f(x,p) - g(x,p) = k\mathcal{M}_{\theta,\phi}(x,p), \qquad (x,p) \in \mathbb{R}^2.
$$

Since f and g are periodic and $\mathcal{M}_{\theta,\phi}$ is not periodic we must have $k = 0$ and therefore $f = g$. Indeed, the weak formulation (1.1) applied to $f - g$ implies that

$$
k \int_{\mathbb{T}}\int_{\mathbb{R}} \mathcal{M}_{\theta,\phi}(x,p)(v(p)\partial_x\varphi + F(x)\partial_p\varphi)\,\mathrm{d}p\,\mathrm{d}x = 0, \qquad \varphi \in C_b^1(\mathbb{T}\times\mathbb{R})
$$

and after integration by parts we get

$$
k \int_{\mathbb{R}} v(p)\varphi(0,p)(\mathcal{M}_{\theta,\phi}(L,p) - \mathcal{M}_{\theta,\phi}(0,p))\,\mathrm{d}p = 0.
$$

Since the potential is not periodic we obtain

$$k \int_{\mathbb{R}} v(p)\varphi(0,p)M_\theta(p)\ \mathrm{d}p = 0, \qquad \varphi \in C_b^1(\mathbb{T}\times\mathbb{R}).$$

In particular taking $\varphi(x,p) = v(p)$, $(x,p) \in \mathbb{T}\times\mathbb{R}$ yields

$$k \int_{\mathbb{R}} |v(p)|^2 M_\theta(p)\ \mathrm{d}p = 0,$$

saying that $k = 0$. □

3. Existence of periodic weak solution

In order to construct a periodic solution for the linear Boltzmann equation we appeal to perturbation techniques. We start by considering smooth electric fields; the regularity will be removed by a standard approximation argument. We consider the periodic equation with a damping term and a source:

$$(3.1) \qquad \alpha f(x,p) + v(p)\partial_x f + F(x)\partial_p f = Q(f) + S(x,p), \qquad (x,p)\in\mathbb{T}\times\mathbb{R},$$

where $\alpha > 0$. To this end, we adapt Definition 1.1 as follows.

DEFINITION 3.1. Let $F \in L^\infty(\mathbb{T}), S \in L^1(\mathbb{T}\times\mathbb{R})$ and $\alpha > 0$. We say that $f \in L^1(\mathbb{T}\times\mathbb{R})$ is a periodic weak solution for (3.1) iff

$$(3.2) \qquad -\int_{\mathbb{T}}\int_{\mathbb{R}} f(x,p)(-\alpha\varphi(x,p) + v(p)\partial_x\varphi + F(x)\partial_p\varphi)\ \mathrm{d}p\ \mathrm{d}x$$
$$= \int_{\mathbb{T}}\int_{\mathbb{R}} Q(f)\varphi(x,p)\ \mathrm{d}p\ \mathrm{d}x + \int_{\mathbb{T}}\int_{\mathbb{R}} S(x,p)\varphi(x,p)\ \mathrm{d}p\ \mathrm{d}x$$

for any function $\varphi \in C_c^1(\mathbb{T}\times\mathbb{R})$.

We check that the above formulation holds true for any $\varphi \in C_b^1(\mathbb{T}\times\mathbb{R})$ as well. Furthermore the following identity holds by integrating (3.1) with respect to the momentum variable p:

$$(3.3) \qquad \alpha\int_{\mathbb{R}} f(x,p)\ \mathrm{d}p + \frac{\mathrm{d}}{\mathrm{d}x}\int_{\mathbb{R}} v(p)f(x,p)\ \mathrm{d}p = \int_{\mathbb{R}} S(x,p)\ \mathrm{d}p, \qquad x\in\mathbb{T}\ .$$

The basis of the existence proof relies on the following claim, where the problem with source term in $L^1(\mathbb{T}\times\mathbb{R})$ is investigated. The solution is shown to satisfy estimates uniformly with respect to the penalization parameter $\alpha > 0$.

PROPOSITION 3.2. *Assume that* $S \in L^1(\mathbb{T}\times\mathbb{R}), E \in W^{1,\infty}(\mathbb{T})$ *and* $\alpha > 0$. *Then there is a unique periodic solution of* (3.1) *satisfying*

$$(3.4) \qquad \|f\|_{L^1(\mathbb{T}\times\mathbb{R})} \le \frac{1}{\alpha}\|S\|_{L^1(\mathbb{T}\times\mathbb{R})}.$$

Moreover the following properties hold:

a) *if* $S \ge 0$, *then* $f \ge 0$,

b) *if* $S \in L^\infty(\mathbb{T}; L^1(\mathbb{R}))$, *then*

$$\left\| \int_{\mathbb{R}} v(p)|f(\cdot,p)|\ \mathrm{d}p \right\|_{L^\infty(\mathbb{T})} \le \frac{c_0}{\alpha}\|S\|_{L^\infty(\mathbb{T};L^1(\mathbb{R}))},$$

c) *if* $S \ge 0$, $S \in L^\infty(\mathbb{T}\times\mathbb{R})$ *and* $\int_{\mathbb{R}} v(p)pf(\cdot,p)\ \mathrm{d}p \in L^\infty(\mathbb{T})$, *then* $f \in L^\infty(\mathbb{T}\times\mathbb{R})$, $\langle f\rangle \in L^\infty(\mathbb{T})$.

This statement is obtained as a consequence of the following well-posedness result.

LEMMA 3.3. *Let $E \in W^{1,\infty}(\mathbb{T}), S \in L^1(\mathbb{T}\times\mathbb{R})$ and $\alpha > 0$. Then there is a unique periodic weak solution $f \in L^1(\mathbb{T}\times\mathbb{R})$ of the problem*

$$(\alpha+\sigma(p))f(x,p)+v(p)\partial_x f+F(x)\partial_p f = S(x,p), \qquad (x,p)\in\mathbb{T}\times\mathbb{R} \tag{3.5}$$

satisfying

$$\|f\|_{L^1(\mathbb{T}\times\mathbb{R})} \le \frac{1}{\alpha+s_0}\|S\|_{L^1(\mathbb{T}\times\mathbb{R})}. \tag{3.6}$$

If $S \ge 0$, then $f \ge 0$, and more generally if $S \in L^\infty(\mathbb{T}\times\mathbb{R})$, then

$$-\frac{1}{\alpha+s_0}\|S_-\|_{L^\infty(\mathbb{T}\times\mathbb{R})} \le f(x,p) \le \frac{1}{\alpha+s_0}\|S_+\|_{L^\infty(\mathbb{T}\times\mathbb{R})}, \qquad \text{a.e. } (x,p)\in\mathbb{T}\times\mathbb{R}, \tag{3.7}$$

where $S_\pm = \max(0,\pm S)$. Finally, if $S \in L^\infty(\mathbb{T};L^1(\mathbb{R}))$, then

$$\left\|\int_{\mathbb{R}} v(p)|f(\cdot,p)|\,\mathrm{d}p\right\|_{L^\infty(\mathbb{T})} \le \frac{c_0}{\alpha+s_0}\|S\|_{L^\infty(\mathbb{T};L^1(\mathbb{R}))}. \tag{3.8}$$

PROOF. We start by proving the uniqueness: for any two solutions f, g we have

$$(\alpha+\sigma(p))|f-g|+v(p)\partial_x|f-g|+F(x)\partial_p|f-g| = 0$$

and therefore

$$(\alpha+s_0)\int_{\mathbb{T}}\int_{\mathbb{R}}|f-g|\,\mathrm{d}p\,\mathrm{d}x \le \int_{\mathbb{T}}\int_{\mathbb{R}}(\alpha+\sigma(p))|f-g|\,\mathrm{d}p\,\mathrm{d}x = 0,$$

implying that $f = g$.

Owing to the regularity of the electric field, let us consider the characteristics (X,P) defined by

$$\frac{\mathrm{d}X}{\mathrm{d}s} = v(P(s;x,p)), \qquad \frac{\mathrm{d}P}{ds} = qE(X(s;x,p)),$$

with the conditions

$$X(0;x,p) = x, \qquad P(0;x,p) = p.$$

Integrating equation (3.1) along the characteristics yields that

$$f(x,p) = \int_{-\infty}^{0} e^{\int_0^s\{\alpha+\sigma(P(\tau;x,p))\}\,\mathrm{d}\tau}S(X(s;x,p),P(s;x,p))\,\mathrm{d}s \tag{3.9}$$

is a weak solution for (3.5). Since E is periodic, we have

$$X(s;x+L,p) = X(s;x,p)+L, \qquad P(s;x+L,p) = P(s;x,p)$$

and therefore f is also L-periodic.

The L^1 bound (3.6) follows by using $\sigma(p) \ge s_0$ and integrating over $\mathbb{T}\times\mathbb{R}$ the inequality

$$(\alpha+\sigma(p))|f(x,p)|+v(p)\partial_x|f|+F(x)\partial_p|f| \le |S(x,p)|.$$

The L^∞ bounds (3.7) follow immediately from the explicit formula (3.9).

It remains to justify the estimate on the current (3.8). Assume that $S \in L^\infty(\mathbb{T};L^1(\mathbb{R}))$ and observe that

$$(\alpha+s_0)\int_{\mathbb{R}}|f(x,p)|\,\mathrm{d}p+\frac{\mathrm{d}}{\mathrm{d}x}\int_{\mathbb{R}}v(p)|f(x,p)|\,\mathrm{d}p \le \int_{\mathbb{R}}|S(x,p)|\,\mathrm{d}p.$$

We set $u(x) = \int_{\mathbb{R}} v(p)|f(x,p)|\,dp$. Then, the relativistic bound (0.5) implies that $c_0 \int_{\mathbb{R}} |f(x,p)|\,dp \geq |u(x)|$, so one gets

$$\frac{\pm(\alpha + s_0)}{c_0} u(x) + u'(x) \leq \int_{\mathbb{R}} |S(x,p)|\,dp, \qquad x \in \mathbb{R}.$$

With the + sign, we integrate between $x - nL$ and x, with $n \in \mathbb{N}$, and we deduce that

$$\begin{aligned} u(x) \;&\leq\; u(x-nL)\exp\left(-\frac{(\alpha+s_0)nL}{c_0}\right) \\ &\quad + \|S\|_{L^\infty(\mathbb{T};L^1(\mathbb{R}))} \int_{x-nL}^{x} \exp\left(-\frac{(\alpha+s_0)(x-y)}{c_0}\right)\,dy. \end{aligned}$$

Letting $n \to +\infty$ we obtain $u(x) \leq \frac{c_0}{\alpha+s_0}\|S\|_{L^\infty(\mathbb{T};L^1(\mathbb{R}))}$ (and clearly we can dominate it by $\frac{c_0}{s_0}\|S\|_{L^\infty(\mathbb{T};L^1(\mathbb{R}))}$, which provides an estimate uniform with respect to α). Proceeding similarly with the $-$ sign, we deduce that $u(x) \geq -\frac{c_0}{\alpha+s_0}\|S\|_{L^\infty(\mathbb{T};L^1(\mathbb{R}))}$ $\geq -\frac{c_0}{s_0}\|S\|_{L^\infty(\mathbb{T};L^1(\mathbb{R}))}$. □

Proof of Proposition 3.2. We consider the sequence of periodic weak solutions $(f_\pm^{(n)})_{n\in\mathbb{N}}$ defined by

$$\sigma(p) f_\pm^{(0)}(x,p) + \alpha f_\pm^{(0)}(x,p) + v(p)\partial_x f_\pm^{(0)} + F(x)\partial_p f_\pm^{(0)} = S_\pm(x,p), \qquad (x,p) \in \mathbb{T}\times\mathbb{R}$$

and for any $n \in \mathbb{N}$,

$$\sigma(p) f_\pm^{(n+1)} + \alpha f_\pm^{(n+1)} + v(p)\partial_x f_\pm^{(n+1)} + F(x)\partial_p f_\pm^{(n+1)} = M_\theta \langle f_\pm^{(n)}\rangle_s + S_\pm, \qquad (x,p) \in \mathbb{T}\times\mathbb{R}, \tag{3.10}$$

where $S_\pm$ are the positive/negative parts of S. Thanks to Lemma 3.3 the sequence $(f_\pm^{(n)})_{n\in\mathbb{N}}$ is well defined. We have $f_\pm^{(0)} \geq 0$ and we check recursively that $0 \leq f_\pm^{(n)} \leq f_\pm^{(n+1)}$ for any $n \in \mathbb{N}$. Integrating over $\mathbb{T}\times\mathbb{R}$ we get

$$\begin{aligned} \int_{\mathbb{T}}\int_{\mathbb{R}} (\alpha + \sigma(p))\, f_\pm^{(n+1)}(x,p)\,dp\,dx &= \int_{\mathbb{T}}\int_{\mathbb{R}} M_\theta(p)\langle f_\pm^{(n)}\rangle_s\,dp\,dx + \int_{\mathbb{T}}\int_{\mathbb{R}} S_\pm\,dp\,dx \\ &= \int_{\mathbb{T}}\int_{\mathbb{R}} \sigma(p) f_\pm^{(n)}\,dp\,dx + \int_{\mathbb{T}}\int_{\mathbb{R}} S_\pm\,dp\,dx \\ &\leq \int_{\mathbb{T}}\int_{\mathbb{R}} \sigma(p) f_\pm^{(n+1)}\,dp\,dx + \int_{\mathbb{T}}\int_{\mathbb{R}} S_\pm\,dp\,dx, \end{aligned}$$

implying that $\sup_{n\in\mathbb{N}} \int_{\mathbb{T}}\int_{\mathbb{R}} f_\pm^{(n)}\,dp\,dx \leq \alpha^{-1}\int_{\mathbb{T}}\int_{\mathbb{R}} S_\pm\,dp\,dx$. By the monotone convergence theorem we deduce that $(f_\pm^{(n)})_n$ converges in $L^1(\mathbb{T}\times\mathbb{R})$. Let $f_\pm = \lim_{n\to\infty} f_\pm^{(n)} \geq 0$. Passing to the limit for $n \to +\infty$ we deduce that $f_\pm \geq 0$ are periodic weak solutions of

$$\alpha f_\pm + v(p)\partial_x f_\pm + F(x)\partial_p f_\pm = Q(f_\pm) + S_\pm, \qquad (x,p) \in \mathbb{T}\times\mathbb{R}$$

satisfying

$$\int_{\mathbb{T}}\int_{\mathbb{R}} f_\pm\,dp\,dx = \alpha^{-1}\int_{\mathbb{T}}\int_{\mathbb{R}} S_\pm\,dp\,dx.$$

Therefore $f = f_+ - f_-$ is a periodic weak solution of (3.1) satisfying (3.4).

Assume now that S belongs to $L^\infty(\mathbb{T}; L^1(\mathbb{R}))$. The estimate in b) follows exactly as in the proof of Lemma 3.3 since we have

$$\alpha \int_{\mathbb{R}} |f| \, \mathrm{d}p + \frac{\mathrm{d}}{\mathrm{d}x} \int_{\mathbb{R}} v(p)|f| \, \mathrm{d}p = \int_{\mathbb{R}} Q(f)\mathrm{sgn}(f) \, \mathrm{d}p + \int_{\mathbb{R}} S \, \mathrm{sgn}(f) \, \mathrm{d}p \leq \int_{\mathbb{R}} |S| \, \mathrm{d}p.$$

The final step consists in proving the L^∞ estimate. Let $S \in L^\infty(\mathbb{T} \times \mathbb{R}), S \geq 0$ and assume that there is $\mathcal{K} > 0$ such that

$$\int_{\mathbb{R}} v(p) p f_+^{(n)} \, \mathrm{d}p \leq \int_{\mathbb{R}} v(p) p f_+ \, \mathrm{d}p = \int_{\mathbb{R}} v(p) p f \, \mathrm{d}p \leq \mathcal{K}, \qquad \text{a.e. } x \in \mathbb{T}.$$

Let $\tilde{\mathcal{K}} > 1$, which will be defined later on, such that $\|S\|_{L^\infty(\mathbb{T}\times\mathbb{R})} \leq s_0 \tilde{\mathcal{K}}/2$.

By Lemma 3.3 we know that

$$\|f_+^{(0)}\|_{L^\infty(\mathbb{T}\times\mathbb{R})} \leq \frac{1}{s_0 + \alpha} \|S\|_{L^\infty(\mathbb{T}\times\mathbb{R})} \leq \frac{\|S\|_{L^\infty(\mathbb{T}\times\mathbb{R})}}{s_0} \leq \tilde{\mathcal{K}}.$$

Then we wish to find some $\tilde{\mathcal{K}}$ such that the property extends to the whole sequence.

Assume that $f_+^{(n)} \in L^\infty(\mathbb{T} \times \mathbb{R})$ with $\|f_+^{(n)}\|_{L^\infty(\mathbb{T}\times\mathbb{R})} \leq \tilde{\mathcal{K}}$. Then, on the one hand, we can estimate the integral of $f_+^{(n)}$ with respect to the momentum p by splitting the domain of integration as

$$\begin{aligned} \langle f_+^{(n)} \rangle &= \int_{\mathbb{R}} f_+^{(n)} \mathbf{1}_{\{|p|<R\}} \, \mathrm{d}p + \int_{\mathbb{R}} f_+^{(n)} \mathbf{1}_{\{|p|\geq R\}} \, \mathrm{d}p \\ &\leq 2R \|f_+^{(n)}\|_{L^\infty(\mathbb{T}\times\mathbb{R})} + \frac{1}{Rv(R)} \int_{\mathbb{R}} v(p) p f_+^{(n)} \mathbf{1}_{\{|p|\geq R\}} \, \mathrm{d}p \\ &\leq 2R \|f_+^{(n)}\|_{L^\infty(\mathbb{T}\times\mathbb{R})} + \frac{1}{Rv(R)} \mathcal{K} \\ &\leq 2R\tilde{\mathcal{K}} + \frac{1}{Rv(R)} \mathcal{K}. \end{aligned}$$

Hence, we define $R = R(\tilde{\mathcal{K}})$ such that $2R\tilde{\mathcal{K}} = \frac{\mathcal{K}}{Rv(R)}$ to obtain the inequality

$$\|\langle f_+^{(n)} \rangle\|_{L^\infty(\mathbb{T})} \leq 4R(\tilde{\mathcal{K}})\tilde{\mathcal{K}}.$$

Since $v(R)/R \to 1/m$ as $R \to 0$, we observe that the relation $R(\tilde{\mathcal{K}})^2 v(R(\tilde{\mathcal{K}})) = \mathcal{K}/2\tilde{\mathcal{K}}$ implies that $R(\tilde{\mathcal{K}})$ tends to 0 as $\tilde{\mathcal{K}} \to \infty$ (it behaves like $\tilde{\mathcal{K}}^{-1/3}$).

On the other hand, a direct application of Lemma 3.3, see (3.7), leads to

$$\begin{aligned} \|f_+^{(n+1)}\|_{L^\infty(\mathbb{T}\times\mathbb{R})} &\leq \frac{s_1}{s_0} \left(\int_{\mathbb{R}} \exp\left(-\frac{\mathcal{E}(p')}{\theta} \right) \mathrm{d}p' \right)^{-1} \|\langle f_+^{(n)} \rangle\|_{L^\infty(\mathbb{T})} + \frac{1}{s_0} \|S\|_{L^\infty(\mathbb{T}\times\mathbb{R})} \\ &\leq \frac{s_1}{s_0} 4R(\tilde{\mathcal{K}})\tilde{\mathcal{K}} \left(\int_{\mathbb{R}} \exp\left(-\frac{\mathcal{E}(p')}{\theta} \right) \mathrm{d}p' \right)^{-1} + \frac{1}{s_0} \|S\|_{L^\infty(\mathbb{T}\times\mathbb{R})}. \end{aligned}$$

Therefore, the uniform estimates rely on the possibility of finding $\tilde{\mathcal{K}}$ such that $\|S\|_{L^\infty(\mathbb{T}\times\mathbb{R})} \leq s_0 \tilde{\mathcal{K}}/2$ and

$$\frac{s_1}{s_0} 4R(\tilde{\mathcal{K}})\tilde{\mathcal{K}} \left(\int_{\mathbb{R}} \exp\left(-\frac{\mathcal{E}(p')}{\theta} \right) \mathrm{d}p' \right)^{-1} + \frac{1}{s_0} \|S\|_{L^\infty(\mathbb{T}\times\mathbb{R})} \leq \tilde{\mathcal{K}}. \tag{3.11}$$

This is obviously possible for $\tilde{\mathcal{K}}$ large enough since $\lim_{\tilde{\mathcal{K}}\to+\infty} R(\tilde{\mathcal{K}}) = 0$.

Accordingly, we have obtained the estimates

$$\|f_+^{(n)}\|_{L^\infty(\mathbb{T}\times\mathbb{R})} \leq \tilde{\mathcal{K}}, \qquad \|\langle f_+^{(n)} \rangle\|_{L^\infty(\mathbb{T})} \leq 4R(\tilde{\mathcal{K}})\tilde{\mathcal{K}}, \qquad n \in \mathbb{N}.$$

Since $\lim_{n\to\infty} f_+^{(n)}(x,p) = f_+(x,p) = f(x,p)$ for a.a. $(x,p) \in \mathbb{T}\times\mathbb{R}$, we deduce that $\|f\|_{L^\infty(\mathbb{T}\times\mathbb{R})} \le \tilde{\mathcal{K}}$. The monotone convergence theorem also proves that $\lim_{n\to\infty}\langle f_+^{(n)}\rangle(x) = \langle f\rangle(x)$ for a.a. $x\in\mathbb{T}$ and thus

$$\|\langle f\rangle\|_{L^\infty(\mathbb{T})} \le 4R(\tilde{\mathcal{K}})\tilde{\mathcal{K}}. \qquad \square$$

By Proposition 3.2 we know that for any $\alpha > 0$ there is a unique periodic weak solution f_α for the problem

$$\alpha f_\alpha(x,p) + v(p)\partial_x f_\alpha + F(x)\partial_p f_\alpha = Q(f_\alpha) + \alpha M_\theta(p), \qquad (x,p)\in\mathbb{T}\times\mathbb{R}. \tag{3.12}$$

These solutions are nonnegative and satisfy for any $\alpha > 0$,

$$\begin{aligned} &\int_{\mathbb{T}}\int_{\mathbb{R}} f_\alpha(x,p)\,\mathrm{d}p\,\mathrm{d}x = \int_{\mathbb{T}}\int_{\mathbb{R}} M_\theta(p)\,\mathrm{d}p\,\mathrm{d}x = L,\\ &\left\|\int_{\mathbb{R}} v(p)f_\alpha(\cdot,p)\,\mathrm{d}p\right\|_{L^\infty(\mathbb{T})} \le c_0. \end{aligned} \tag{3.13}$$

We split the end of the proof of Theorem 0.1 into several steps.

Step 1: Existence of a nontrivial solution

By applying the weak formulation of (3.12) to the test function $\mathcal{E}(p) + q\phi(x)$ one gets

$$\begin{aligned} \alpha\int_{\mathbb{T}}\int_{\mathbb{R}} f_\alpha(\mathcal{E}(p)+q\phi(x))\,\mathrm{d}p\,\mathrm{d}x + q\phi(L)\int_{\mathbb{R}} v(p)f_\alpha(L,p)\,\mathrm{d}p - q\phi(0)\int_{\mathbb{R}} v(p)f_\alpha(0,p)\,\mathrm{d}p\\ + \int_{\mathbb{T}}\int_{\mathbb{R}}(\sigma(p)f_\alpha - M_\theta(p)\langle f_\alpha\rangle_s)(\mathcal{E}(p)+q\phi(x))\,\mathrm{d}p\,\mathrm{d}x\\ = \alpha\int_{\mathbb{T}}\int_{\mathbb{R}} M_\theta(p)(\mathcal{E}(p)+q\phi(x))\,\mathrm{d}p\,\mathrm{d}x. \end{aligned}$$

Therefore by taking into account (3.13) and

$$\int_{\mathbb{T}}\int_{\mathbb{R}} M_\theta(p)\langle f_\alpha\rangle_s\mathcal{E}(p)\,\mathrm{d}p\,\mathrm{d}x \le s_1\int_{\mathbb{T}}\langle f_\alpha\rangle\,\mathrm{d}x\int_{\mathbb{R}} M_\theta(p)\mathcal{E}(p)\,\mathrm{d}p = s_1 L\int_{\mathbb{R}} M_\theta(p)\mathcal{E}(p)\,\mathrm{d}p$$

we deduce that

$$\sup_{0<\alpha<1}\int_{\mathbb{T}}\int_{\mathbb{R}} f_\alpha(x,p)\mathcal{E}(p)\,\mathrm{d}p\,\mathrm{d}x < +\infty.$$

Now multiplying (3.12) by p one gets

$$\int_{\mathbb{R}}(\alpha+\sigma(p))pf_\alpha\,\mathrm{d}p + \frac{\mathrm{d}}{\mathrm{d}x}\int_{\mathbb{R}} v(p)pf_\alpha(x,p)\,\mathrm{d}p = F(x)\langle f_\alpha\rangle + \int_{\mathbb{R}} pM_\theta(p)\langle f_\alpha\rangle_s\,\mathrm{d}p. \tag{3.14}$$

This relation proves that $x\mapsto\int_{\mathbb{R}} v(p)pf_\alpha(x,p)\,\mathrm{d}p$ is continuous since $\int_{\mathbb{R}}(\alpha+\sigma(p))pf_\alpha\,\mathrm{d}p$ and $F(x)\langle f_\alpha\rangle + \int_{\mathbb{R}} pM_\theta(p)\langle f_\alpha\rangle_s\,\mathrm{d}p$ are integrable on $\mathbb{T}$, uniformly with respect to α. Then (0.6) leads to

$$\sup_{0<\alpha<1}\int_{\mathbb{T}}\int_{\mathbb{R}} v(p)pf_\alpha(x,p)\,\mathrm{d}p\,\mathrm{d}x \le \sup_{0<\alpha<1}\int_{\mathbb{T}}\int_{\mathbb{R}}(mc_0^2+\mathcal{E}(p))f_\alpha(x,p)\,\mathrm{d}p\,\mathrm{d}x < +\infty.$$

Therefore there exists $x_\alpha\in\mathbb{T}$ such that

$$\sup_{0<\alpha<1}\int_{\mathbb{R}} v(p)pf_\alpha(x_\alpha,p)\,\mathrm{d}p \le \frac{1}{L}\sup_{0<\alpha<1}\int_{\mathbb{T}}\int_{\mathbb{R}} v(p)pf_\alpha(x,p)\,\mathrm{d}p\,\mathrm{d}x < +\infty.$$

We now integrate (3.14) between x_α and x for any $x \in [x_\alpha, x_\alpha + L]$. Using (0.6) again we deduce that there is a constant C such that

$$\sup_{0<\alpha<1} \left\| \int_{\mathbb{R}} v(p) p f_\alpha(\cdot, p) \, dp \right\|_{L^\infty(\mathbb{T})} \leq C,$$
$$\sup_{0<\alpha<1} \int_{\mathbb{T}} \int_{\mathbb{R}} (mc_0^2 + \mathcal{E}(p)) f_\alpha(x,p) \, dp \, dx < +\infty,$$

and thus applying Proposition 3.2-c) we obtain uniform L^∞ bounds for f_α and $\langle f_\alpha \rangle$. Therefore there is a sequence $(\alpha_n)_{n\in\mathbb{N}}$ converging towards zero such that

$$\lim_{n\to\infty} f_{\alpha_n} = f \qquad \text{weakly-}\star \text{ in } L^\infty(\mathbb{T}\times\mathbb{R}),$$
$$\lim_{n\to\infty} \langle f_{\alpha_n} \rangle = \rho \qquad \text{weakly-}\star \text{ in } L^\infty(\mathbb{T}),$$
$$\lim_{n\to\infty} \int_{\mathbb{R}} v(p) f_{\alpha_n} \, dp = j \qquad \text{weakly-}\star \text{ in } L^\infty(\mathbb{T}).$$

It is easily seen that

$$(1+\mathcal{E}(p))f \in L^1(\mathbb{T}\times\mathbb{R}),$$
$$f \in L^\infty(\mathbb{T}\times\mathbb{R}), \qquad f \geq 0, \qquad \int_{\mathbb{T}}\int_{\mathbb{R}} f \, dp \, dx = L,$$
$$\rho \in L^1(\mathbb{T}) \cap L^\infty(\mathbb{T}),$$

and we check immediately that $\rho = \langle f \rangle$ and $j = \int_{\mathbb{R}} v(p) f \, dp \in [-c_0, c_0]$. Now passing to the limit for $n \to +\infty$ in the weak formulation of (3.1) we deduce that f is a periodic weak solution of (0.1). This ends the proof of b) in Theorem 0.1 at least for smooth electric fields.

Step 2: Extension to rough electric fields
Notice that all the bounds of the above solution depend on $m, c_0, q, s_0, s_1, \theta$ and $\|E\|_{L^\infty(\mathbb{T})}$ but not on $\|E'\|_{L^\infty(\mathbb{T})}$. Therefore, we can consider $(E_n)_{n\in\mathbb{N}}$ to be a sequence of smooth fields, for each n, $E_n \in W^{1,\infty}(\mathbb{T})$, which converges a.e. towards $E \in L^\infty(\mathbb{T})$, with $\|E_n\|_{L^\infty(\mathbb{T})} \leq \|E\|_{L^\infty(\mathbb{T})}$. Let f_n be the associated sequence of periodic weak solutions, as given by Step 1. We have the uniform bounds

$$\sup_{n\in\mathbb{N}} \|f_n\|_{L^\infty(\mathbb{T}\times\mathbb{R})} + \sup_{n\in\mathbb{N}} \int_{\mathbb{T}}\int_{\mathbb{R}} (1 + v(p)p) f_n(x,p) \, dp \, dx < +\infty,$$

$$\sup_{n\in\mathbb{N}} \|\langle f_n \rangle\|_{L^\infty(\mathbb{T})} < +\infty, \qquad \int_{\mathbb{T}}\int_{\mathbb{R}} f_n(x,p) \, dp \, dx = L, \qquad \left| \int_{\mathbb{R}} v(p) f_n \, dp \right| \leq c_0.$$

Therefore, we can extract a weakly-$\star$ convergent subsequence $(f_{n_k})_{k\in\mathbb{N}}$ in $L^\infty(\mathbb{T}\times\mathbb{R})$, and letting k go to ∞ proves part b) of Theorem 0.1 for any bounded electric field.

Step 3: Properties of the current
Let us discuss the sign of the current of the solution obtained in b), which we denote by f_* from now on. To this end we use another entropy dissipation relation. We combine the equalities

$$(v(p)\partial_x + F(x)\partial_p)(f_* \ln f_*) = Q(f_*)(1 + \ln f_*),$$
$$\frac{1}{\theta}(v(p)\partial_x + F(x)\partial_p)(f_*(\mathcal{E}(p) + q\phi(x))) = \frac{1}{\theta} Q(f_*)(\mathcal{E}(p) + q\phi(x)),$$

and we deduce that

$$(v(p)\partial_x+F\partial_p)(f_*(\ln f_*+\theta^{-1}(\mathcal{E}(p)+q\phi(x)))) = Q(f_*)(1+\ln f_*+\theta^{-1}(\mathcal{E}(p)+q\phi(x))).$$

Taking into account the periodicity of f_* and the fact that $\int_{\mathbb{R}} Q(f_*)\,\mathrm{d}p = 0$, integration with respect to $(x,p) \in \mathbb{T}\times\mathbb{R}$ yields

$$\frac{q}{\theta}(\phi(L)-\phi(0))\int_{\mathbb{R}} v(p)f_*\,\mathrm{d}p - \int_{\mathbb{T}}\int_{\mathbb{R}} Q(f_*)\ln\frac{f_*}{M_\theta(p)}\,\mathrm{d}p\,\mathrm{d}x = 0. \tag{3.15}$$

Now using the monotonicity of the ln function we have

$$\begin{aligned}
-\int_{\mathbb{R}} Q(f_*)\ln\frac{f_*}{M_\theta}\,\mathrm{d}p &= -\frac{1}{2}\int_{\mathbb{R}}\int_{\mathbb{R}} s(p,p')\{M_\theta(p)f_*(x,p')-M_\theta(p')f_*(x,p)\} \\
&\quad\times\left\{\ln\frac{f_*(x,p)}{M_\theta(p)}-\ln\frac{f_*(x,p')}{M_\theta(p')}\right\}\,\mathrm{d}p'\,\mathrm{d}p \\
&= \frac{1}{2}\int_{\mathbb{R}}\int_{\mathbb{R}} s(p,p')M_\theta(p)M_\theta(p')\left\{\frac{f_*(x,p')}{M_\theta(p')}-\frac{f_*(x,p)}{M_\theta(p)}\right\} \\
&\quad\times\left\{\ln\frac{f_*(x,p')}{M_\theta(p')}-\ln\frac{f_*(x,p)}{M_\theta(p)}\right\}\,\mathrm{d}p'\,\mathrm{d}p \\
&\geq 0 \qquad (3.16)
\end{aligned}$$

and therefore $q\int_{\mathbb{R}} v(p)f_*\,\mathrm{d}p\int_{\mathbb{T}} E\,\mathrm{d}x \geq 0$. It is worth pointing out that, in the specific case of a constant scattering function, (3.15) leads to the following remarkable identity:

$$\frac{q}{\theta}(\phi(L)-\phi(0))\int_{\mathbb{R}} v(p)f_*\,\mathrm{d}p + \frac{1}{\tau}\int_{\mathbb{T}}\int_{\mathbb{R}}(f_*-\langle f_*\rangle M_\theta(p))\ln\frac{f_*}{\langle f_*\rangle M_\theta(p)}\,\mathrm{d}p\,\mathrm{d}x = 0.$$

Let us now show that

$$\int_{\mathbb{R}} v(p)f_*\,\mathrm{d}p \neq 0.$$

We argue by contradiction. If the current of f_* vanishes, then by (3.15), (3.16) we know that f_*/M_θ do not depend on p, implying that $f_* = \langle f_*\rangle M_\theta$ and therefore

$$(v(p)\partial_x + F(x)\partial_p)(\langle f_*\rangle M_\theta) = Q(\langle f_*\rangle M_\theta) = 0.$$

Hence, we have $f_* = \langle f_*\rangle M_\theta = k\mathcal{M}_{\theta,\phi}$, which is periodic only if $k=0$ (since $\int_{\mathbb{T}} E\,\mathrm{d}x \neq 0$). But it would contradict $\int_{\mathbb{T}}\int_{\mathbb{R}} f_*\,\mathrm{d}p\,\mathrm{d}x = L$.

Step 4: Solution with a given current

It remains to prove the existence of a periodic solution for any given current $j\in\mathbb{R}$, when the electric field has a nonvanishing average.

Given $j\in\mathbb{R}$, we can consider

$$f = \frac{j}{q\int_{\mathbb{R}} v(p)f_*\,\mathrm{d}p}\,f_*,$$

which defines a periodic weak solution with current equal to j. Observe that $\mathrm{sgn} f = \mathrm{sgn}\big(\frac{j}{q\int_{\mathbb{R}} v(p)f_*\,\mathrm{d}p}\big)$. Since $q\int_{\mathbb{R}} v(p)f_*\,\mathrm{d}p\int_{\mathbb{T}} E\,\mathrm{d}x > 0$, we conclude that $\mathrm{sgn} f = \mathrm{sgn}\big(\frac{j}{\int_{\mathbb{T}} E\,\mathrm{d}x}\big)$. The proof of Theorem 0.1 is now completed. □

Acknowledgements

I. Gamba acknowledges support from NSF under grants FRG-DMS-0757450 and DMS 0807712; M. Bostan and T. Goudon acknowledge support from the INRIA program AEN-Fusion. Support from the Institute for Computational Engineering and Sciences at the University of Texas at Austin is also gratefully acknowledged.

References

1. N. Ben Abdallah and J. Dolbeault, *Relative entropies for the Vlasov-Poisson system in bounded domains*, C. R. Acad. Sci. Paris Sér. I Math. **330** (2000), 867–872. MR1771949 (2001m:76112)
2. N. Ben Abdallah, I. M. Gamba, and A. Klar, *The Milne problem for high field kinetic equations*, SIAM J. Appl. Math. **64** (2004), 1709–1736. MR2084207 (2005g:82050)
3. M. Bostan, I.M. Gamba, T. Goudon, and A. Vasseur, *Boundary value problems for the linear Boltzmann equation involving a variable force field*, Preprint.
4. P. Markowich, C. Ringhofer, C. Schmeiser, *Semiconductor equations*, Springer-Verlag, Vienna, 1990. MR1063852 (91j:78011)
5. F. Poupaud, *Runaway phenomena and fluid aproximation under high fields in semiconductor kinetic theory*, Z. Angew. Math. Mech. **72** (1992), 359–372. MR1178932 (93h:82074)
6. ———, *Boundary value problems for the stationary Vlasov-Maxwell system*, Forum Math. **4** (1992), 499–527. MR1176884 (93i:35144)

Laboratoire de Mathématiques de Besançon, UMR CNRS 6623, Université de Franche-Comté, 16 route de Gray, 25030 Besançon cedex, France

E-mail address: mbostan@univ-fcomte.fr

ICES & Department of Mathematics, University of Texas at Austin, Texas 78712 USA

E-mail address: gamba@math.utexas.edu

Project-Team SIMPAF–INRIA Lille Nord Europe Research Centre & Laboratoire Paul Painlevé, UMR CNRS 8524-USTL, Park Plazza 40, avenue Halley F-59650 Villeneuve d'Ascq cedex, France

E-mail address: thierry.goudon@inria.fr

Amer. Math. Soc. Transl.
(2) Vol. **229**, 2010

Smooth Approximations of Solutions to Nonconvex Fully Nonlinear Elliptic Equations

Luis Caffarelli and Luis Silvestre

To Nina Uraltseva with our admiration.

ABSTRACT. We show that fully nonlinear elliptic PDEs (which may not have classical solutions) can be approximated with integro-differential equations which have $C^{2,\alpha}$ solutions. For these approximated equations we prove a uniform $C^{1,\alpha}$ estimate. We also study the rate of convergence.

1. Introduction

Consider a Dirichlet problem for a nonconvex fully nonlinear elliptic equation

$$\begin{aligned} F(D^2u(x)) &= f(x) \quad \text{in } \Omega, \\ u(x) &= g(x) \quad \text{on } \partial\Omega. \end{aligned}$$

In the case of F convex or concave, the Dirchlet problem is known to have classical solutions provided that $\partial\Omega$, f and g are smooth enough. In the nonconvex case, we only have solutions in the viscosity sense, which may not be C^2 even if f is very smooth.

The possibility of approximating solutions of nonconvex fully nonlinear equations by classical solutions has been a long standing issue. The recent work [**3**] provided such an approximation, but it required considerably technical arguments. The ideas in this article are based on the work [**4**] on integral fully nonlinear equations.

Let φ be a compactly supported, smooth, symmetric, probability density and given a smooth function u, let

$$I_\varepsilon(x) = \int_{\mathbb{R}^n} (u(x+y) - u(x)) \frac{1}{\varepsilon^{n+2}} \varphi\left(\frac{y}{\varepsilon}\right) \, \mathrm{d}y.$$

If we let $\varepsilon \to 0$, I_ε converges to an integral on the unit sphere

$$I_0(x) = \int_{S_1} u_{\sigma\sigma}(x) \Phi(\sigma) \, \mathrm{d}\sigma,$$

where the weight $\Phi(\sigma)$ depends on the integral of φ in the direction σ:

$$\Phi(\sigma) = \int_0^\infty t^{3-n}\varphi(t\sigma)\, dt. \tag{1.1}$$

In turn, $u_{\sigma\sigma} = \sum \sigma_i\sigma_j\partial_{ij}u$ and thus

$$I_0(x) = \sum a_{ij}\partial_{ij}u,$$

where the coefficients a_{ij} are

$$a_{ij} = \int_{S_1} \sigma_i\sigma_j\Phi(\sigma)\, d\sigma.$$

Conversely, given an elliptic (positive definite) matrix $\{a_{ij}\} = B^2 = B^tB$, if φ is a radially symmetric probability density so that $\int_{\mathbb{R}^n} |x|^2\varphi(x)\, dx = 2n$ and we define

$$\tilde{\varphi}(x) = \frac{1}{\det B}\varphi(B^{-1}x),$$

then the resulting operator of the limiting process with $\tilde{\varphi}$ is indeed $a_{ij}\partial_{ij}u$.

Consider now the Isaacs fully nonlinear equation

$$\begin{aligned} &F(D^2u) = f(x) \quad \text{in } \Omega, \\ &\text{where } F(D^2u) = \inf_\beta \sup_\gamma a_{ij}^{\gamma\beta}\partial_{ij}u = \inf_\beta \sup_\gamma L_{\gamma\beta}u. \end{aligned} \tag{1.2}$$

Assume that the equation is uniformly elliptic: there exist constants $0 < \lambda \le \Lambda$ such that for every index α, β, $\lambda I \le \{a_{ij}^{\alpha\beta}\} \le \Lambda I$. Thus, we can write every linear operator as

$$a_{ij}^{\alpha\beta}\partial_{ij}u = \frac{\lambda}{2}\triangle u + \lim_{\varepsilon\to 0} I_\varepsilon^{\alpha\beta}(u),$$

where

$$I_\varepsilon^{\alpha\beta}u = \int_{\mathbb{R}^n} (u(x+y) - u(x))\frac{1}{\varepsilon^{n+2}\det B_{\alpha\beta}}\varphi\left(B_{\alpha\beta}^{-1}\frac{y}{\varepsilon}\right)\, dy,$$

and $B_{\alpha\beta}^2 = \{a_{ij}^{\alpha\beta}\} - \frac{\lambda}{2}I$. The function φ is a compactly supported radially symmetric probability density with second moment equal to one in any direction.

The above discussion suggests that solutions v of the equation

$$E_\varepsilon(v) = \frac{\lambda}{2}\triangle v + \inf_\beta \sup_\alpha I_\varepsilon^{\alpha\beta}(v) = f(x) \tag{1.3}$$

would be, as ε goes to zero, approximations of the solution u to the original fully nonlinear elliptic PDE (1.2). The interesting fact about these approximations is that for every $\varepsilon > 0$ their solutions v are $C^{2,\alpha}$ functions (with estimates depending on ε) if $f \in C^\alpha$. Moreover, since the first term $\frac{\lambda}{2}\triangle v$ is fixed and the second term has some type of ellipticity, it is possible to obtain some a priori estimates independent of ε which coincide, in fact, with known regularity results for fully nonlinear elliptic PDE without convexity assumptions.

In this note we intend to prove existence, smoothness, and rate of convergence to the solution as $\varepsilon \to 0$. We will prove that the problem is well posed for every $\varepsilon > 0$ once the boundary value g is extended in a neighborhood of Ω. Then we will show that the equation has an interior $C^{1,\alpha}$ estimate independent of ε (corresponding to the interior $C^{1,\alpha}$ estimate for fully nonlinear PDEs). In the last section, we will show that the rate of convergence of the solutions to our approximated problem to the solution to (1.2) is of the order ε^α for some $\alpha > 0$.

2. Some simple properties of the approximate equation

In this section we construct the approximate equation and observe some elementary properties.

To fix ideas, let us choose a smooth, radially symmetric function φ that is supported in B_2, $0 \le \varphi \le 1$ in its support, and

$$\int_{\mathbb{R}^n} |x|^2 \varphi(x)\, dx = 2n.$$

This choice is made so that for any C^2 function u,

$$\lim_{\varepsilon \to 0} \int_{\mathbb{R}^n} (u(x+y) - u(x)) \frac{1}{\varepsilon^{n+2}} \varphi\left(\frac{y}{\varepsilon}\right)\, dy = \triangle u(x).$$

Also,

$$\lim_{\varepsilon \to 0} \int_{\mathbb{R}^n} (u(x+y) - u(x)) \frac{1}{\varepsilon^{n+2} \det(B)} \varphi\left(B^{-1} \frac{y}{\varepsilon}\right)\, dy = a_{ij} \partial_{ij} u(x),$$

where $B^2 = \{a_{ij}\}$.

By construction all matrices $B_{\alpha\beta}$ in the definition of E_ε have their eigenvalues bounded above and below, so the support of $\frac{1}{\det B} \varphi(B^{-1}x)$ will always be contained in a ball B_Q, where Q is a universal constant given by $2/\sqrt{\lambda}$. Therefore, the value of $E_\varepsilon u(x)$ depends only on the values of u in $B_{Q\varepsilon}(x)$. This universal constant Q will appear in several estimates in this paper.

Let g be a continuous function defined on $\partial\Omega$. We want to solve the approximated problem $E_\varepsilon u = f$ in Ω and $u = g$ on $\partial\Omega$. However, in order for every $I_\varepsilon^{\alpha\beta}$ to be well defined, we need to extend g to a neighborhood of Ω of size $Q\varepsilon$.

A natural continuous extension of g to $\mathcal{C}\Omega$ can be done by a *sup-convolution*:

$$\tilde{g}(x) = \sup_{x^* \in \partial\Omega} g(x^*) - |x - x^*|^2.$$

For the existence of the solution of the equation for a positive value ε, any continuous extension would work. For computing the rates of convergence, it is better to extend g in a way that the extension is as regular as the original function g. If g is $C^{1,1}$ and Ω has the exterior ball condition, then $\tilde{g}$ will be $C^{1,1}$ in a neighborhood of $\partial\Omega$. The approximated problem that we will study is

$$\begin{aligned} E_\varepsilon u &= f(x) \quad \text{in } \Omega, \\ u &= \tilde{g} \quad \text{in } (\partial\Omega)_{Q\varepsilon}, \end{aligned} \tag{2.1}$$

where $(\partial\Omega)_{Q\varepsilon} = \{x \in \mathcal{C}\Omega : \operatorname{dist}(x, \partial\Omega) < Q\varepsilon\}$ is a $Q\varepsilon$ neighborhood of $\partial\Omega$.

We point out the scaling of the equation. If u is a solution of $E_\varepsilon u = f$, then $\bar{u}(x) = su(tx)$ is a solution of $\bar{E}_{t\varepsilon}(u) = st^2 f$, where $\bar{E}_{t\varepsilon}$ is also an operator as in (1.3) but with $t\varepsilon$ instead of ε.

We note that the equation is constructed so that $E_\varepsilon P = F(D^2 P)$ for any quadratic polynomial P, where F is the original fully nonlinear equation. The following lemmas provide more precise statements.

LEMMA 2.1. *Assume φ is radially symmetric, nonnegative and supported in a ball B_R. If u is superharmonic in $B_R(x)$, then*

$$Iu(x) = \int_{\mathbb{R}^n} (u(x+y) - u(x)) \varphi(y)\, dy \le 0.$$

PROOF. It is a simple average of the mean value theorem. □

COROLLARY 2.2. *Let u be a function such that for a fixed matrix a_{ij},*

$$a_{ij}\partial_{ij}u \leq 0 \quad \textit{in a ball } B_{r+Q\varepsilon} \quad (\textit{resp.} \geq 0).$$

Then also

$$\int_{B_{Q\varepsilon}} (u(x+y) - u(x)) \frac{1}{\det B} \varphi(B^{-1}y) \, dy \leq 0 \quad \textit{in } B_r \quad (\textit{resp.} \geq 0).$$

PROOF. We apply Lemma 2.1 to $\tilde{u}(x) = u(Bx)$. □

COROLLARY 2.3. *Let λ_{min} and λ_{max} be the smallest and largest eigenvalues of D^2u. Assume that in a ball $B_{Q\varepsilon}$, $\lambda_{min} \leq 0 \leq \lambda_{max}$ and $-\lambda_{min} \geq c(\lambda, \Lambda)\lambda_{max}$, φ as above. Then*

$$\int_{B_{Q\varepsilon}} (u(x+y) - u(x)) \frac{1}{\det B} \varphi(B^{-1}y) \, dy \leq 0.$$

Here B is a positive definite matrix, and λ and Λ above refer to the ellipticity constants.

PROOF. Under the change of variables $y^* = B^{-1}y$, u remains superharmonic if $C(\lambda, \Lambda)$ is large enough. □

Recalling that E_ε is given by the Laplacian plus an inf-sup of integral operators as in Corollary 2.3, we have the following corollary.

COROLLARY 2.4. *If λ_{min} and λ_{max} are the smallest and largest eigenvalues of D^2u, $\lambda_{min} \leq 0 \leq \lambda_{max}$ and $-\lambda_{min} \geq c(\lambda, \Lambda)\lambda_{max}$, φ as above, then $E_\varepsilon u \leq 0$.*

3. Subsolutions, supersolutions, and Perron's method

In this section we will prove that the approximated problem has a unique solution using Perron's method. The smoothness of the integral part of the equation allows us to use very classical arguments.

We will assume that the domain Ω is connected and the classical Laplace equation is solvable for any continuous Dirichlet data. We will use this to show that the solution to our approximated problem also achieves its boundary data g continuously. For example we can consider a connected domain Ω with a uniform external ball condition: there exists a ρ_0 such that for every point $x \in \partial\Omega$, there exists a ball $B_{\rho_0}(y)$ contained in $\mathcal{C}\Omega$ such that $x \in \partial B_{\rho_0}(y)$.

Let us analyze the structure of the approximated equation E_ε. For each pair of indexes α and β,

$$\begin{aligned} I_\varepsilon^{\alpha\beta} &= \int_{\mathbb{R}^n} (u(x+y) - u(x)) \frac{\varphi(B_{\alpha\beta}^{-1}y)}{\varepsilon^{n+2}\det(B_{\alpha\beta})} \, dy \\ &= \int_{\mathbb{R}^n} u(x+y) \frac{\varphi(B_{\alpha\beta}^{-1}y)}{\varepsilon^{n+2}\det(B_{\alpha\beta})} \, dy - \frac{1}{\varepsilon^2} \left(\int \varphi(y) \, dy \right) u(x). \end{aligned}$$

Thus, we can write

$$E_\varepsilon(x) = \frac{\lambda}{2} \triangle u - \frac{C}{\varepsilon^2} u + H_\varepsilon u,$$

where $C = \int \varphi$ and H_ε is the inf-sup of convolutions of u with smooth functions

$$H_\varepsilon u(x) = \inf_\alpha \sup_\beta \int_{\mathbb{R}^n} u(x+y) \frac{\varphi(B_{\alpha\beta}^{-1}y)}{\varepsilon^{n+2}\det(B_{\alpha\beta})} \, dy.$$

Since we are using a φ that is smooth and compactly supported, for every index α and β, the integrals above are uniformly smooth (for given $\varepsilon > 0$) even if u is only L^1_{loc}. Therefore, if $u \in L^1_{loc}$, the function $H_\varepsilon u$ will be locally Lipschitz. This allows us to understand the notion of subsolutions and supersolutions in a classical distributional sense. Even for $u \in L^1_{loc}$, we can make sense of

$$\frac{\lambda}{2}\triangle u - \frac{C}{\varepsilon^2}u + H_\varepsilon u \geq f(x)$$

in the sense of distributions since the third term is just a Lipschitz function.

The first two terms $\frac{\lambda}{2}\triangle u - \frac{C}{\varepsilon^2}u$ correspond to a linear elliptic operator with constant coefficients. In the following proposition, we list some classical properties of linear ellipic equations with constant coefficients that will be useful in this section of the paper.

PROPOSITION 3.1. (a) *If the function u satisfies $\frac{\lambda}{2}\triangle u - \frac{C}{\varepsilon^2}u = f$ for some constant C in an open set Ω, then u has two derivatives more than f. More precisely, if $f \in C^\alpha$, then $u \in C^{2,\alpha}$ in the interior of Ω.*

(b) *The Dirichlet problem*

$$\begin{aligned} \frac{\lambda}{2}\triangle u - \frac{C}{\varepsilon^2}u &= -f && \text{in } B_r, \\ u &= v && \text{on } \partial B_r \end{aligned}$$

has nonnegative Green and Poisson functions G_r and P_r such that

$$u(x) = \int_{B_r} G_r(x,y)f(y)\,\mathrm{d}y + \int_{\partial B_r} P_r(x,y)v(y)\,\mathrm{d}\sigma(y).$$

(c) *The inequality $\frac{\lambda}{2}\triangle u - \frac{C}{\varepsilon^2}u \leq D$ in an open set Ω is equivalent to the fact that for every $x \in \Omega$ and every $B_r(x) \subset \Omega$, the following inequality holds:*

$$u(x) \geq \int_{B_r} w(|y|)u(y)\,\mathrm{d}y - g(r)D, \tag{3.1}$$

where $w(r)$ and $g(r)$ are the nonnegative functions given by

$$\begin{aligned} w(|y|) &= \frac{1}{r}P(0,y), \\ g(r) &= \frac{1}{r}\int_0^r \int_{B_s} G_s(0,y)\,\mathrm{d}y\,\mathrm{d}s. \end{aligned}$$

PROOF. Parts (a) and (b) are very classical and hold for any operator of the form $a\triangle u - bu$ for any $a > 0$ and $b \geq 0$. Part (c) is just the result of integrating part (b) and taking the average. Note that $P(0,y)$ depends only on $|y|$ because the problem is radially symmetric. □

Note that since the function $u \equiv 1$ satisfies $\frac{\lambda}{2}\triangle u - \frac{C}{\varepsilon^2}u = -\frac{C}{\varepsilon^2}$, then we have the identity

$$\int_{B_r} w(|y|)\,\mathrm{d}y + \frac{C}{\varepsilon^2}g(r) = 1. \tag{3.2}$$

It is also classical that the inequality $\frac{\lambda}{2}\triangle u - \frac{C}{\varepsilon^2}u \leq D$ in the weak sense is equivalent to u being lower semicontinuous and satisfying (3.1).

LEMMA 3.2. *If $u \in L^1_{loc}$ is a supersolution (resp. subsolution) of the equation*

$$E_\varepsilon u = \frac{\lambda}{2}\triangle u - \frac{C}{\varepsilon^2}u + H_\varepsilon u \leq f(x) \quad (\textit{resp. } \geq f(x)) \ \textit{ in } \Omega,$$

then u is lower (resp. upper) semicontinuous.

PROOF. From the equation we see that locally $\frac{\lambda}{2}\triangle u - \frac{C}{\varepsilon^2}u$ is bounded above. This is a standard linear elliptic equation with constant coefficients. It is a classical result that this implies that u has a lower semicontinuous representative. □

We now prove the strong comparison principle.

LEMMA 3.3 (Comparison principle). *Assume Ω is a connected domain and f is continuous. Let u and v be a subsolution and a supersolution respectively. If $u - v$ has an absolute maximum at some interior point in Ω, then $u - v$ is constant in Ω.*

PROOF. From Lemma 3.2, u is upper semicontinuous and v is lower semicontinuous. Thus $u - v$ is upper semicontinuous and it achieves its maximum in $\overline{\Omega}$.

Assume $u - v$ assumes a maximum at some interior point $x \in \Omega$. For any indexes α, β, we would have

$$I_\varepsilon^{\alpha\beta}(u-v)(x) = \frac{1}{\varepsilon^{n+2}\det B_{\alpha\beta}} \int ((u-v)(x+y) - (u-v)(x))\varphi\left(B_{\alpha\beta}^{-1}\frac{y}{\varepsilon}\right)\, dy.$$

We see that since $u-v$ achieves its maximum at x, then $(u-v)(x+y) \leq (u-v)(x)$ for every y small. Moreover, either $u = v$ in a neighborhood of x or $I_\varepsilon^{\alpha\beta}[u-v](x) \leq -\eta_0$ for any α, β (with η_0 depending on the function $u - v$ and the ellipticity constants λ and Λ). Assume the latter, for all indexes α and β, $I_\varepsilon^{\alpha\beta}u(x) \leq I_\varepsilon^{\alpha\beta}v(x) - \eta_0$. Therefore

$$-\frac{C}{\varepsilon^2}u(x)+H_\varepsilon u(x) = \inf\sup I_\varepsilon^{\alpha\beta}u(x) \leq \inf\sup I_\varepsilon^{\alpha\beta}v(x) - \eta_0 = -\frac{C}{\varepsilon^2}v(x)+H_\varepsilon v(x)-\eta_0.$$

Let $h = v - u$. Regrouping the terms, the inequality above says that

$$h(x) \leq \frac{\varepsilon^2}{C}(H_\varepsilon v(x) - H_\varepsilon u(x) - \eta_0).$$

So, by (3.2),

$$h(x) \leq \int_{B_r} w(|y|)h(x)\, dy + g(r)(H_\varepsilon v(x) - H_\varepsilon u(x) - \eta_0). \tag{3.3}$$

On the other hand, h satisfies the equation

$$\frac{\lambda}{2}\triangle h - \frac{C}{\varepsilon^2}h = H_\varepsilon u - H_\varepsilon v.$$

The functions $H_\varepsilon u$ and $H_\varepsilon v$ are Lipschitz. Thus, for every y in a neighborhood of x, $|H_\varepsilon u(y) - H_\varepsilon u(x)| \leq \eta_0/4$ and $|H_\varepsilon v(y) - H_\varepsilon v(x)| \leq \eta_0/4$. Then, in a neighborhood of x we have

$$\frac{\lambda}{2}\triangle h - \frac{C}{\varepsilon^2}h \leq H_\varepsilon u(x) - H_\varepsilon v(x) + \frac{1}{2}\eta_0.$$

So, from Proposition 3.1, for r small we have

$$\begin{aligned} h(x) &\geq \int_{B_r} w(|y|)h(y)\, dy - g(r)(H_\varepsilon u(x) - H_\varepsilon v(x) + \frac{1}{2}\eta_0) \\ &\geq \int_{B_r} w(|y|)h(x)\, dy + g(r)(H_\varepsilon v(x) - H_\varepsilon u(x) - \frac{1}{2}\eta_0). \end{aligned}$$

But this is a contradiction with (3.3).

Thus we conclude that u must be equal to v in a neighborhood of x. By the classical connectedness argument, $u = v$ in the whole domain. □

The uniqueness of the solution to the equation (2.1) is an immediate consequence of the comparison principle. We will also use it to prove the existence of a solution by following a more or less classical Perron's method approach.

Let u be the infimum of all supersolutions

$$\begin{aligned} E_\varepsilon v &\leq f \quad \text{in } \Omega, \\ v &\geq g \quad \text{in } (\partial\Omega)_{Q\varepsilon}. \end{aligned}$$

We will prove that u is the solution to equation (2.1) provided that f and g are continuous. As usual for Perron's method, we divide the proof in showing that u is a solution to the equation inside the domain Ω and showing that u achieves continuously the boundary values on $\partial\Omega$.

First of all, we must make sure that the infimum u is well defined.

LEMMA 3.4. *Assume that g and f are bounded continuous functions. The function u above is well defined and bounded.*

PROOF. We must find an upper and lower bound.

Let $M = \sup |f|$ and $N = \sup |g|$. Let R be a large enough radius such that the ball B_R contains an ε neighborhood of Ω. Then it is an elementary computation that the functions

$$b^\pm(x) = M(R^2 - |x|^2)^\pm \pm N$$

are respectively a supersolution and a subsolution of equation (2.1) since they are second-order polynomials and thus the value of $E_\varepsilon b^\pm$ coincides with $F(D^2 b^\pm)$.

By the comparison principle (Lemma 3.3), every supersolution v is larger than b^- in the whole domain Ω.

On the other hand, for every supersolution v, the function $\min v, b^+$ is also a supersolution. So u is equal to the infimum of all functions $\min v, b^+$, which are all bounded below and above by b^- and b^+. Thus $b^- < u < b^+$. □

THEOREM 3.5. *Let f and g be continuous functions and Ω be a domain for which the classical Dirichlet problem for the Laplace equation has a continuous solution up to the boundary. The function u above satisfies $E_\varepsilon u = f$ in Ω, $u = g$ in $(\partial\Omega)_{Q\varepsilon}$ and it is continuous.*

PROOF. We can consider only supersolutions v that are bounded above and below by b^+ and b^- as in the proof of Lemma 3.4. Let M be the maximum value of $|b^\pm|$, so that all v satisfy $|v| \leq M$.

By the boundedness of v and the smoothness of φ, we know that $H_\varepsilon v$ is uniformly bounded and Lipschitz. In particular, there is a constant C such that for every supersolution v,

$$\frac{\lambda}{2}\Delta v \leq f + \frac{C}{\varepsilon^2} v - H_\varepsilon v \leq C.$$

It follows from Green's formula that a constant upper bound on the Laplacian is equivalent to the inequality

$$v(x) \geq \frac{1}{|B_r|} \int_{B_r(x)} v(y)\, \mathrm{d}y - C\omega_n r^2$$

for any ball $B_r(x) \subset \Omega$, where ω_n is a dimensional constant and C is the same constant as in the inequality above.

This condition is clearly preserved by taking the infimum, so $\triangle u$ is also bounded above by the same constant and consequently u is lower semicontinuous.

We will prove now that $E_\varepsilon u \leq f$ in Ω.

Let v_k be a sequence of supersolutions such that at a given point x, $v_k(x) \to u(x)$.

Let $D = \liminf_{j\to\infty} H_\varepsilon v_j(x)$.

For any $\eta > 0$, let us pick a large k_0 such that

$$H_\varepsilon v_k(x) \geq D - \eta \quad \text{for all } k \geq k_0.$$

Since u is lower semicontinuous, $u(y) \geq u(x) - \eta$ in a neighborhood of x. Since $v_k \geq u$ everywhere, then $v_k(y) \geq u(y) \geq u(x) - \eta$ in a neighborhood of x.

Since the functions $H_\varepsilon v_j$ are uniformly Lipschitz and f is continuous. We can consider a smaller neighborhood if necessary so that for every y in that neighborhood, $|H_\varepsilon v_k(y) - H_\varepsilon v_k(x)| \leq \eta$ and $|f(y) - f(x)| \leq \eta$. Therefore, for every y in a neighborhood of x and $k \geq k_0$,

$$\begin{aligned}\frac{\lambda}{2}\triangle v_k(y) - \frac{C}{\varepsilon^2} v_k(y) &= f(y) - H_\varepsilon v_k(y)\\ &\leq f(x) - H_\varepsilon v_k(x) + 2\eta\\ &\leq f(x) - D + 3\eta.\end{aligned}$$

We can then find a small radius $r_0 > 0$ such that $\frac{\lambda}{2}\triangle v_k(y) - \frac{C}{\varepsilon^2} v_k(y) \leq f(x) - D + 3\eta$ in $B_{r_0}(x)$. From Proposition 3.1, this is equivalent to the fact that for $r < r_0$,

$$v_k(x) \geq \frac{1}{|B_r|}\int_{B_r(x)} w(|y|) v_k(y)\, \mathrm{d}y - (2\frac{f(x) - D + 3\eta}{\lambda})g(r).$$

Since $v_k(y) \geq u(y)$ and $H_\varepsilon u(x) \leq D$,

$$v_k(x) \geq \frac{1}{|B_r|}\int_{B_r(x)} w(|x|) u(y)\, \mathrm{d}y - (2\frac{f(x) - H_\varepsilon(x) + 3\eta}{\lambda})g(r).$$

Taking the limit in the right-hand side as $k \to +\infty$, we have

$$u(x) \geq \frac{1}{|B_r|}\int_{B_r(x)} w(|x|) u(y)\, \mathrm{d}y - (2\frac{f(x) - H_\varepsilon(x) + 3\eta}{\lambda})g(r).$$

This holds for every $x \in \Omega$ and r small enough (depending on η and x). This is saying that $\frac{\lambda}{2}\triangle u - \frac{C}{\varepsilon^2} u \leq -H_\varepsilon u + f + 4\eta$ for any η. Therefore $E_\varepsilon u \leq f$ in Ω and u is a supersolution.

Now we have to show that u is also a subsolution by proving the opposite inequality. Since u is a supersolution of (2.1), we know that $f - H_\varepsilon u + \frac{C}{\varepsilon^2} u - \frac{\lambda}{2}\triangle u$ is a nonnegative measure μ. Assume μ is nonzero. So there must be some point x_0 and $\eta > 0$ where $\mu(B_r(x_0)) \geq \eta r^n$ if r is small enough.

Since $H_\varepsilon u$ is a Lipschitz function and f is continuous, we can find an $r > 0$ such that

$$\sup_{B_r(x_0)} |-H_\varepsilon u + f + H_\varepsilon u(x_0) - f(x_0)| < c\eta$$

for c arbitrarily small.

Let us substitute u in a tiny ball $B_\rho(x)$ by the solution to

$$-\frac{C}{\varepsilon^2}\bar{u} + \frac{\lambda}{2}\triangle\bar{u} = -H_\varepsilon u(x_0) + f(x_0) - 2c\eta \qquad \text{in } B_\rho(x_0), \tag{3.4}$$

$$\bar{u} = u \qquad \text{on } \partial B_\rho(x_0), \tag{3.5}$$

so that in B_ρ, $\frac{C}{\varepsilon^2}(\bar{u} - u) - \frac{\lambda}{2}\triangle(\bar{u} - u) \geq \mu - 3c\eta$ with $\mu(B_{\rho/2}(x_0)) \geq \eta\rho^n/2^n$. If we choose c small enough (by making ρ small) this implies that $\bar{u} < u$ in $B_\rho(x_0)$ (strict inequality).

Since $\bar{u} \leq u$ inside $B_\rho(x_0)$, $\bar{u} = u$ outside $B_\rho(x_0)$, and $E_\varepsilon u \leq f$, then also $E_\varepsilon\bar{u} \leq f$ outside $B_\rho(x_0)$.

On the other hand, the value of the difference of the integral terms $|H_\varepsilon u - H_\varepsilon\bar{u}|$ can be made arbitrarily small by taking ρ small enough ($\rho \ll \varepsilon$). In particular, it can be made smaller than $c\eta$. Therefore $\bar{u}$ will also be a supersolution of (2.1) inside $B_\rho(x_0)$ if ρ is very small. But then $\bar{u} \geq u$ and this is a contradiction.

The contradiction came from assuming that $f - \inf\sup I_\varepsilon^{\alpha\beta}u - \frac{\lambda}{2}\triangle u$ was nonzero. So u must be a solution in Ω.

By construction $u \geq g$ in $(\partial\Omega)_{Q\varepsilon}$. Let us show that $u = g$ in $(\partial\Omega)_{Q\varepsilon}$ and u is continuous on $\partial\Omega$.

Since all v are uniformly bounded, then every integral operator $I_\varepsilon^{\alpha\beta}v$ is uniformly bounded for each $\varepsilon > 0$ (depending on ε). That means that we can find a simple supersolution and a subsolution by solving the following problems:

$$\triangle s^{\pm}(x) = \mp C \quad \text{in } \Omega,$$
$$s = g \quad \text{in } (\partial\Omega)_{Q\varepsilon},$$

where C is the upper bound for all $|I_\varepsilon^{\alpha\beta}v|$. Therefore the functions $\min(s^+, M)$ and $\max(s^-, -M)$ (recall $|v| \leq M$) are a supersolution and a subsolution, respectively, which have the same value on $(\partial\Omega)_{Q\varepsilon}$ and are continuous on $\partial\Omega$. Therefore u must be in between the two, which implies that $u = g$ in $(\partial\Omega)_{Q\varepsilon}$ and u is continuous on $\partial\Omega$. □

4. Smoothness

We start this section by pointing out that if $f \in C^\alpha$, for every $\varepsilon > 0$, the solution u of

$$E_\varepsilon u = f \quad \text{in } \Omega,$$
$$u = \tilde{g} \quad \text{in } (\partial\Omega)_{Q\varepsilon}$$

is a C^2 function in the interior of Ω. Indeed, from construction it is a continuous function. But the integral term $H_\varepsilon u$ is Lipschitz. Since u satisfies the equation

$$\frac{\lambda}{2}\triangle u - \frac{C}{\varepsilon^2}u = f(x) - H_\varepsilon u,$$

the right-hand side is C^α and the right-hand side is a linear elliptic operator with constant coefficients that has $C^{2,\alpha}$ estimates. Thus $u \in C^{2,\alpha}$ by the classical estimates for the linear elliptic equations with constant coefficients.

Of course the estimate above on the C^2 norm of u depends on the value of ε. In this section we will prove an interior $C^{1,\alpha}$ independent of ε. The fact that $u \in C^2$ for every $\varepsilon > 0$ means that we are only dealing with classical solutions. In this section we assume that $f \in C^\alpha$ for convenience, but the estimates depend on $\|f\|_{L^\infty}$.

Since the $C^{1,\alpha}$ estimate obtained in this section is independent of ε, it becomes a regularity result for fully nonlinear equations as $\varepsilon \to 0$. The results in this section correspond to the results in chapter 5 in [**2**] for fully nonlinear elliptic equations. On the other hand, one advantage of dealing with approximate equations as in this paper is that since all solutions are classical, many technicalities in the proofs are not necessary.

We obtain a $C^{1,\alpha}$ estimate by applying the following proposition to the first derivatives of the solution of (2.1).

LEMMA 4.1. *Let $0 < \lambda \leq \Lambda$ and $\varepsilon > 0$. Let u be a solution to the following equation:*

$$\frac{\lambda}{2}\triangle u + \int_{\mathbb{R}^n} (u(x+y) - u(x)) \frac{1}{\varepsilon^{n+2}\det B(x)} \varphi\left(B(x)^{-1}\frac{y}{\varepsilon}\right) \,\mathrm{d}y = f(x) \quad \text{in } B_1,$$

where $B : B_1 \to \mathbb{R}^{n\times n}$ is a matrix-valued function such that for every x, $\sqrt{\lambda} I \leq B(x) \leq \sqrt{\Lambda} I$ and f is a bounded function. Then u satisfies the estimate

$$||u||_{C^\alpha(B_{1/2})} \leq C(||u||_{L^\infty(B_{1+Q\varepsilon})} + ||f||_{L^\infty(\Omega)}),$$

where C depends on λ, Λ and n, but not on ε or on any modulus of continuity of B.

Note that the matrices $B(x)$ can change from point to point. The estimate in Lemma 4.1 does not depend on any modulus of continuity of $B(x)$. This is crucial to obtain $C^{1,\alpha}$ estimates for fully nonlinear equations since we do not have any control a priori of the continuity of the coefficients of the linearized equation for the derivatives of the solution.

Instead of proving Lemma 4.1, we will prove a more general result. Because of the bounds from above and below for the eigenvalues of $B(x)$, for every x the function $\varphi\left(B(x)^{-1}\frac{y}{\varepsilon}\right)$ (as a function of y) is nonnegative and supported in some ball $B_{Q\varepsilon}$. So Lemma 4.1 is a particular case of the following lemma.

THEOREM 4.2. *Let $Q > 0$ and $\varepsilon > 0$. Let u be a (classical) solution to the following equation:*

$$L_\varepsilon u := \triangle u + \int_{B_{Q\varepsilon}} (u(x+y) - u(x)) k(x,y) \,\mathrm{d}y = f(x) \quad \text{in } B_1, \tag{4.1}$$

where $k(x,y)$ is a nonnegative function, symmetric in y, such that $k(x,y) \leq \varepsilon^{-n-2}$ if $|y| < Q\varepsilon$ and f is a bounded function. Then u satisfies the estimate

$$||u||_{C^\alpha(B_{1/2})} \leq C(||u||_{L^\infty(1+Q\varepsilon)} + ||f||_{L^\infty(\Omega)}),$$

where C depends on Q and n, but not on ε or on any modulus of continuity of k.

The proof of Theorem 4.2 uses the classical idea of showing that the oscillation in dyadic balls decreases geometrically. For that we will show a growth lemma, whose proof depends on the scale even though the estimate is uniform in scale at the end.

We recall the scaling of the equation. If u is a solution of $L_\varepsilon u = f$ for some operator L_ε as in Theorem 4.2, then $\bar{u}(x) = su(tx)$ is a solution of $\bar{L}_{t\varepsilon}(u) = st^2 f$, where $\bar{L}_{t\varepsilon}$ is also an operator as in Theorem 4.2 but with $t\varepsilon$ instead of ε.

There are two different scales in this problem. When looking at a scale larger than ε, the ellipticity of the integral part of L plays a role. When looking at a finer scale than ε, then the integral term in the equation can be considered just a smooth perturbation for the Laplace equation.

If we want to prove a Hölder continuity result, we must be able to show that the oscillation of the function decreases at all scales. When looking at a fine scale, we must consider rescalings of the original function of the form $\rho^\alpha u(x/\rho)$, which will solve an equation for an operator $L_{\varepsilon/\rho}$ with ε/ρ large if ρ is smaller than ε.

In the next few lemmas, we write e instead of ε to stress that we will apply the lemmas at different scales. If we apply it at scale ρ, we would need to consider an operator $L_{\varepsilon/\rho}$ as above for which $e = \varepsilon/\rho$ may be large.

LEMMA 4.3. *Assume $e > e_0$ (for a large e_0). There exists an $\eta_0 > 0$, $0 < \mu < 1$ and $M > 1$ depending only on Q and n such that if*

$$\Delta u + \int_{B_{Qe}} (u(x+y) - u(x))k(x,y)\, dy \leq \eta_0 \quad \textit{in } B_1,$$

$$u \geq 0 \quad \textit{in } B_{1+Qe},$$

$$\inf_{B_{1/2}} u \leq 1,$$

then $|\{u > M\} \cap B_{1/4}| \geq \mu$.

PROOF. Let $v = \min(u/M, 1)$. Since $e > e_0$, we can control the L^∞ norm of the integral term

$$\int_{B_{Qe}} (u(x+y) - u(x))k(x,y)\, dy \geq -\int k(x,y)\, dy \geq -Ce_0^{-2} \geq -\eta_0$$

if e_0 is large enough. Thus, we obtain an estimate for the plain Laplacian of the function

$$\Delta v \leq 2\eta_0 \text{ in } B_1.$$

On the other hand, since $\inf_{B_{1/2}} u \leq 1$, then $\inf_{b_{1/2}} v \leq 1/M$. Therefore the set $\{v \geq M\}$ cannot cover a large portion of $\bigcap B_{1/4}$ if η_0 is small, i.e.

$$|\{v < M\} \cap B_{1/4}| \geq \mu. \qquad \square$$

LEMMA 4.4. *Assume $e < e_0$, where e_0 is the one from Lemma 4.3. There exists an $\eta_0 > 0$, $0 < \mu < 1$ and $M > 1$ depending only on Q and n such that if*

$$\Delta u + \int_{B_{Qe}} (u(x+y) - u(x))k(x,y)\, dy \leq f(x) \quad \textit{in } B_1,$$

$$u \geq 0 \quad \textit{in } B_{1+Qe},$$

$$\inf_{B_{1/2}} u \leq 1, \quad \textit{and}$$

$$||f||_{L^n(B_1)} \leq \eta_0,$$

then $|\{u < M\} \cap B_{1/4}| \geq \mu$.

In order to prove the lemma above, we prove the following version of the Alexandroff-Backelman-Pucci estimate at a coarse scale.

LEMMA 4.5 (coarse ABP estimate). *Assume $e \leq e_0$. Let $u : B_1 \to \mathbb{R}$ be a function such that $u \geq 0$ in $(\partial B_1)_{Qe_0} = B_{1+Qe_0} \setminus B_1$ and*

$$\Delta u + \int_{B_{eR}} (u(x+y) - u(x))k(x,y)\, dy \leq f(x) \quad \textit{in } B_1$$

for some nonnegative function f with the same assumption in k as in Theorem 4.2.

Let us extend u as zero outside B_1 and let Γ be the convex envelope of u in B_{1+Qe_0}. Then the classical ABP estimate holds:

$$-\min_{B_1} u \leq C \left(\int_{\{u=\Gamma\}} f(x)^n \, dx \right)^{1/n}.$$

PROOF. As in the classical proof of the ABP estimate,

$$-\min_{B_1} u \leq C|\nabla\Gamma(\{u=\Gamma\})|^{1/n} = C \left(\int_{\{u=\Gamma\}} \det(D^2\Gamma)^n \, dx \right)^{1/n}.$$

For every point $x \in \{u = \Gamma\}$, the integral term in the equation is nonnegative,

$$\int_{B_{eR}} (u(x+y) - u(x))k(x,y) \, dy \geq 0,$$

since all incremental quotients are nonnegative if $u(x) = \Gamma(x)$ (because $e < e_0$). Therefore we have $\triangle\Gamma(x) \leq \triangle u(x) \leq f(x)$.

On the other hand, since $D^2\Gamma(x)$ cannot have a negative eigenvalue, then by the arithmetic-geometric mean inequality, $\triangle\Gamma(x)/n \geq \det(D^2\Gamma)^{1/n}$. Thus

$$-\min_{B_1} u \leq C \left(\int_{\{u=\Gamma\}} \det(D^2\Gamma)^n \, dx \right)^{1/n} \leq C \left(\int_{\{u=\Gamma\}} f(x)^n \, dx \right)^{1/n}. \qquad \square$$

PROOF OF LEMMA 4.4. We observe that for large p, a smooth function $b(x)$ given by $(|x|^{-p} - 1)^+$ outside of $B_{1/8}$ and some smooth extension inside $B_{1/8}$ is a subsolution $Lb \geq 0$ outside $B_{1/4}$, and Lb is bounded independently of e ($e > e_0$) inside $B_{1/4}$.

Now we apply ABP to $u - b$ and we proceed as in the proof in [**2**], chapter 4 (Lemma 4.5). $\square$

By combining Lemmas 4.4 and 4.3, we have that the pointwise estimate holds at every scale. We have the corollary that holds for any value of e.

COROLLARY 4.6. *There exists an $\eta_0 > 0$, $0 < \mu < 1$ and $M > 1$ depending only on Q and n such that for any $e > 0$, if*

$$\triangle u + \int_{B_{Qe}} (u(x+y) - u(x))k(x,y) \, dy \leq f(x) \quad \text{in } B_1,$$

$$u \geq 0 \quad \text{in } B_{1+Qe},$$

$$\inf_{B_{1/2}} u \leq 1, \quad \text{and}$$

$$||f||_{L^\infty(B_1)} \leq \eta_0,$$

then $|\{u < M\} \cap B_{1/4}| \geq \mu$.

PROOF. We apply either Lemma 4.4 or Lemma 4.3 depending on whether $e \geq e_0$ or $e < e_0$. $\square$

The previous result implies the L^δ estimate.

COROLLARY 4.7. *There exists an $\eta_0 > 0$, $\delta > 0$ and C depending only on Q and n such that for any $e > 0$, if*

$$\triangle u + \int_{B_{Qe}} (u(x+y) - u(x))k(x,y)\, dy \le f(x) \quad \text{in } B_1,$$

$$u \ge 0 \quad \text{in } B_{1+Q\varepsilon}, \quad \text{and}$$

$$||f||_{L^\infty(B_1)} \le \eta_0,$$

then

$$|\{u < t\} \cup B_{1/4}| \le Ct^{-\delta} \inf_{B_{1/2}} u$$

for some constant C depending only on Q and n.

PROOF. We follow the proof in [**2**], chapter 4. The L^δ estimate is proved using only an estimate such as Corollary 4.6 at every scale. □

We are now ready to prove Theorem 4.2.

PROOF OF THEOREM 4.2. First of all we point out that we can rescale the equation to make $||f||_{L^\infty}$ as small as we wish, and this estimate will be preserved by the C^α scaling throughout the proof.

We will prove a decay in the oscillation of balls around the origin,

$$\operatorname*{osc}_{B_{4^{-k}}(0)} u \le C(1-\theta)^k ||u||_{L^\infty(B_{1/2+Q\varepsilon})} \tag{4.2}$$

for a universal $\theta > 0$, which immediately implies the result with $\alpha = -\log(1-\theta)/\log 4$.

We prove (4.2) by induction. For $k = 0$ it is true with $C = 1$.

Assume it is true for some $k \in \mathbb{N}$ with $C = 1$. We consider two cases: either $4^k Q\varepsilon < 1/8$ or $4^k Q\varepsilon \ge 1/2$.

Let us first discuss the case $4^k Q\varepsilon < 1/2$.

We use a classical idea of De Giorgi. The values of u remain in an interval $[a, b]$ for $x \in B_{2^{-k}}$ with $b - a \le (1-\theta)^k ||u||_{L^\infty(B_{1/2+Q\varepsilon})}$. For every $x \in B_{4^{-k-1}}$, u is either above or below $(a+b)/2$. So in at least half of the points (in measure), u will be one side of $(a+b)/2$. Without loss of generality, let us say that it stays above in at least half of the ball:

$$\{u \ge (a+b)/2\} \cap B_{4^{-k-1}}| \ge \frac{1}{2}|B_{4^{-k-1}}|.$$

So, we rescale by considering

$$v = \frac{2}{a+b}(u(2^{-2k-1}x) - a)$$

so that v solves an equation such as (4.1) but with $2^{2k+1}\varepsilon$ instead of ε and $v \ge 0$ in B_2. In this case, $2^{2k+1}Q\varepsilon < 1$, so $B_{1+2^{2k+1}Q\varepsilon} \supset B_2$. Then we can apply Corollary 4.7 and obtain

$$\inf_{B_{1/2}} v \ge c|\{v \ge 1\} \cap B_{1/4}| \ge c$$

for some universal constant c. Scaling back to u, this means that $u \ge a + c\frac{a+b}{2}$ in $B_{4^{k+1}}$, so the inductive step is proved with $\theta = c/2$ and $C = 1$.

The previous iteration will continue for as long as $4^k Q\varepsilon < 1/2$. Let k be the smallest integer such that $4^k Q\varepsilon \geq 1/2$. The previous iteration process will reach k, so that we have

$$\operatorname*{osc}_{B^{4^{-k}}(0)} u \leq (1-\theta)^k ||u||_{L^\infty(B_{1/2+Q\varepsilon})}.$$

Let $v = (1-\theta)^{-k} u(4^{-k}x)$, so that v satisfies

$$\begin{aligned} \operatorname*{osc}_{B_1} v &\leq 1, \\ \operatorname*{osc}_{B_4} v &\leq (1-\theta)^{-1} < 2, \\ \triangle v + \int_{B_1} (v(x+y) - v(x)) \tilde{k}(x,y) \, dy \quad \text{in } B_1 &= 0, \end{aligned}$$

where $\tilde{k}(x,y) = k(4^k x, 4^k y) \leq Q^{-n-2}$ if $|y| \leq 1$ and zero otherwise. But then the integral term in the equation is bounded by a universal constant C (depending only on Q and n; recall that θ is also universal). So $|\triangle v| \leq C$ in B_1.

Therefore, by the C^α estimates of the Laplace equation, there is a universal constant C such that

$$\operatorname*{osc}_{B_r} v \leq C r^\alpha.$$

Scaling back, (4.2) holds for some universal constant C for all positive values of k. □

We now state the $C^{1,\alpha}$ estimate.

THEOREM 4.8. *Let u be a solution of*

$$\begin{aligned} E_\varepsilon u &= f \quad \text{in } B_1, \\ u &= g \quad \text{in } B_{1+Q\varepsilon} \setminus B_1, \end{aligned}$$

where g is a bounded function. Then u satisfies the estimate:

$$||u||_{C^{1,\alpha}(B_{1/2})} \leq C \left(||g||_{L^\infty} + ||f||_{L^\infty} \right),$$

where α and C are universal constants (they depend on λ, Λ and n, but not on ε).

Note that the above result can be scaled to obtain that if $E_\varepsilon u = f$ in B_r, then

$$||u||_{C^{1,\alpha}(B_{r/2})} \leq C \left(\frac{1}{r^{1+\alpha}} ||u||_{L^\infty(B_{r+Q\varepsilon})} + r^{1-\alpha} ||f||_{L^\infty}(B_r) \right).$$

PROOF. For zero right-hand side, Theorem 4.8 is a standard consequence of Lemma 4.1 (see Corollary 5.7 in [**2**]). The main point of the proof is that if $E_\varepsilon u_1 = 0$ and $E_\varepsilon u_2 = 0$, then $L_\varepsilon(u_1 - u_2) = 0$ for some operator L_ε as in Lemma 4.1. Thus we can apply Lemma 4.1 to incremental quotients of u. First we apply Lemma 4.1 to obtain an estimate for u in C^α. Then we can iteratively obtain estimates with higher exponents by applying Lemma 4.1 to incremental quotients of the form $v(x) = (u(x+he) - u(x))/h^\beta$ for any unit vector e and $h > 0$. In this way we can pass for an estimate in C^β to an estimate in $C^{\beta+\alpha}$ as long as $\beta + \alpha \leq 1$. Thus, after a finite number of steps we obtain an estimate of the Lipschitz norm of u, and finally we apply 4.1 to all directional derivatives u_e and finish the proof.

For nonzero right-hand side, the estimate is a consequence of the zero right-hand side estimate and the scaling of the equation (see chapter 8 in [**2**] and also [**1**] for an application of the same method to integro-differential equations). □

5. A Lipschitz estimate almost up to the boundary

In this section we obtain a uniform Lipschitz estimate in the points inside Ω whose distance to the boundary is at least of order ε. This would become an *up to the boundary* regularity estimate as $\varepsilon \to 0$.

In order to obtain this estimate, we construct barriers to be used in domains with the exterior ball condition. In order to get regularity estimates almost up to the boundary that are uniform in ε, we would need to construct barriers that work for every ε (small enough). This is the purpose of this section.

We recall that Ω has the uniform external ball condition if there exists a ρ_0 such that for every point $x \in \partial\Omega$, there exists a ball $B_{\rho_0}(y)$ contained in $\mathcal{C}\Omega$ such that $x \in \partial B_{\rho_0}(y)$.

If Ω has the external ball condition for a radius $\rho_0 > 0$, then when we consider the ε neighborhood $(\partial\Omega)_{Q\varepsilon}$, it also has the exterior ball condition if ε is small, since the exterior boundary of $(\partial\Omega)_{Q\varepsilon}$ is the parallel surface of $\partial\Omega$ at distance $Q\varepsilon$ which has the exterior ball condition with radius $\rho_0 - Q\varepsilon$.

We apply Corollary 2.4 to the function $v = -|x|^{-p}$ with p a large universal constant. If $|x| > Q\varepsilon$, we obtain that $E_\varepsilon v(x) \geq 0$. We will use this fact to create a barrier of the form $v(x) = a - b|x - x_0|^{-p}$ outside of a ball $B_\rho(x_0)$ which touches $(\partial\Omega)_{Q\varepsilon}$ from the exterior. Naturally this is possible assuming that $\rho \leq \rho_0 - Q\varepsilon$ and $\rho > Q\varepsilon$. So, let us say that $\rho = \rho_0/2$ and ε is small enough. Adding an extra quadratic term, we can also make barriers with a nonzero right-hand side:

$$E_\varepsilon \left[a - b|x - x_0|^{-p} - \frac{c}{\lambda}|x|^2\right] \leq -c \quad \text{outside } B_\rho(x_0).$$

We apply this barrier function to prove the following lemma.

LEMMA 5.1. *Let u be a solution to (2.1). Assume Ω has a uniform external ball condition and $g \in C^{1,1}$.*

There is a small universal ε_0 such that if $\varepsilon < \varepsilon_0$ and $x, y \in \Omega$ be such that $\operatorname{dist}(x, \partial\Omega) \leq 2d$ *and* $\operatorname{dist}(x, y) \leq d$, *then $|u(x) - u(y)| \leq C(d + \varepsilon)$ for a universal constant C.*

PROOF. Let z_0 be the closest point to x on the exterior boundary of $(\partial\Omega)_{Q\varepsilon}$. $z_0 \in \mathcal{C}\Omega$ and $\operatorname{dist}(x, z_0) \leq d + \varepsilon$. Since Ω has the exterior ball condition (and thus also does the exterior boundary of $(\partial\Omega)_{Q\varepsilon}$) there is a ball $B_\rho(x_0)$ tangent to $\partial(\partial\Omega)_{Q\varepsilon}$ from the outside. The functions

$$A(x) = g(z_0) + b\rho^{-p} - b|x - x_0|^{-p} - \frac{c}{\lambda}|x|^2,$$
$$B(x) = g(z_0) - b\rho^{-p} + b|x - x_0|^{-p} + \frac{c}{\lambda}|x|^2$$

satisfy $E_\varepsilon A \leq -c$ and $E_\varepsilon B \geq c$ in Ω. So if we choose $c = \max|f|$ and b depending on the $C^{1,1}$ norm of g, then A will be a supersolution and B a subsolution to the problem (2.1). Thus $B \leq u \leq A$. But the oscillation of A and B as well as $|A - B|$ in $B_d(x)$ is bounded by $C(d + \varepsilon)$, where C is a constant depending on λ, Λ, n and ρ. Therefore $|u(x) - u(y)| \leq C(d + \varepsilon)$. □

THEOREM 5.2. *Let u be a solution to (2.1). Assume Ω has a uniform external ball condition and $g \in C^{1,1}$.*

There is a small universal ε_0 such that if $\varepsilon < \varepsilon_0$ and $x \in \Omega$ such that $\operatorname{dist}(x, \Omega) \geq Q\varepsilon$, *then $|\nabla u| \leq C$ for a constant C depending on λ, Λ, n and ρ_0.*

PROOF. Let $d = \text{dist}(x, \partial\Omega)/2$. From the assumptions, we know that $Q\varepsilon < d < \text{diam}(\Omega)$. From Lemma 5.1, $\text{osc}_{B_d(x)} u \leq Cd$. Let us consider the function $\bar{u}(z) = \frac{1}{d}u(x + dz)$. Then $\bar{u}$ satisfies

$$E_{\varepsilon/d}\bar{u}(z) = \frac{\lambda}{2}\Delta\bar{u} + \inf_{\beta}\sup_{\alpha} I^{\alpha\beta}_{\varepsilon/d}(\bar{u}) = df(x + dz),$$
$$\operatorname*{osc}_{B_{2Q}} \bar{u} \leq C.$$

We can apply Theorem 4.8 and from the interior estimate on the gradient conclude that $|\nabla\bar{u}(0)| \leq C$ for some constant C depending on λ, Λ, n and Ω but not on ε. But that implies that $|\nabla u(x)| \leq C$, which finishes the proof. □

6. Rate of convergence

In this section we prove that the solution u_ε to the approximate problem (2.1) approaches the solution u to the original equation (1.2) uniformly with a rate of the form $C\varepsilon^\alpha$ for some small $\alpha > 0$. We state this in the following theorem.

THEOREM 6.1. *Assume $g \in C^{1,1}$ and f is a Lipschitz function. There exists a universal constant C and $\alpha > 0$ (depending only on λ, Λ, n and the exterior ball condition ρ_0 of the domain) such that*

$$||u_\varepsilon - u||_{L^\infty} \leq C\varepsilon^\alpha(||g||_{C^{1,1}} + ||f||_{Lip}).$$

This result can be proved as an application of a general result from [**3**]. We start by recalling the notion of δ-solutions.

DEFINITION 6.2. *Fix $\delta > 0$. We say that a continuous function v is a δ-supersolution (resp. δ-subsolution) of (1.2) in Ω if, for all $x_0 \in \Omega$ such that $B_\delta(x_0) \subset \Omega$, a polynomial P such that $|P| \leq C\delta^{-\sigma}$, for some universal C, $\sigma > 0$, and $P \leq v$ (resp. $P \geq v$) in $B_\delta(x_0)$ can touch v from below (resp. above) at x_0, i.e. $P(x_0) = v(x_0)$, only if $F(D^2P) \leq 0$ (resp. $F(D^2P) \geq 0$). Finally, a continuous function v is a δ-solution if it is both a δ-supersolution and a δ-subsolution.*

This definition is relevant since the solution to our approximated equation (2.1) is a $Q\varepsilon$-solution. We prove that in the following lemma.

LEMMA 6.3. *If u solves (2.1), then u is a $Q\varepsilon$-solution of (1.2).*

PROOF. If a quadratic polynomial P touches u from above at a point x, then on the one hand, $\triangle P \geq \Delta u$. On the other hand, if $P \geq u$ in $B_{Q\varepsilon}(x)$, then $P \geq u$ in the full domain of integration of every integral, so $I^{\alpha\beta}_\varepsilon P \geq I^{\alpha\beta}_\varepsilon u$ for every α, β. Therefore $E_\varepsilon P \geq E_\varepsilon u$.

Since P is simply a quadratic polynomial, the value of $E_\varepsilon P$ coincides with the value of the original second-order elliptic operator $\inf\sup a^{\alpha\beta}_{ij}\partial_{ij}P$. Thus u is a $Q\varepsilon$-solution. □

The following theorem is proved in [**3**].

THEOREM 6.4. *Let Ω be an open subset of $\mathbb{R}^n$ with regular boundary and consider a solution $u \in C^{0,1}(\overline{\Omega})$ of (1.2). Assume that $v^+ \in C^\gamma(\overline{\Omega})$ (resp. $v^- \in C^\gamma(\overline{\Omega})$) is a δ-subsolution (resp. δ-supersolution) of (1.2) for some fixed $\gamma \in (0,1)$. If $v^+ \geq u + c\delta^{\bar{\alpha}}$ (resp. $v^- \leq u - c\delta^{\bar{\alpha}}$) on $\partial\Omega$ for some positive constants c and α,* *then there exist uniform constants $C > 0$ and $\alpha \in (0, \bar{\alpha})$ such that, for δ sufficiently small,*

$$v^+ \leq u + C\delta^\alpha \quad (\textit{resp. } v^- \geq u - C\delta^\alpha).$$

Combining Theorem 6.4 with Theorem 5.2, we can prove Theorem 6.1. Nevertheless, for completeness, we will provide a detailed sketch of the proof of Theorem 6.1 using the ideas from [**3**]. The proof uses somewhat sophisticated regularity results for fully nonlinear elliptic equations, which can be found in [**2**]. Since we do not aim at the level of generality as in Theorem 6.4 but only to the particular case of the approximated solutions of this paper, we are able to simplify a few steps in the proof.

Proof of Theorem 6.1. By multiplying f, g and u by an appropriate constant, we can assume that $||g||_{C^{1,1}} + ||f||_{Lip} = 1$. We have already shown that the approximation u_ε is uniformly Lipschitz $Q\varepsilon$ away from $\partial\Omega$.

The solution u separates from the boundary value g linearly from the boundary (depending on the exterior ball condition). From Lemma 5.1, the approximation u_ε separates from the boundary value less than $C\varepsilon$ in a $Q\varepsilon$ neighborhood of $\partial\Omega$. So $|u_\varepsilon - u| \leq C\varepsilon$ for all points $x \in \Omega$ such that $\operatorname{dist}(x, \partial\Omega) \leq Q\varepsilon$.

Since u is a solution of the fully nonlinear uniformly elliptic equation (1.2), all first derivatives $u_i = \partial_i u$ are in the class of solutions to equations with *measurable coefficients*. More precisely, for every index i, we have $M^+(D^2 u_i) \geq 0$ and $M^-(D^2 u_i) \leq 0$ (where M^+ and M^- are the extremal Pucci operators). There is a result saying that the Hessians of functions in such a class are in L^θ for some small $\theta > 0$ (see [**5**] or [**2**], Proposition 7.4). In other words, $u_i \in W^{2,\theta}(B_1)$; more precisely, for every $t > 0$, every derivative u_i has a paraboloid of opening t tangent from above (or below) except in a set of measure $t^{-\theta}$.

In terms of the value of the function u itself, this means that for every $t > 0$, except in a singular set of measure $t^{-\theta}$, the function u has a second-order Taylor expansion, meaning that there is some second-order polynomial P_x (depending on the point x) such that $||P||_{C^{1,1}} \leq t$ and

$$|u(x + y) - P_x(y)| \leq Ct|y|^3.$$

In this regular set, which we will call R, we can obtain an estimate of $E_\varepsilon u$. Recall that $E_\varepsilon P = F(D^2 P) = 0$. So the error comes from the integral term in E_ε applied to the cubic part, which is of order $Ct\varepsilon$. Thus $|E_\varepsilon u(x)| \leq Ct\varepsilon$ except in a set of measure $t^{-\theta}$.

Let us choose $t = \varepsilon^{\alpha-1}$ for some $\alpha \in (0, 1)$ to be determined later. So we have $|E_\varepsilon u(x) - f(x)| \leq C\varepsilon^\alpha$ except in a set of measure $\varepsilon^{\theta(1-\alpha)}$.

So we have that $|E_\varepsilon u(x)|$ is small except in a set of small measure. The question is how to fill that gap. We will use a sup-convolution to regularize the solution u and from the regularity estimates in u we will estimate its difference with the sup-convolution u^*.

Let u^* be the sup-convolution of u:

$$u^*(x) = \max_{y \in \overline{B_1}} u(y) - \varepsilon^{-\alpha}|x - y|^2.$$

Since u is a Lipschitz function, for every $x \in \Omega$, the maximum in the sup-convolution is achieved at some $y_* \in \overline{\Omega}$, i.e. $u^*(x) = u(y_*) - \varepsilon^{-\alpha}|x - y_*|^2$, for which $|x - y_*| < C\varepsilon^\alpha$. Moreover, $u_* \leq u + C\varepsilon^\alpha$. On the other hand, it is clear by definition that $u^* \geq u$ since $y = x$ is a candidate for the maximum.

Let $M \subset \Omega$ be the set of all points $y \in \Omega$ such that for some $x \in \Omega$, $u^*(x) = u(y) - \varepsilon^{-\alpha}|x-y|^2$. From this definition, at every such point y the graph of u has a tangent paraboloid from above with opening $\varepsilon^{-\alpha}$. Since u is a solution to the uniformly elliptic equation (1.2), the Harnack inequality implies that it also has a tangent paraboloid from below with opening $-C\varepsilon^{-\alpha}$ for a universal constant C. Therefore u is differentiable at every point $y \in M$, and ∇u is Lipschitz in M with Lipschitz constant $C\varepsilon^{-\alpha}$. From the fact that y is the point where the maximum in the definition is achieved, the gradient must satisfy $\nabla u(y) = -2\varepsilon^{-\alpha}(y-x)$. Therefore the map $X(y) := x$ is well defined and Lipschitz, with Lipschitz constant bounded above by a universal C.

We will estimate the minimum of the function $v = C\varepsilon^\alpha + u_\varepsilon - u^*$. By choosing C appropriately we can make sure that $v \geq 0$ in a $Q\varepsilon$ neighborhood of ∂Q. For the interior, we will use the ABP estimate from Lemma 4.5.

For every $x \in \Omega$, $x = X(y)$ for some $y \in M$, and $E_\varepsilon u_*(x) \geq E_\varepsilon u(y)$ in the sense that there is a translation of the graph of u around the point y which is tangent from below to the graph of u_* at the point x.

We will estimate $E_\varepsilon u_*(x)$ depending on whether x is the image by $X(y)$ of a point y in the regular set R or not.

If x is the image $X(y)$ of some point $y \in R$, then $E_\varepsilon u_*(x) \geq E_\varepsilon u(y) \geq f(y) - C\varepsilon^\alpha \geq f(x) - C\varepsilon^\alpha$ (using that f is Lipschitz).

If x is any generic point in Ω (not necessarily the image by $X(y)$ of a regular point), then just by the definition of the sup-convolution, u_* has a tangent paraboloid from below with opening $C\varepsilon^{-\alpha}$ and thus $E_\varepsilon u_*(x) \geq -C\varepsilon^{-\alpha}$.

Therefore, the sup-convolution u_* satisfies the following equation in Ω:

$$E_\varepsilon u_* \geq \begin{cases} f(x) - C\varepsilon^\alpha & \text{in } R, \\ -C\varepsilon^{-\alpha} & \text{outside } R. \end{cases}$$

Therefore the function v is a subsolution of the linearized equation

$$\frac{\lambda}{2}\Delta v + \int_{\mathbb{R}^n} (v(x+y) - v(x)) \frac{1}{\varepsilon^{n+2} \det B(x)} \varphi\left(B(x)^{-1}\frac{y}{\varepsilon}\right) \, dy \leq \begin{cases} C\varepsilon^\alpha & \text{in } R, \\ C\varepsilon^{-\alpha} & \text{outside } R, \end{cases}$$

where the matrix $B(x)$ satisfies point-wise the ellipticity estimates $\sqrt{\lambda} I \leq B(x) \leq \sqrt{\Lambda} I$ but may be discontinuous with respect to x.

We apply Lemma 4.5 in the set $\Omega_0 = \{x \in \Omega : \operatorname{dist}(x, \partial\Omega) \geq Q\varepsilon\}$ and obtain

$$\min -v \leq C\left(\int_R \varepsilon^{\alpha n} \, dx + \int_{\Omega_0 - R} \varepsilon^{-\alpha n} \, dx\right)^{1/n} \leq C(\varepsilon^{\alpha n} + \varepsilon^{\theta(1-\alpha) - \alpha n})^{1/n}.$$

We can choose $\alpha = \theta/(2n+\theta)$ and obtain $v \geq -C\varepsilon^\alpha$. But this implies that $u_* - u_\varepsilon \leq C\varepsilon^\alpha$, which in turn implies that $u_\varepsilon - u \leq C\varepsilon^\alpha$ since $|u_* - u| \leq C\varepsilon^\alpha$.

We finished the proof that $u_\varepsilon - u \leq C\varepsilon^\alpha$. The other inequality follows in the same way. □

We note that even in the case when the solution u to the limiting problem has $C^{2,\delta}$ estimates for some small $\delta > 0$ (as in the convex case) we cannot expect a much better rate of convergence. Indeed, from $u \in C^{2,\delta}(\overline{\Omega})$, we could estimate $u - u_\varepsilon$ at every point $x \in \Omega$. We would have a second-order polynomial P_x such that

$$|u(y) - P_x(y)| \leq C|x-y|^{2+\delta}.$$

Therefore $|E_\varepsilon u(x) - F(D^2u(x))| \leq C\varepsilon^\delta$ for every point x. But from here we would only obtain $|u - u_\varepsilon| \leq \varepsilon^\delta$. On the other hand, in the convex case, if F is smooth ($C^{1,\alpha}$), then u is $C^{3,\alpha}$ (from Schauder estimates on the first derivatives) and we may gain a factor of ε in the rate of convergence in a smooth domain after using this extra regularity.

Acknowledgments

Both authors were partially supported by NSF grants.

References

[1] L. Caffarelli and L. Silvestre. *Regularity results for nonlocal equations by approximation.* Preprint, ArXiv:0902.4030, 2009.

[2] L. A. Caffarelli and X. Cabre. *Fully nonlinear elliptic equations.* Amer. Math. Soc., Providence, RI, 1995. MR1351007 (96h:35046)

[3] L. A. Caffarelli and P. E. Souganidis. *Rates of convergence for the homogenization of uniformly elliptic pde in strongly mixing random media.* Preprint, 2008.

[4] Luis Caffarelli and Luis Silvestre. *The Evans-Krylov theorem for non local fully non linear equations.* Preprint.

[5] Fang-Hua Lin. *Second derivative L^p-estimates for elliptic equations of nondivergent type.* Proc. Amer. Math. Soc., **96** (1986), no. 3, 447–451. MR822437 (88b:35058)

Amer. Math. Soc. Transl.
(2) Vol. **229**, 2010

Hölder Continuity of Solutions of 2D Navier-Stokes Equations with Singular Forcing

Peter Constantin and Gregory Seregin

Dedicated to Nina Nikolaevna Uraltseva

ABSTRACT. We discuss the regularity of solutions of 2D incompressible Navier-Stokes equations forced by singular forces. The problem is motivated by the study of complex fluids modeled by the Navier-Stokes equations coupled to a nonlinear Fokker-Planck equation describing microscopic corpora embedded in the fluid. This leads naturally to bounded added stress and hence to $W^{-1,\infty}$ forcing of the Navier-Stokes equations.

1. Introduction

We discuss the regularity of solutions of 2D incompressible Navier-Stokes equations forced by singular forces. The problem is motivated by the study of complex fluids modeled by the Navier-Stokes equations coupled to a nonlinear Fokker-Planck equation describing microscopic corpora embedded in the fluid. This leads naturally to bounded added stress and hence to $W^{-1,\infty}$ forcing of the Navier-Stokes equations. A more detailed description of the problem in question, together with an application of the results in the present paper can be found in our forthcoming paper [**2**].

In this paper, we focus on the 2D Navier-Stokes issues. The global existence of energy solutions and their uniqueness are well known as classical results of J. Leray for the Cauchy problem and O. Ladyzhenskaya for initial boundary value problems in bounded domains. These results remain to be true for singular forces as well.

The regularity of energy solutions with relatively smooth forces is also known. Regularity can be established, for instance, by scalar multiplication of the Navier-Stokes equation by the Stokes operator of the velocity field, integration by parts, and application of Ladyzhenskaya's inequality

$$\|u\|^2_{L^4(\mathbb{R}^2)} \leq \sqrt{2}\|u\|_{L^2(\mathbb{R}^2)}\|\nabla u\|_{L^2(\mathbb{R}^2)}, \qquad \forall u \in C_0^\infty(\mathbb{R}^2).$$

This procedure yields summability of the second spatial derivatives. Further regularity can be obtained perturbatively, with the help of the linear theory.

2000 *Mathematics Subject Classification.* Primary 35Kxx, 76Dxx.
Key words and phrases. Navier-Stokes equations, Hölder continuity, singular forcing.

The regularity of energy solutions with singular forcing is limited. The best one can expect is Hölder continuity of the velocity field. We prove Hölder continuity at a local level, in both space and in time. We assume that our local solution has finite energy and that the pressure field is in L^2. This latter assumption seems restrictive: we are not able to justify it for general initial boundary value problems with reasonable singular forcing. The assumption is however satisfied in the absence of boundaries, i.e., for the Cauchy problem in the whole space and for the initial value problem on the torus. We briefly explain in this paper how the local regularity results can be applied to the Cauchy problem in the whole space.

In our proof, the Hölder continuity of the velocity field depends quantitatively on the modulus of continuity of the function $\omega \mapsto \int_\omega |u|^4 dz$. In order to be able to apply this regularity result to coupled systems or to families of Navier-Stokes systems, this modulus of continuity needs to be a priori uniformly controlled. We achieve this in the absence of boundaries by obtaining higher integrability of the velocity, $u \in L^\infty(dt; L^r(dx))$, $r \geq 4$. In order to obtain the higher integrability we prove the generalized Ladyzhenskaya inequality, which reads

$$\|u\|^2_{L^{2r}(\mathbb{R}^2)} \leq \frac{r}{\sqrt{2}} \|u\|_{L^r(\mathbb{R}^2)} \|\nabla u\|_{L^2(\mathbb{R}^2)}, \qquad \forall u \in C_0^\infty(\mathbb{R}^2)$$

for $r \geq 2$. The proof is elementary and can be found in the Appendix.

2. Notation and Local Regularity Result

We assume that Ω and Ω_1 are domains in $\mathbb{R}^2$ such that $\Omega_1 \Subset \Omega$ and $0 < T_1 < T$, and let

$$Q = \Omega \times (-T, 0), \qquad Q_1 = \Omega_1 \times (-T_1, 0).$$

Parabolic balls will be denoted as $Q(z_0, R) = B(x_0, R) \times (t_0 - R^2, t_0)$, where $z_0 = (x_0, t_0)$, $x_0 \in \mathbb{R}^2$, $t_0 \in \mathbb{R}$, and $B(x_0, R)$ is an open disk in $\mathbb{R}^2$ having radius R and centered at the point x_0.

We use the following notation for mean values:

$$(f)_{z_0,R} = \frac{1}{|Q(z_0,R)|} \int\limits_{Q(z_0,R)} f(z)dz, \qquad [p]_{x_0,R} = \frac{1}{|B(x_0,R)|} \int\limits_{B(x_0,R)} p(x)dx.$$

$L^p(\Omega)$ and $W^{l,p}(\Omega)$ stand for the usual Lebesgue and Sobolev spaces of functions defined on Ω, and the norm of the Lebesgue space is denoted by $\|\cdot\|_{m,\Omega}$. For the forcing we are going to use a functional space $M_{2,\gamma}(Q)$ with parameter $0 \leq \gamma < 1$ and seminorm

$$\|f\|_{M_{2,\gamma}(Q)} = \sup_{Q(z_0,R) \subset Q} R^{1-\gamma} \Big(\frac{1}{|Q(z_0,R)|} \int\limits_{Q(z_0,R)} |f(z) - (f)_{z_0,R}|^2 dz \Big)^{\frac{1}{2}} < \infty.$$

We denote by c all positive universal constants. Our regularity result can be formulated as follows.

THEOREM 2.1. *Assume that we are given functions*

$$u \in L^4(Q; \mathbb{R}^2), \quad p \in L^2(Q), \quad F \in M_{2,\gamma}(Q; \mathbb{M}^{2\times 2}) \tag{2.1}$$

with $0 \leq \gamma < 1$, *satisfying the Navier-Stokes equations*

$$\partial_t u + u \cdot \nabla u - \Delta u + \nabla p = -\operatorname{div} F, \quad \operatorname{div} u = 0 \tag{2.2}$$

in Q *in the sense of distributions.*

Then

$$u \in C^{\gamma}(\overline{Q}_1) \tag{2.3}$$

if $0<\gamma<1$ *and*

$$u \in BMO(Q_1) \tag{2.4}$$

if $\gamma = 0$.

REMARK 2.2. *The Hölder continuity and the BMO space are defined with respect to parabolic metrics.*

REMARK 2.3. *The corresponding norms are estimated in terms of the quantities* $\|u\|_{4,Q}$, $\|p\|_{2,Q}$, $\|F\|_{M_{2,\gamma}(Q)}$, $\mathrm{dist}(\Omega_1, \partial\Omega)$, $T - T_1$, *and the modulus of continuity of the function* $\omega \mapsto \int\limits_{\omega} |u|^4 dz$.

Several additional results can be proved by means of minor modifications of the proof of Theorem 2.1. Before stating one of them, we define the usual energy spaces for the 2D Navier-Stokes equations. Let H and V be completions of the set of all divergence-free vector fields from $C_0^\infty(\mathbb{R}^2;\mathbb{R}^2)$ with respect to the L^2 norm and the Dirichlet integral, respectively.

PROPOSITION 2.4. *Let* $u \in L^\infty(0,T;H)\cap L^2(0,T;V)$, $p \in L^2(0,T;L^2(\mathbb{R}^2))$ *be a solution of the Cauchy problem*

$$\partial_t u + u\cdot\nabla u - \Delta u + \nabla p = -\mathrm{div}\, F, \qquad \mathrm{div}\, u = 0, \tag{2.5}$$

$$v(\cdot,0) = a(\cdot) \in H, \tag{2.6}$$

where $F \in L^q(Q_T;\mathbb{M}^{2\times 2}) \cap L^2(Q_T;\mathbb{M}^{2\times 2})$ *with* $q>4$ *and* $Q_T = \mathbb{R}^2 \times (0,T)$.

Then, given $0 < s \le T$, *there exists a constant* C *depending only on* s, *the norms of* F *in* $L^q(Q_T;\mathbb{M}^{2\times 2})$ *and in* $L^2(Q_T;\mathbb{M}^{2\times 2})$ *and the norm of* a *in* H *such that*

$$\|u\|_{L^\infty(\mathbb{R}^2\times(s,T))} \le C. \tag{2.7}$$

Moreover, the function u *is Hölder continuous in* $\mathbb{R}^2\times[s,T]$ *with exponent* $\gamma = 1-\frac{4}{q}$.

REMARK 2.5. *The existence and uniqueness of a solution to the Cauchy problems (2.5) and (2.6) with the above properties is well known; see* [**5**].

REMARK 2.6. *The same statement is valid in the case of periodic boundary conditions. More generally, it is true as long as the pressure field is in* L^2.

3. Proof of Theorem 2.1

We are going to analyze differentiability properties of the velocity field u in terms of the following functionals:

$$\Phi(u;z_0,\varrho) = \Big(\int\limits_{Q(z_0,\varrho)} |u-(u)_{z_0,\varrho}|^4 dz\Big)^{\frac12}, \quad \Psi(u;z_0,\varrho) = \Big(\int\limits_{Q(z_0,\varrho)} |u|^4 dz\Big)^{\frac12},$$

$$D(p;z_0,\varrho) = \int\limits_{Q(z_0,\varrho)} |p-[p]_{x_0,\varrho}|^2 dz.$$

The following two statements are well known.

LEMMA 3.1. *Let the function* $v \in L^4(Q(z_0,R))$ *satisfy the heat equation*

$$\partial_t v - \Delta v = 0$$

in $Q(z_0,R)$. *Then*

$$\Phi(v;z_0,\varrho) \le c\Big(\frac{\varrho}{R}\Big)^4 \Phi(v;z_0,R) \tag{3.1}$$

for all $0<\varrho\le R$.

LEMMA 3.2. *Given* $G \in L^2(Q(z_0,R);\mathbb{M}^{2\times 2})$, *there exists a unique function*

$$w \in C([t_0-R^2,t_0];L^2(B(x_0,R);\mathbb{R}^2)) \cap L^2([t_0-R^2,t_0];W^{1,2}(B(x_0,R);\mathbb{R}^2))$$

such that

$$\partial_t w - \Delta w = -\operatorname{div} G$$

in $Q(z_0,R)$ *and*

$$w = 0$$

on the parabolic boundary of $Q(z_0,R)$. *Moreover, the function* w *satisfies the estimates:*

$$|w|^2_{2,Q(z_0,R)} \equiv \sup_{t_0-R^2<t<t_0} \|w(\cdot,t)\|^2_{2,B(x_0,R)} + \|\nabla w\|^2_{2,Q(z_0,R)}$$

$$\le 2\|G\|^2_{2,Q(z_0,R)}, \tag{3.2}$$

$$\Phi(w;z_0,R) \le c|w|^2_{2,Q(z_0,R)}. \tag{3.3}$$

The next couple of statements are about some properties of the solutions of the system (2.2).

LEMMA 3.3. *Under the assumptions of Theorem 2.1 we have*

$$\Phi(u;z_0,\varrho) \le c\Big\{\Big[\Big(\frac{\varrho}{R}\Big)^4 + \Psi(u;z_0,R)\Big]\Phi(u;z_0,R)$$

$$+D(p;z_0,R) + MR^{2+2\gamma}\Big\} \tag{3.4}$$

whenever $Q(z_0,R) \subset Q$ *and* $0<\varrho\le R$. *Here,* $M = \|F\|^2_{M_{2,\gamma}(Q)}$.

PROOF. Setting

$$G = F - (F)_{z_0,R} + (p - [p]_{x_0,R})\mathbb{I} + (u-(u)_{z_0,R})\otimes u$$

in Lemma 3.2, we get the following estimate for w:

$$\Phi(w;z_0,R) \le c\int_{Q(z_0,R)} \Big[|F-(F)_{z_0,R}|^2 + |p-[p]_{x_0,R}|^2 + |u-(u)_{z_0,R}|^2|u|^2\Big]dz$$

$$\le c\Big[MR^{2+2\gamma} + D(p;z_0,R) + \Psi(u;z_0,R)\Phi(u;z_0,R)\Big]. \tag{3.5}$$

Obviously, the function $v = u - w$ satisfies the heat equation. Then, applying Lemma 3.1, we find

$$\Phi(u-w;z_0,\varrho) \le c\Big(\frac{\varrho}{R}\Big)^4\Phi(u-w;z_0,R).$$

The latter inequality gives us:

$$\Phi(u;z_0,\varrho) \le c\Big[\Big(\frac{\varrho}{R}\Big)^4\Phi(u;z_0,R) + \Phi(w;z_0,R)\Big]. \tag{3.6}$$

Combining (3.5) and (3.6), we arrive at (3.4), and thus Lemma 3.3 is proved.

LEMMA 3.4. *Under the assumptions of Theorem 2.1, we have the estimate*

$$D(p;z_0,\varrho)\le c\Big[\Big(\frac{\varrho}{R}\Big)^4D(p;z_0,R)+\Psi(u;z_0,R)\Phi(u;z_0,R)+MR^{2+2\gamma}\Big] \tag{3.7}$$

whenever $Q(z_0,R)\subset Q$ *and* $0<\varrho\le R$.

PROOF. The crucial part of the proof is the pressure decomposition

$$p=p_1+p_2,$$

where the first component p_1 satisfies the identity

$$\int\limits_{B(x_0,R)} p_1\Delta\varphi dx=-\int\limits_{B(x_0,R)}\Big((u-(u)_{z_0,R})\otimes u+F-(F)_{z_0,R}\Big):\nabla^2\varphi dx,$$

where the test function $\varphi\in W^2_2(B(x_0,R))$ is subject to the Dirichlet boundary condition: $\varphi=0$ on $\partial B(x_0,R)$. It is not difficult to show that such a function p_1 exists and obeys the estimate

$$\int\limits_{Q(z_0,R)}|p_1-[p_1]_{x_0,R}|^2dz\le c\int\limits_{Q(z_0,R)}|p_1|^2dz$$

$$\le c\Big[\Psi(u;z_0,R)\Phi(u;z_0,R)+MR^{2+2\gamma}\Big]. \tag{3.8}$$

The second counterpart of the pressure p_2 is a harmonic function and thus satisfies the estimate

$$\int\limits_{B(x_0,\varrho)}|p_2-[p_2]_{x_0,\varrho}|^2dx\le c\Big(\frac{\varrho}{R}\Big)^4\int\limits_{B(x_0,R)}|p_2-[p_2]_{x_0,R}|^2dx$$

$$\le c\Big(\frac{\varrho}{R}\Big)^4\int\limits_{B(x_0,R)}|p-[p]_{x_0,R}|^2dx+c\int\limits_{B(x_0,R)}|p_1|^2dx$$

for any $0<\varrho\le R$. Hence, by (3.8), we show

$$D(p_2;z_0,\varrho)\le c\Big[\Big(\frac{\varrho}{R}\Big)^4D(p,z_0,R)+\Psi(u;z_0,R)\Phi(u;z_0,R)+MR^{2+2\gamma}\Big]. \tag{3.9}$$

Taking into account the simple inequality

$$D(p;z_0,\varrho)\le 2D(p_1;z_0,\varrho)+2D(p_2;z_0,\varrho)$$
$$\le\int\limits_{Q(z_0,R)}|p_1|^2dz+2D(p_2;z_0,\varrho),$$

we deduce the estimate (3.7) from (3.8) and (3.9). Lemma 3.4 is proved.

Now, we pass to the proof of Theorem 2.1. Assuming that $Q(z_0,R)\subset Q$ and $\tau\in(0,1)$, we find from (3.4) and (3.7) two inequalities:

$$\Phi(u;z_0,\tau^2R)\le c(\tau^4+\Psi(u;z_0,\tau R))\Phi(u;z_0,\tau R)+cD(p;z_0,\tau R)+cM(\tau R)^{2+2\gamma}$$

and

$$(1+c)D(p;z_0,\tau R)\le(1+c)c\Big\{\tau^4D(p;bz_0,R)+\Psi(u;z_0,R)\Phi(u;z_0,R)$$
$$+MR^{2+2\gamma}\Big\}.$$

Adding the latter inequalities and introducing the new functional

$$\Theta(z_0, R) = \Phi(u; z_0, \tau R) + D(p; z_0, R),$$

we arrive at the basic estimate

$$\begin{aligned}\Theta(z_0, \tau R) &\leq c(\tau^4 + \Psi(u; z_0, \tau R))\Theta(z_0, R)\\ &+c\Psi(u; z_0, R)\Phi(u; z_0, R) + cMR^{2+2\gamma}.\end{aligned} \tag{3.10}$$

It is not difficult to the check validity of the following inequality:

$$\Phi(u; z_0, \tau R) \leq c\Phi(u; z_0, R)$$

for any $\tau \in (0, 1)$ and any $R > 0$. Then from (3.10) it follows that

$$\Theta(z_0, \tau R) \leq c(\tau^4 + \Psi(u; z_0, R))\Theta(z_0, R/\tau) + cMR^{2+2\gamma} \tag{3.11}$$

under the assumption that $Q(z_0, R/\tau) \subset Q$. Letting $\gamma_1 = (1 + \gamma)/2$ and choosing $\tau = \tau(\gamma) \in (0, 1)$ so that

$$c\tau^{2-2\gamma_1} \leq 1/2, \tag{3.12}$$

we can choose $R_0 < \tau \min\{\mathrm{dist}(\Omega_1, \partial\Omega), \sqrt{T - T_1}\}$ so that

$$\Psi(u; z_0, R) < \tau^4 \tag{3.13}$$

for all $z_0 \in Q_1$ and all $0 < R \leq R_0$. It is here that the modulus of continuity of $\omega \mapsto \int_\omega |u|^4 dxdt$ is used. So, summarizing all the above, we have

$$\Theta(z_0, \tau R) \leq \tau^{2+2\gamma_1}\Theta(z_0, R/\tau) + cMR^{2+2\gamma} \tag{3.14}$$

for all $z_0 \in Q_1$ and all $0 < R \leq R_0$. To reduce (3.14) to a known iterative procedure, we let $\varrho = R/\tau$ and $\vartheta = \tau^2$. As a result, we find

$$\Theta(z_0, \vartheta\varrho) \leq \vartheta^{1+\gamma_1}\Theta(z_0, \varrho) + cM\vartheta^{1+\gamma}\varrho^{2+2\gamma} \tag{3.15}$$

for all $z_0 \in Q_1$ and all $0 < \rho \leq R_0/\tau$. The inequality (3.15) can be easily iterated; see [**1**]:

$$\Theta(z_0, \vartheta^k R_0/\tau) \leq \vartheta^{k(1+\gamma)}\Big(\Theta(z_0, R_0/\tau) + c_1 MR_0^{2+2\gamma}\Big)$$

for any $k \in \mathbb{N}$. Here and in what follows, all positive constants depending only on γ are denoted by c_1. The latter inequality implies that

$$\Phi(u; z_0, \vartheta^k R_0) \leq \vartheta^{k(1+\gamma)}\Big(\Theta(z_0, R_0/\tau) + c_1 MR_0^{2+2\gamma}\Big)$$

for any $k \in \mathbb{N}$ and, hence,

$$\Phi(u; z_0, \varrho) \leq \varrho^{1+\gamma}H \tag{3.16}$$

for any $0 < \varrho \leq R_0$, where

$$H = c_1\Big(\frac{1}{R_0}\Big)^{1+\gamma}\Big[\Theta(z_0, R_0/\tau) + MR_0^{2+2\gamma}\Big].$$

Obviously, H is a function of R_0, $\|u\|_{4,Q}$, $\|p\|_{2,Q}$, M, $\mathrm{dist}(\Omega_1, \partial\Omega)$, $T - T_1$, and γ.

Now, our next step is to figure out how does $\Psi(u; z_0, R)$ depend on R. By (3.16), we have

$$|(u)_{z_0,\varrho/2} - (u)_{z_0,\varrho}| \leq \frac{c}{\varrho}\Phi^{1/2}(u; z_0, \varrho) \leq cH^{1/2}\varrho^{-(1-\gamma)/2}$$

for any $0 < \varrho \leq R_0$. Therefore,

$$|(u)_{z_0,R_0/2^k} - (u)_{z_0,R_0}| \leq cH^{1/2}\sum_{i=0}^{k-1}\Big(\frac{R_0}{2^i}\Big)^{-(1-\gamma)/2}$$

$$= cH^{1/2}\Big(\frac{R_0}{2^k}\Big)^{-(1-\gamma)/2}\sum_{i=0}^{k-1}\Big(\frac{1}{2^{k-i}}\Big)^{(1-\gamma)/2} \leq c_1H^{1/2}\Big(\frac{R_0}{2^k}\Big)^{-(1-\gamma)/2}$$

for any $k \in \mathbb{N}$, or

$$|(u)_{z_0,\varrho} - (u)_{z_0,R_0}| \leq c_1H^{1/2}\frac{1}{\varrho^{(1-\gamma)/2}} \tag{3.17}$$

for any $0 < \varrho \leq R_0$. Proceeding and making use of (3.17), we find

$$\Psi^{1/2}(u; z_0, \varrho) \leq \Phi^{1/2}(u; z_0, \varrho) + c\varrho|(u)_{z_0,\varrho}| \leq c_1\varrho^{(1+\gamma)/2}H^{1/2} + c\varrho|(u)_{z_0,R_0}|.$$

The latter implies that

$$\Psi(u; z_0, \varrho) \leq \varrho^{(1+\gamma)}H_1 \tag{3.18}$$

for any $0 < \varrho \leq R_0$, where H_1 depends on the same arguments as H.

Coming back to the basic estimate (3.10) and taking into account (3.12), (3.13), (3.16), and (3.18),

$$\Theta(z_0, \tau R) \leq \tau^{2+2\gamma}\Theta(z_0, R) + c(M + HH_1)R^{2+2\gamma}$$

for any $0 < R \leq R_0$. After iterations of the latter inequality, we show that

$$\Theta(z_0, \tau^k R) \leq \tau^{k(2+2\gamma)}\Big(\Theta(z_0, R_0) + c(M + HH_1)R_0^{2+2\gamma}\Big)$$

for any $k \in \mathbb{N}$. Consequently, we obtain:

$$\Phi(u; z_0, \varrho) \leq \varrho^{2+2\gamma}H_2 \tag{3.19}$$

for any $z_0 \in Q_1$ and for any $0 < \varrho \leq R_0$ with H_2 depending on the same arguments as H and H_1. Finally, the Hölder continuity of u subject to (3.19) follows from known considerations; see, for example, [**3**] or [**4**]. Theorem 2.1 is proved.

Proof of Proposition 2.4. Our first remark is that

$$L^q(Q_T) \subset M_{2,\gamma}(Q_T) \tag{3.20}$$

if $\gamma = 1 - \frac{4}{q}$. The second remark is that

$$u \in L^4(Q_T) \tag{3.21}$$

and the corresponding norm is bounded by $\|a\|_{2,\mathbb{R}^2} + \|F\|_{2,Q_T}$.

In order to handle the modulus of continuity of the function $\omega \mapsto \int\limits_\omega |u|^4dz$, we are going to show that, for any $0 < s < T$,

$$\sup_{s<t<T}\|u(\cdot,t)\|_{4,\mathbb{R}^2} \leq C(s, \|F\|_{4,Q_T}, \|a\|_{2,\mathbb{R}^2}). \tag{3.22}$$

Assume that (3.22) has already been proved. Let the number $\tau \in (0, 1)$ be defined by (3.12). Then by (3.22) we can find a number $0 < R_1 < \frac{1}{2}\sqrt{s}$ such that

$$\Big(\int\limits_{Q(z_0,R_1)} |u|^4dz\Big)^{\frac{1}{2}} < R_1 \sup_{s/2<t<T}\|u(\cdot,t)\|_{4,\mathbb{R}^2}$$

$$< R_1C(s/2, \|F\|_{4,Q_T}, \|a\|_{2,\mathbb{R}^2}) < \tau^4 \tag{3.23}$$

for any $z_0=(x_0,t_0)$ such that $t_0>s$. Obviously, R_1 depends only on s, $\|F\|_{q,Q_T}$, $\|F\|_{2,Q_T}$, and $\|a\|_{2,\mathbb{R}^2}$ as, by interpolation, $\|F\|_{4,Q_T}$ is estimated by $\|F\|_{q,Q_T}$ and $\|F\|_{2,Q_T}$. Then we repeat the proof of Theorem 2.1, replacing Q with $Q(z_0,R_1)$ and Q_1 with $Q(z_0,R_1/2)$, and establish $u\in L^\infty(Q(z_0,R_1/2))$ with a uniform estimate with respect to z_0 having $t_0>s$. Thus, the solution is Hölder continuous.

So, let us prove (3.22). To this end, we test the Navier-Stokes equations by $|u|^2u$. As a result we have

$$\frac{1}{4}\partial_t\int_{\mathbb{R}^2}|u|^4dx+\int_{\mathbb{R}^2}\Big(|u|^2|\nabla u|^2dx+2|u|^2|\nabla|u||^2\Big)dx$$
$$=\int_{\mathbb{R}^2}\Big(2|u|pu\cdot\nabla|u|+F:(|u|^2\nabla u+2|u|u\otimes\nabla|u|)\Big)dx.$$

After an application of the Cauchy inequality with a suitable weight, we find the following estimate:

$$\partial_t\|u\|^4_{4,\mathbb{R}^2}\le c\int_{\mathbb{R}^2}\Big(p^2|u|^2+|F|^2|u|^2\Big)dx$$
$$\le c\Big(\|p\|^2_{4,\mathbb{R}^2}+\|F\|^2_{4,\mathbb{R}^2}\Big)\|u\|^2_{4,\mathbb{R}^2}.$$

It remains to make use of the pressure equation which, in the case of the Cauchy problem, gives us

$$\partial_t\|u(\cdot,t)\|^4_{4,\mathbb{R}^2}\le c\Big(\|u(\cdot,t)\|^4_{8,\mathbb{R}^2}+\|F(\cdot,t)\|^2_{4,\mathbb{R}^2}\Big)\|u(\cdot,t)\|^2_{4,\mathbb{R}^2}. \tag{3.24}$$

To evaluate the right-hand side of (3.24), we are going to use a particular case of the generalized Ladyzhenskaya inequality

$$\|u(\cdot,t)\|^2_{8,\mathbb{R}^2}\le c\|u(\cdot,t)\|_{4,\mathbb{R}^2}\|\nabla u\|_{2,\mathbb{R}^2}.$$

The proof of the generalized Ladyzhenskaya inequality is given in the Appendix. But then (3.24) can be reduced to the form

$$\partial_t\|u(\cdot,t)\|^4_{4,\mathbb{R}^2}\le c\Big(\|u(\cdot,t)\|^4_{4,\mathbb{R}^2}\|\nabla u\|^2_{2,\mathbb{R}^2}+\|F(\cdot,t)\|^4_{4,\mathbb{R}^2}+\|u(\cdot,t)\|^4_{4,\mathbb{R}^2}\Big).$$

Multiplying the last inequality by a suitable cutoff function in t, keeping in mind that our solution has the finite energy bounded by $\|a\|^2_{2,\mathbb{R}^2}+\|F\|^2_{2,Q_T}$ from above, and using Gronwall's lemma, we prove (3.22). Proposition 2.4 is proved.

4. Appendix: Generalized Ladyzhenskaya Inequality

The inequality

$$\|u\|^2_{4,\mathbb{R}^2}\le\sqrt{2}\|u\|_{2,\mathbb{R}^2}\|\nabla u\|_{2,\mathbb{R}^2},\qquad\forall u\in C^\infty_0(\mathbb{R}^2), \tag{4.1}$$

was used by Ladyzhenskaya in [**5**] to prove the unique global solvability initial boundary value problem for the Navier-Stokes equations in bounded domains of $\mathbb{R}^2$. The generalized version of the Ladyzhenskaya inequality is as follows:

$$\|u\|^2_{2r,\mathbb{R}^2}\le\frac{r}{\sqrt{2}}\|u\|_{r,\mathbb{R}^2}\|\nabla u\|_{2,\mathbb{R}^2},\qquad\forall u\in C^\infty_0(\mathbb{R}^2) \tag{4.2}$$

for $r\ge 2$. The proof of (4.2) is essentially the same as the proof of (4.1). The main ingredient of it is the following identity:

$$|u|^r(x_1,x_2)=r\int_{-\infty}^{x_1}|u|^{r-1}(t,x_2)(|u|)_{,1}(t,x_2)dtdx_2.$$

Using the same identity with respect to x_2, we find

$$\int_{\mathbb{R}^2} |u|^{2r} dx_1 dx_2 \le r^2 \int_{\mathbb{R}^2} |u|^{r-1} |u_{,1}| dx_1 dx_2 \int_{\mathbb{R}^2} |u|^{r-1} |u_{,2}| dx_1 dx_2$$
$$\le \frac{r^2}{2} \int_{\mathbb{R}^2} |u|^{2(r-1)} dx_1 dx_2 \int_{\mathbb{R}^2} |\nabla u|^2 dx_1 dx_2.$$

By interpolation,

$$\int_{\mathbb{R}^2} |u|^{2(r-1)} dx_1 dx_2 \le \Big(\int_{\mathbb{R}^2} |u|^{2r} dx_1 dx_2 \Big)^{\frac{r-2}{r}} \Big(\int_{\mathbb{R}^2} |u|^{r} dx_1 dx_2 \Big)^{\frac{2}{r}}.$$

Now we deduce (4.2) from the latter inequalities.

Acknowledment. P.C.'s research was partially sponsored by NSF grant DMS-0804380. G. S.'s research was partially supported by the RFFI grant 08-01-00372-a. P.C. gratefully acknowledges the hospitality of Oxford University's OxPDE Center.

References

[1] Campanato, S., *Equazioni paraboliche del secondo ordine e spazi* $\mathcal{L}^{2,\theta}(\Omega,\delta)$. Ann. Mat. Pura Appl. **73**(1966), 55–102. MR0213737 (35:4595)
[2] Constantin, P., Seregin, G., *Global regularity of solutions of coupled Navier-Stokes equations and nonlinear Fokker-Planck equations*, to appear.
[3] Da Prato, G., *Spazi* $\mathcal{L}^{2,\theta}(\Omega,\delta)$ *e loro proprieta.* Ann. Mat. Pura Appl. **69**(1965), 383–392. MR0192330 (33:556)
[4] Giaquinta, M., Struwe, M., *On the partial regularity of weak solutions of nonlinear parabolic systems.* Math. Z. **179**(1982), 437–451. MR652852 (83f:35062)
[5] Ladyzhenskaya, O. A., *Global solvability of a boundary value problem for the Navier-Stokes equations in the case of two spatial variables*, Doklady AN USSR, **123**(1958), 427–429.

UNIVERSITY OF CHICAGO

ST.PETERSBURG DEPARTMENT OF STEKLOV INSTITUTE AND OXFORD UNIVERSITY

Amer. Math. Soc. Transl.
(2) Vol. **229**, 2010

Currents and Curvature Varifolds in Continuum Mechanics

Mariano Giaquinta, Paolo Maria Mariano, Giuseppe Modica, and Domenico Mucci

Dedicated to Nina Nikolaevna Uraltseva.

ABSTRACT. The use in continuum mechanics of weak diffeomorphisms, Cartesian currents and curvature k-varifolds with boundary is discussed. The attention is focused on the analysis of the existence of minimizers for the energies of elastic complex bodies and bodies admitting cracks. The interplay between the analytical features and the physical needs is stressed.

1. Mechanical reasons and some historical remarks

The scenario opens on the problem of finding minimizers (ground states) of the energy of elastic simple Cauchy bodies undergoing large deformations. Ingredients are: (i) a minimalist description of the morphology of a body obtained only through its gross place in the physical space, a geometrical picture defining the so-called Cauchy bodies; (ii) an energy varying along transplacements describing changes of places. Convenience suggests to manage two (isomorphic) copies of the physical ambient space, say $\mathbb{R}^d$ and $\hat{\mathbb{R}}^d$. The first copy hosts a place $\mathcal{B}$, an open, connected set with Lipschitz boundary, that can be in principle occupied by the body and is taken as a reference. One-to-one, differentiable, orientation-preserving maps

$$x \longmapsto u := u(x) \in \hat{\mathbb{R}}^d, \quad x \in \mathcal{B},$$

allow one to reach new places $u(\mathcal{B})$.

The constitutive assumption of a conservative setting for both material properties and external actions justifies the presumption of the existence of an energy $\mathcal{E}(u, \mathcal{B})$. The adjective 'simple', used in specifying the bodies under scrutiny, underlines that the bulk elastic energy depends on u through its spatial first derivative only. The total energy to be analyzed is a 3-form over the first jet of the trivial fiber bundle $\mathcal{B} \times \hat{\mathbb{R}}^d$ with sections defined by the transplacement maps $x \longmapsto u(x)$. It reads

$$\mathcal{E}(u, \mathcal{B}) := \int_{\mathcal{B}} e(x, Du(x))\, dx + \int_{\mathcal{B}} \hat{e}(u(x))\, dx,$$

2000 *Mathematics Subject Classification.* Primary 49J45, 49Q15, 74A30, 74A45, 74A60.
Key words and phrases. Graphs, Cartesian currents, curvature varifolds, mechanics of complex bodies, continuum mechanics.

where $e(\cdot)$ is the *bulk elastic energy* and $\hat{e}(\cdot)$ the *potential of external actions*. Boundary conditions of Dirichlet type are presumed here for the map $x \longmapsto u(x)$. The prescription of invariance with respect to isometric changes in observers, that is, invariance with respect to the action of the Euclidean group $\hat{\mathbb{R}}^d \ltimes SO(d)$ over $\hat{\mathbb{R}}^d$, implies that $e(\cdot)$ cannot be convex in $Du(x)$, the deformation gradient. Such a result (see [**48**]) opened the problem of finding minimal physically significant requirements allowing the existence of minimizers of $\mathcal{E}(\cdot, \mathcal{B})$. The notion of polyconvexity of $e(\cdot)$ [**39**] has been the decisive ingredient for proving the existence of minimizers in nonlinear elasticity (see [**4**]). Minimizers have been found to be only locally weak invertible maps. Global invertibility was established later under appropriate Dirichlet boundary conditions [**5**]. Further results in that spirit are in [**46, 47, 42, 41**]. A critical review of the subsequent work can be found in [**6**]. More precisely, $W^{1,p}(\mathcal{B}, \mathbb{R}^3)$ is the Sobolev space hosting naturally the minimizers of $\mathcal{E}(\cdot, \mathcal{B})$. Their existence accrues from classical results, e.g. De Giorgi's semicontinuity result [**20**], once the weak compactness with respect to sequences of minors of the gradient of deformation Du is established. The condition is assured straight away when deformations admit bounded energy in $W^{1,p}$, $p > 3$. In the case $p = 3$, it has been found in [**40**] by using the orientation-preserving property of the deformations. When $p < 3$ another problem arises: in fact, nonnegligible is the presence of transplacement maps with graphs admitting boundaries with projections into the interior of $\mathcal{B}$. Such boundaries describe the formation of 'holes' and/or open 'fractures' of various natures. The need of avoiding the presence of singular sets, a need dictated by the will of describing *purely* elastic deformations, which are deformations with no material threshold allowing a phase transition which might determine plastic flows or rupture of material bonds, opened a research program leading to the notion of *weak diffeomorphisms* defined in [**24**]. Such maps are in essence transplacements satisfying a special condition, the so-called *condition of zero boundary*, which prevents the formation of undesired 'holes' or open 'fractures'. Weak diffeomorphisms are, in fact, orientation-preserving maps which allow the frictionless contact of parts of the external boundary of the body and prevent the self-penetration of the matter. The existence of minimizers of $\mathcal{E}(\cdot, \mathcal{B})$ in the class of weak diffeomorphisms has been established in [**24**].

The definition of weak diffeomorphisms is based, amid other things, on properties of the d-current integrations (*currents*, for short) over the graphs of them. The current associated with the map $x \longmapsto u(x)$ is, in essence, a linear functional, indicated by G_u, over smooth d-forms with compact support in $\mathcal{B} \times \hat{\mathbb{R}}^d$. The boundary ∂G_u of G_u is defined by duality: $\partial G_u(\omega) = G_u(d\omega)$, where ω is a generic $(d-1)$-form with compact support over $\mathcal{B} \times \hat{\mathbb{R}}^d$, provided that $\partial G_u(\omega) = 0$ on $\mathcal{B} \times \hat{\mathbb{R}}^d$. The condition of zero boundary appearing in the definition of weak diffeomorphisms is thus $\partial G_u(\omega) = 0$. The analysis in terms of currents provides us with the weak compactness with respect to the sequences of minors of Du for $p \leq 3$ (see also [**26**]). For $p > 3$, the condition $\partial G_u(\omega) = 0$ is obviously satisfied.

Roughly speaking, the essential idea followed in defining currents is to control the properties of the graphs of the maps under scrutiny by duality, that is, through the properties of functionals involving the graphs themselves and forms over the product space in which the graphs are defined. The technique is of interest per se in the calculus of variations (see the treatise [**27**]). It is useful for the analysis of a number of problems in different settings.

In dealing with nonlinear elasticity of simple bodies, a problem is the possible appearance of a gap phenomenon (see related discussions in [**18**]). Results describing the approximation of weak diffeomorphisms with regular maps are, in fact, not available. A gap phenomenon appears in the case of energies constructed over general manifold-valued maps. The physical setting falling within this scheme is, for example, the one of spin structures admitting Dirichlet energy (also called the Frank energy) when they self-organize and do not undergo gross deformation. The mathematical scenario has deep features. The first result mentioned here dates back to 1964: Every smooth map $\nu : \mathfrak{X} \longrightarrow \mathcal{Y}$ between two compact, differentiable Riemannian manifolds can be deformed in its own homotopy class into a minimizer of the Dirichlet energy

$$\frac{1}{2}\int_{\mathfrak{X}} |D\nu|^2 \; dvol_{\mathfrak{X}},$$

provided that the sectional curvature, $\sigma_{\mathcal{Y}}$, of $\mathcal{Y}$ is nonnegative [**16**]. A distinguished representation of the homotopy class can be constructed this way. The question was whether the condition $\sigma_{\mathcal{Y}} \leq 0$ could be dropped or weakened.

Consider a geodesic ball $B_R(p)$ in $\mathcal{Y}$ and suppose that (i) $B_R(p)$ is disjoint from the cut locus of its centre p, (ii) the radius R satisfies the inequality

$$R < \frac{\pi}{2\sqrt{k}}.$$

The scalar k is determined by the supremum of the sectional curvature of $\mathcal{Y}$ in $B_R(p)$. When such a supremum is negative, k is set equal to 0. A class C_φ of maps $\nu : B^n \longrightarrow B_R(p)$ is defined over the unit ball B^n in $\mathbb{R}^n$. The index φ indicates that the members of C_φ agree with a boundary datum φ over ∂B_R. Results in [**31**] show the existence of a unique harmonic map ν in C_φ which is a minimizer of the Dirichlet energy and is smooth. When $\mathcal{Y}$ is the standard n-dimensional unit sphere S^n in $\mathbb{R}^{n+1}$, minimizers of the Dirichlet energy amid maps with range in a fixed hemisphere, say the upper hemisphere, are smooth. Moreover, if one considers differentiable maps $\nu : B^n \to S^n$, the *equator map*

$$\left(0, \frac{x}{|x|}\right)$$

is a weak harmonic map. It is stable if $n \geq 8$. In fact, every stationary harmonic map from B^n into S^n is regular up to dimension 7 [**29, 44**] according to the fact that the normal vector field to the minimal surface defines a harmonic map. If one considers maps $\nu : B^n \to S^2$, with $n \geq 3$ and no further restriction, the map $\nu(x) = \frac{x}{|x|}$ is also a minimizer of the Dirichlet energy [**9**].

The dimension of the base and the target manifold play a role in the analysis. Consider, for example, differentiable maps $\nu : B^3 \to S^2$ and their associated Dirichlet energies. Take a Lipschitz map $\varphi : \partial B^3 \to S^2$ and define $W^{1,2}_\varphi(B^3, S^2)$ to be the space of the $W^{1,2}(B^3, S^2)$-map with traces agreeing with φ on ∂B^3. It is not empty because at least $\nu(x) := \varphi(x/|x|)$ belongs to it. A gap phenomenon occurs for the Dirichlet energy, namely

$$\inf_{W^{1,2}_\varphi(B^3,S^2)} \int_{B^3} |D\nu(x)|^2 \; dx < \inf_{W^{1,2}_\varphi(B^3,S^2)\cap C^0(B^3,S^2)} \int_{B^3} |D\nu(x)|^2 \; dx,$$

even when $\deg\varphi = 0$ [**30**]. Notice that φ has no smooth extension to B^3 with values in S^2 when $\deg\varphi \neq 0$.

A question is the evaluation of the relaxed energy, that is, the limit value of the Dirichlet energy obtained by using special sequences of maps, selected in some way. Two strategies have been followed. The first one involves sequences $\{\nu_k\}$ in $C^1\left(B^3, S^2\right)$ converging weakly in $W^{1,2}$, namely $\nu_k \rightharpoonup \nu$ in $W^{1,2}\left(B^3, S^2\right)$. In evaluating the relaxed Dirichlet energy as $k \to \infty$, one finds an energy involving the occurrence of point charges with degree equal to ± 1 [**9**]. The second strategy is on the geometric side: sequences $\{\nu_k\}$ in $C^1\left(B^3, S^2\right)$ are once more considered, but their convergence is evaluated in terms of currents. A generic sequence $\{\nu_k\}$ is said to be convergent when the associated sequence of currents converges weakly, that is, when $G_{\nu_k} \rightharpoonup T$. In order to evaluate the relaxed energy one has first to write the energy in terms of currents, precisely in terms of the 3-vector orienting the graph of the generic ν, a 3-vector given by the Radon-Nikodým derivative of the current associated with ν with respect to its total variation. The new representation of the Dirichlet energy is called the *parametric extension.* A *strong density result* in [**25**] shows that the relaxed Dirichlet energy evaluated as before over sequences converging in terms of currents, namely for $G_{\nu_k} \rightharpoonup T$, coincides with its parametric extension to T. Even in this case singularities appear. They are charges connected by line defects where 'fusion' of the spin structures occurs: the spin field becomes multi-valued over the line, at every x there taking as value the entire S^2. A preference between the two methods summarized above is addressed primarily by physical instances. In fact, the selection of a functional space in which analyses are developed is essentially a constitutive prescription when one manages models with some physical meaning.

From the abstract side, refinements and extensions of the theory of the currents, namely the definition and the analysis of entities called *semi-currents* (see [**3**]) and *Cartesian currents* in $cart^{2,1}\left(B^n \times \mathcal{Y}\right)$ ([**23**]), are tools for the analysis of Dirichlet energies over differentiable maps $\nu : B^n \to \mathcal{Y}$ and $\nu : \mathfrak{X} \to \mathcal{Y}$, where $\mathfrak{X}$ and $\mathcal{Y}$ are as above. Precisely, semi-currents over the trivial bundle $B^n \times \mathcal{Y}$ with basis B^n are currents evaluated by testing the graphs of the maps under analysis over a compactly supported smooth form having at most two differentials in the fiber $\mathcal{Y}$. They represent limits of sequences of smooth maps $\nu : B^n \to \mathcal{Y}$ with equibounded Dirichlet energies. When such currents have (i) no boundary, (ii) finite Dirichlet energy and (iii) they admit a special representation of the weak limit of sequences of graphs of smooth maps with bounded Dirichlet energies, they are called Cartesian currents in $cart^{2,1}\left(B^n \times \mathcal{Y}\right)$ (see [**23**]). Precisely, consider smooth maps $\nu_k : B^n \to \mathcal{Y}$ and related bounded Dirichlet energies. Denote by $\gamma_1 ... \gamma_R$ a basis of integral cycles in the spherical subgroup $H_2^{sph}\left(\mathcal{Y}\right)$, by $\mathbb{L}_r\left(T\right)$ an $(n-2)$-dimensional current in B^n and by S_T a current which does not vanish only on forms which have nonzero differential of the component of the forms with exactly two differentials over $\mathcal{Y}$. The statement that the limit T of G_{ν_k}, as $k \to \infty$, is in $cart^{2,1}\left(B^n \times \mathcal{Y}\right)$ means that there exists a $W^{1,2}\left(B^n, \mathcal{Y}\right)$-map ν_T such that

$$T = G_{\nu_T} + \sum_{r=1}^{R} \mathbb{L}_r\left(T\right) \times \gamma_r + S_T,$$

when the second integral homology group $H_2\left(\mathcal{Y}\right)$ of $\mathcal{Y}$ is torsion-free. The map ν_T defines uniquely the representation of T through its essential features. In such a representation, the geometrical properties of the target manifold $\mathcal{Y}$ and the dimension of B have an influence. In fact, when $n \geq 3$, an additional requirement is that for any generic 2-dimensional subspace K of $\mathbb{R}^n$, the restriction of T to $K \times \mathcal{Y}$ is a 2-dimensional current in $cart^{2,1}((B^n \cap K) \times \mathcal{Y})$.

Weak closedness of $cart^{2,1}(B^n \times \mathcal{Y})$ can then be proven (see the details of the proof in [**23**]). Extensions of the strong density result mentioned above can be achieved in terms of Cartesian currents (see [**27, 28**] and the references therein). The final result (see [**28**] and the references therein) is that the relaxed Dirichlet energy evaluated on sequences $\{\nu_k\}$ in $C^1(B^n, \mathcal{Y})$ converging weakly (write $\nu_k \rightharpoonup \nu$) in $W^{1,2}(B^n, \mathcal{Y})$ equals the infimum of the parametric extension of the energy to Cartesian currents with characteristic map ν_T equal to the limit ν of ν_k.

Additional aspects of the analysis of energies associated with maps between manifolds opened further problems. For example, when $\mathfrak{X}$ is the unit disc and $\mathcal{Y}$ a compact manifold, lack of (sequential) compactness appears as a consequence of invariance with respect to horizontal conformal changes, which are conformal changes of $\mathfrak{X}$ [**44, 49**].

The nontrivial homotopy of the target manifold $\mathcal{Y}$ plays a role in the analyses. Consider $\mathcal{Y}$ to be simply connected and such that its second homotopy group $\pi_2(\mathcal{Y})$ is trivial, namely $\pi_2(\mathcal{Y}) = 0$. A problem is to check whether minimizers of the Dirichlet energy on maps $\nu : B^1 \longrightarrow \mathcal{Y}$ are regular. The positive answer was presented in [**33**]: stationary points are regular; the result is a consequence of the absence of the so-called harmonic spheres, which are forbidden by the condition $\pi_2(\mathcal{Y}) = 0$. Stationary harmonic maps have singular sets of dimension at most $n-2$ [**7**]. Nonstationary harmonic maps display moreover scattered discontinuities: in 1995, Rivière found harmonic maps that are everywhere singular [**43**].

The attention then shifted from the Dirichlet energy

$$\int_{\mathfrak{X}} |D\nu|^2 \, dvol_{\mathfrak{X}},$$

to more complicated quadratic energies such as, for example, Oseen-Frank energy for liquid crystals, the density of which is

$$\begin{aligned} e(\nu, D\nu) &= k_1 (\text{div } \nu)^2 + k_2 (\nu \cdot \text{curl } \nu + q) \\ &\quad + k_3 |\nu \times \text{curl } \nu|^2 + \gamma \left(\text{tr} (D\nu)^2 - (\text{div } \nu)^2 \right), \end{aligned}$$

which satisfies the head-to-tail symmetry $\nu \to -\nu$ for the map $\nu : \mathcal{B} \to P^2$, with P^2 the projective plane, and is objective in the sense that it is invariant with respect to the action of the special orthogonal group over P^2. Intrinsic in this last remark is a refinement of the standard notion of observer arising in the structure of the mechanics of complex bodies (see [**36**]). Of course, since the macroscopic deformation is not accounted for in the expression of $e(\nu, D\nu)$, actual and reference shapes are identified through an isomorphism $i : \mathbb{R}^d \to \hat{\mathbb{R}}^d$. For this reason, the differential operator D can be considered as a spatial derivative with respect to the actual places of the material elements. Oseen-Frank energy evaluated along maps $\nu : \mathcal{B} \to S^2$ has been analyzed so far as a natural extension of the prototype Dirichlet energy. As mentioned above, candidates to be minimizers of energies with Dirichlet's or Oseen's-Frank's structures display concentrations of energy over co-dimension 2 sets. A natural manner of analyzing the essential characteristics of these concentrations is to evaluate the graphs of the maps under scrutiny. Precisely, one considers the values that graphs take over forms compactly supported over a trivial bundle with basis $\mathcal{B}$ and typical fiber S^2. Cartesian currents come then once more into play. Minimizers of the energies under analysis are then naturally expressed in terms of them (see [**27**]).

The mixture of the result mentioned above has allowed the construction of a tool for the investigation of the existence of ground states in elastic complex bodies. They are bodies with complicated material texture admitting changes (mutations) which have a drastic influence on the overall mechanical behavior. Such an influence is exerted through interactions that can be hardly represented only by means of standard stresses. Quasicrystals, ferroelectrics, polymeric bodies or magnetizable materials are paradigmatic examples. A multifield description of their morphology is called upon. It is intrinsically multi-scale and multidimensional. In fact, besides the standard transplacement maps, describing the macroscopic deformation, a manifold-valued map $\nu : \mathcal{B} \to \mathcal{M}$ comes into play to account for the main features of the material morphology *inside* the generic material element (the essential elements of the mechanics of complex bodies can be found in [**11, 35, 38**]). The existence result of minimizers of the pertinent elastic energy in large deformations has been presented in [**37**]. The essential features of the procedure are summarized below because they are a synthesis of both the results in finite elasticity of simple bodies in terms of Cartesian currents, and the ones on the energies associated with maps between manifolds. What one obtains is not properly the trivial superposition of the two aspects because there is energetic coupling between macroscopic deformation and changes in the material morphology at low scales in space. It is recognized that minimizers are weak diffeomorphisms in terms of transplacements, and Sobolev maps in terms of morphological descriptors. Macroscopic deformations are orientation-preserving maps that avoid the interpenetration of matter and the formation of holes. Weak diffeomorphisms minimizing the energy have graphs without boundary, in fact.

However, fractures occur in nature. When a variational description of cracked bodies is adopted by following [**19**], minimality of the energy at every time among all virtual crack-transplacement pairs at that time is required. When time evolution occurs, energy conservation has to be imposed throughout. The difficulty of managing crack geometries in finding minimizers has suggested also the convenient simplification of identifying cracks with the jump sets of displacement fields. BV functions have then been involved [**1, 14**] as candidates to be minimizers of the elastic energy of simple bodies. A difficulty arises: theorems allowing the selection of fields with discontinuity sets describing reasonable (physically significant) crack patterns seem to be not available, at least in the current literature; a detailed review on this topic can be found in [**8**]. In any case, this kind of approach seems not to be able to account for partially opened cracks. In fact, in this case, the transplacement field is continuous across the closed part of the crack although the material bonds are broken. A way to account for these aspects is to resort to k-dimensional varifolds, which are measures on a fiber bundle based on $\mathcal{B}$, constructed taking as a prototype fiber a Grassmanian of k-planes selected over $\mathcal{B}$. Cracks are then intended as supports of varifolds which describe the lateral surfaces of the cracks and the tips. If the treatment is restricted to the case of simple Cauchy bodies for the sake of simplicity, an extended notion of weak diffeomorphisms has to be used in order to represent the main properties of transplacement maps allowing cracks. The possibility of boundaries in the graphs of such maps must be considered. Each boundary accounts for a jump in the transplacement field. To assure that such jumps occur only within the cracked part of $\mathcal{B}$, the mass of the relevant boundary current is assumed to be bounded by the measure of the support of the varifolds accounting for the potential crack patterns. Energy along the margins of the cracks is considered, in accordance with Griffith's original suggestion. Minimizers of the overall energy; i.e., the sum of bulk, surface and tip contributions, are then pairs of transplacement maps and varifolds. The control of minimizing sequences is assured by presuming that the surface energy depends on the curvature of the crack, rather than the sole area as in Griffith's model. For $\mathcal{B} \subset \mathbb{R}^3$, the resulting energy is then

$$\mathcal{E}(u, V, \mathcal{B}) \quad : \quad = \int_{\mathcal{B}} e(x, Du)\ dx + \sum_{k=1}^{2} \alpha_k \int_{\mathcal{G}_k(\mathcal{B})} \left|\mathbf{A}_{(k)}\right|^{p_k}\ dV_k$$
$$+ \sum_{k=1}^{2} \beta_k \mathbf{M}(V_k) + \gamma \mathbf{M}(\partial V_1),$$

where $e(x, Du)$ is the bulk elastic energy, the potential of external bulk actions, namely $\tilde{e}(u)$, can be added, and the third addendum is the sum of the surface and tip energies which are just proportional to the measure of the lateral surfaces of the crack and the tip. These ingredients correspond to the standard Griffith's approach. The extension with respect to Griffith's approach is constituted by the presence of the second and the third addenda. Precisely, the second addendum accounts for the dependence of the surface and line energies on the curvature, $\mathbf{A}_{(k)}$ is in fact the curvature of the k-dimensional varifold V_k, while the third addendum measures the energy at possible corners along the tip. The functional dependence on $\mathbf{A}_{(k)}$ is intended as a measure of the macroscopic influence of low scale effects in the debonding process. The symbol $\mathbf{M}(\cdot)$ represents the 'mass' of the varifold introduced in parentheses. Existence results of pairs of extended weak diffeomorphisms and varifolds minimizing the energy introduced above have been collected in [**22**]. The minimization process is over a class of bodies. It is just this circumstance that renders compatible the variational description sketched above with the intrinsic dissipative nature of the fracture mechanics. An advantage of the formulation is the possibility of deriving the weak form of balance equations even when one knows only that the fracture set is just a k-dimensional rectifiable set.

Extensions to the case of complex bodies seems to be available under additional conditions on the surface energy, which accounts for the morphological descriptor fields.

The fast bird's eye view of the functional methods to analyze various types of energies describing crucial offsprings of condensed matter physics presented so far has not touched on evolution phenomena. The analysis of the conservation laws arising in the general model-building framework of the mechanics of complex bodies is still an open issue. Open also is the description of the evolution of cracks in terms of varifolds.

2. A first tool: Cartesian currents

The reasons for considering graphs of maps and the associated currents, in analyzing the energies of simple and complex deformable bodies, have been mentioned above. Some details may clarify those remarks.

Let $u : \mathcal{B} \to \hat{\mathbb{R}}^d$ be a member of $W^{1,1}(\mathcal{B}, \hat{\mathbb{R}}^d)$. Denote by $M(F)$ the so-called d-vector collecting all the minors of F (namely, the minors of $Du(x)$). $M(F)$ is then at each $x \in \mathcal{B}$ an element of $\Lambda_d(\mathbb{R}^d \times \hat{\mathbb{R}}^d)$, the space of d-vectors with the associated dual space $\Lambda^d(\mathbb{R}^d \times \hat{\mathbb{R}}^d)$. Forms over $\mathcal{B}$ are then maps $\omega : \mathcal{B} \to \Lambda^d(\mathbb{R}^d \times \hat{\mathbb{R}}^d)$. Their space is indicated by $\mathcal{D}^d(\mathcal{B} \times \hat{\mathbb{R}}^d)$. The map assigning to every point x the d-vector $M(F(x))$ is of course of the type $\mathcal{B} \to \Lambda_d(\mathbb{R}^d \times \hat{\mathbb{R}}^d)$.

It is possible to construct the *d-current integration* G_u over the graph of u, by taking only the rectifiable part of the graph of u (see [**27**]). Precisely, G_u is the linear functional on smooth d-forms ω with compact support in $\mathcal{B} \times \hat{\mathbb{R}}^d$ defined by

$$G_u := \int_{\mathcal{B}} \langle \omega(x, u(x)), M(Du(x)) \rangle \, dx.$$

The *boundary current* associated with G_u is indicated by ∂G_u and defined by $\partial G_u(\omega) := G_u(d\omega)$, $\omega \in \mathcal{D}^{d-1}(\mathcal{B}\times\hat{\mathbb{R}}^d)$ with $\mathcal{D}^{d-1}(\mathcal{B}\times\hat{\mathbb{R}}^d)$ the space of $(d-1)-$forms with compact support in $\mathcal{B} \times \hat{\mathbb{R}}^d$. Details on the nature and the properties of Cartesian currents can be found in [**27**].

Currents, as recalled above, are essential tools to define weak diffeomorphisms, the role of which in finding minimizers for the energy of deformable bodies, even simple nonlinear elastic Cauchy bodies, has been introduced in the previous section. Their formal definition reads as follows:

DEFINITION 1. A map $u \in W^{1,1}(\mathcal{B}, \hat{\mathbb{R}}^d)$ is said to be a **weak diffeomorphism** (in short $u \in dif^{1,1}(\mathcal{B}, \hat{\mathbb{R}}^d)$), when

- **(*i*):** $\det Du(x) > 0$ *for almost every* $x \in \mathcal{B}$,
- **(*ii*):** $|M(Du)| \in L^1(\mathcal{B})$,
- **(*iii*):** $\partial G_u = 0$ *on* $\mathcal{D}^{d-1}(\mathcal{B} \times \hat{\mathbb{R}}^d)$,
- **(*iv*):** *for any* $f \in C_c^\infty(\mathcal{B} \times \hat{\mathbb{R}}^d)$,

$$\int_{\mathcal{B}} f(x, u(x)) \det Du(x) \, dx \le \int_{\hat{\mathbb{R}}^d} \sup_{x\in\mathcal{B}} f(x, w) \, dw.$$

The conditions listed in the definition above have explicit physical meaning which allows weak diffeomorphisms to be candidates for the description of standard deformations. Condition (i) is the standard requirement assuring that the map $x \longmapsto u(x)$ is orientation-preserving. The subsequent requirement prescribes that all the minors of Du are bounded in $L^1(\mathcal{B})$. For example, it implies that the average volume change be not infinite. This way, if one thinks of u as a transplacement mapping, one is prescribing that global extreme deformations are prevented. The third condition imposes that the graph of the map $x \longmapsto u(x)$ has no boundary current inside $\mathcal{B} \times \hat{\mathbb{R}}^d$. In other words, such a condition imposes that fractures do not occur. The condition may appear more clear when one thinks of the standard deformation of a body from its reference macroscopic place $\mathcal{B}$ to the actual place $\mathcal{B}_a := u(\mathcal{B})$ as a d-dimensional surface S in $\mathbb{R}^d \times \hat{\mathbb{R}}^d$, having one-to-one projections into both factors of the cross product space. Elements of the tangent bundle of S then describe infinitesimal deformations. At each point of S, namely the pair (x, y), where now $y = u(x)$, the tangent $d-$vector to S at (x, y) can be written in terms of $Du(x)$, $\operatorname{adj} Du(x)$ and $\det Du(x)$, or, alternatively, in terms of $D\hat{u}(y)$, $\operatorname{adj} D\hat{u}(y)$ and $\det D\hat{u}(y)$, where $\hat{u}$ is the inverse deformation with $x = \hat{u}(y)$. It is then evident that the tangent d-vector to S is associated with $M(Du)$. The current G_u is then a global *dual* way to account for the multilinear algebra over the tangent bundle of S. If the current has a boundary, this means that there are regions of S where all

$(d-1)$-forms are not exact (recall that $\partial G_u := G_u(d\omega)$), which means that there are homological discontinuities in $\mathcal{B}$.

Condition (iv) in Definition 1 has been proposed in [**24**] and is a generalization of another condition suggested in [**13**]. It permits us to move $\mathcal{B}$ along u into a region $\mathcal{B}_a$ in such a way that self-contact between parts of the boundary $\partial\mathcal{B}$ is allowed while self-penetration is excluded.

The structure properties of $dif^{1,1}$ are summarized in the theorem reported without proof below.

THEOREM 1 ([**24, 27**]).

(1) *(Closure) Let $\{u_k\}$ be a sequence with $u_k \in dif^{1,1}(\mathcal{B}, \hat{\mathbb{R}}^d)$ for any k. If*

$$u_k \rightharpoonup u \quad \text{and} \quad M(Du_k) \rightharpoonup v$$

weakly in L^1, then $v = M(Du)$ a.e. and $u \in dif^{1,1}(\mathcal{B}, \hat{\mathbb{R}}^d)$.

(2) *(Compactness) Let $\{u_k\}$ be a sequence with $u_k \in W^{1,r}(\mathcal{B}, \hat{\mathbb{R}}^d)$, $r > 1$. Consider u_k as weak diffeomorphisms. Assume that there exists a constant $C > 0$ and a convex function $\vartheta : [0, +\infty) \to \mathbb{R}^+$ such that $\vartheta(t) \to +\infty$ as $t \to 0^+$, and*

$$\|M(Du_k)\|_{L^r(\mathcal{B})} \leq C, \qquad \int_{\mathcal{B}} \vartheta(\det Du_k(x))\, dx \leq C.$$

By taking subsequences $\{u_j\}$ with $u_j \rightharpoonup u$ in $W^{1,r}(\mathcal{B}, \hat{\mathbb{R}}^d)$, one gets $u_j \to u$ in $L^r(\mathcal{B})$, $M(Du_j) \rightharpoonup M(Du)$ in L^r and $\int_{\mathcal{B}} \vartheta(\det Du(x))\, dx \leq C$. In particular, u is a weak diffeomorphism.

Although weak diffeomorphisms can appropriately describe standard deformations, for technical reasons the attention will be focused in the sequel on a subspace of $dif^{1,1}(\mathcal{B}, \hat{\mathbb{R}}^d)$, namely

$$dif^{r,1}(\mathcal{B}, \hat{\mathbb{R}}^d) := \left\{ u \in dif^{1,1}(\mathcal{B}, \hat{\mathbb{R}}^d) \,|\, |M(Du)| \in L^r(\mathcal{B}) \right\},$$

for some $r > 1$.

As remarked in the introduction, the existence of gap phenomena in relaxing energies, defined over jet bundles on the class of weak diffeomorphisms, is still an open problem.

3. Currents in the mechanics of complex bodies

As mentioned in the initial section, the combined use of the properties of weak diffeomorphisms and Sobolev maps allows one to find minimizers of the elastic energies in complex bodies, simple Cauchy bodies being obtained when the macroscopic influence of the material complexity is neglected.

Bodies are called *complex* when changes in their material texture at various subscales (from nano-to-mesoscopic level) prominently influence the gross behavior through peculiar interactions generated by the mutations of the substructure. Their list includes liquid crystals, bodies with dense distributions of microcracks, quasi-periodic alloys, materials with polarization (ferroelectrics or magnetoelastic bodies), various types of composites and bodies with strain-gradient effects. Microstructures can be exploited, even invented anew, to reach predetermined goals. An essential problem in their description is that of bridging scales even from the atomic to the macroscopic level, translating through the continuum limit the prominent aspects of the microstructural features. Inner dimensions are exploited. They are the dimension of the manifold of substructural shapes which is used to represent the peculiar features of the material microstructure.

The work developed so far on the foundations of the mechanics of complex bodies has underlined a wide family of scientific questions of theoretical and applied character (see [**10, 11, 12, 35, 38, 15, 45, 36**] and the references therein). In classical field theories the starting point is to consider a body as an abstract set, the elements of which are called *material elements.* A model of a body starts then from the attribution of geometrical structure to such a set. Essentially, at least a rough idea of the material element must be at our disposal. The clear definition of the nature of the generic material element is not a trivial task. However, in the standard format of continuum mechanics, the problem is overcome by describing the geometry of every material element only through the place in space it occupies. The description is the minimalist one. One considers the material element as a monad in Leibniz' words, that is, a windowless box. No interest is shown for the geometry of the structure of the material element and its changes. Information on it are known just at the level of constitutive structures.

The analysis of complex bodies alters the standard paradigms of continuum mechanics. Events at low scales influence, in fact, the gross behavior. To take into account these effects the first step is to furnish a more detailed representation of the material elements. They have to be considered as *systems* rather than monads. The geometry of the inner microstructure must be represented. In fact, one selects only some prominent geometric features and some morphological descriptor of the inner geometry. Their choice is a structural part of the modeling process. The description is *multifield,* so it is *intrinsically multiscale* and *multidimensional.* The standard *transplacement field*

$$x \longmapsto u := u\,(x) \in \hat{\mathbb{R}}^d, \quad x \in \mathcal{B},$$

pictures macroscopic *deformations,* while the *morphological descriptor field*

$$x \longmapsto \nu := \nu\,(x) \in \mathcal{M}, \quad x \in \mathcal{B},$$

describes the inner geometry of the microstructure. $\mathcal{M}$ is the so-called *manifold of substructural shapes.*

Standard requirements are assumed to hold: $x \longmapsto u$ is essentially an orientation-preserving piecewise C^1-diffeomorphism, and the region $u\,(\mathcal{B})$ has the same geometrical structural properties of $\mathcal{B}$. The field $x \longmapsto \nu$ is assumed to be piecewise differentiable. In order to construct the essential features of the mechanics of complex bodies, it is not necessary to select some specific manifold $\mathcal{M}$. The only necessary assumption is that $\mathcal{M}$ is just a finite-dimensional differentiable manifold (preferably without boundary).

Here, the discussion of standard and generalized measures of deformations, the representation of macroscopic and microscopic actions through the external power, the generalized notions of observers and their changes (notions which correspond to a nontrivial sliding in the standard paradigm), the invariance requirements of the external power (or the companion relative power) leading to balance equations obtained independently of constitutive prescription are not recalled (relevant comments can be found in [**11, 35, 36**]). The attention is focused only on elastic complex bodies. Their behavior is governed by an energy of the type

$$\mathcal{E}(u,\nu,\mathcal{B}) := \int_{\mathcal{B}} e(x,u(x),Du(x),\nu(x),D\nu(x))\ dx,$$

where

$$e(x,u,Du,\nu,D\nu) = \hat{e}(x,Du,\nu,D\nu) - \hat{w}(u,\nu),$$

with $\hat{e}(x,Du,\nu,D\nu)$ the elastic energy and $\hat{w}(u,\nu)$ the potential of external bulk actions; it admits the additive decomposition $\hat{w}(u,\nu) = \hat{w}_1(u) + \hat{w}_2(\nu)$. Equilibrium states are described by minimizers of such an energy. Conditions assuring their existence are sketched below.

Consider the energy density e as a map

$$e : \mathcal{B} \times \hat{\mathbb{R}}^d \times \mathcal{M} \times M^+_{d\times d} \times M_{N\times d} \to \bar{\mathbb{R}}^+$$

with values $e(x,u,F,\nu,N)$, where $F := Du(x)$, $N := D\nu(x)$, and assume that the properties (H1) and (H2) listed below hold.

(H1): e is such that there exists a Borel function

$$Pe : \mathcal{B} \times \hat{\mathbb{R}}^d \times \mathcal{M} \times \Lambda_d(\mathbb{R}^d \times \hat{\mathbb{R}}^d) \times M_{N\times d} \to \bar{\mathbb{R}}^+,$$

with values $Pe(x,u,\nu,\xi,N)$, which is

(a) l. s. c. in (u,ν,ξ,N) for a.e. $x \in \mathcal{B}$,
(b) convex in (ξ,N) for any (x,u,ν),
(c) such that $Pe(x,u,\nu,M(F),N) = e(x,u,\nu,F,N)$ for any list of entries

$$(x,u,\nu,F,N) \quad \text{with} \quad \det F > 0.$$

In terms of Pe, the energy functional becomes

$$\mathcal{E}(u,\nu,\mathcal{B}) = \int_{\mathcal{B}} Pe(x,u(x),\nu(x),M(F),N)\ dx. \tag{3.1}$$

(H2): The energy density e satisfies the growth condition

$$e(x,u,\nu,F,N) \geq C_1\left(|M(F)|^r + |N|^s\right) + \vartheta(\det F) \tag{3.2}$$

for any (x,u,ν,F,N) with $\det F > 0$, $r,s > 1$, $C_1 > 0$ constants and $\vartheta : (0,+\infty) \to \mathbb{R}^+$ a convex function such that $\vartheta(t) \to +\infty$ as $t \to 0^+$.

In essence, the assumption that Pe is convex in $(M(F),N)$ for any (x,y,ν) is an assumption of stability of the material. It accounts for a possible interplay between the gradient of the gross deformation and the inhomogeneity of the microstructure distribution within the body. In fact, the inhomogeneity, that is, the way in which the microstructure varies from place to place, is measured by the gradient of the morphological descriptor. The growth condition (3.2) has a constitutive nature. It prescribes that the energy admits a polynomial lower bound which has the typical structure of a decomposed energy of Ginzburg-Landau type. It describes only interactions between neighboring material elements and does not account for the energy associated with the self-actions inside every material element. For this reason, with (3.2) the assumption is in a sense that substructural events within the generic material element may only increase the overall energy.

Dirichlet boundary conditions for u and ν are imposed here over portions of the boundary $\partial\mathcal{B}$ indicated by $\partial\mathcal{B}_u$ and $\partial\mathcal{B}_\nu$, respectively. In fact, it is assumed that the field $x \longmapsto u(x)$ takes assigned values $x \longmapsto u_0(x)$ over $\partial\mathcal{B}_u$, while $x \longmapsto \nu(x)$ is prescribed to be $x \longmapsto \nu_0(x)$ over $\partial\mathcal{B}_\nu$.

Under these conditions, existence of minimizers for the energy can be investigated in the space

$$\mathcal{W}_{r,s} := \left\{ (u,\nu) \, | u \in dif^{r,1}(\mathcal{B}, \hat{\mathbb{R}}^d), \ \nu \in W^{1,s}(\mathcal{B}, \mathcal{M}) \right\}.$$

From the closure theorem for weak diffeomorphisms reported above and standard semicontinuity results, an existence theorem can be derived.

THEOREM 2 ([**37**]). *Under the hypotheses (H1) and (H2), if there is a pair* $(u_0, \nu_0) \in \mathcal{W}_{r,s}$ *such that* $\mathcal{E}(u_0, \nu_0) < +\infty$, *the functional* $\mathcal{E}$ *achieves the minimum value in the classes*

$$\mathcal{W}^d_{r,s} := \{(u,\nu) \in \mathcal{W}_{r,s} | u = u_0 \text{ on } \partial\mathcal{B}_u, \nu = \nu_0 \text{ on } \partial\mathcal{B}_\nu\}$$

and

$$\mathcal{W}^c_{r,s} := \left\{ (u,\nu) \in \mathcal{W}_{r,s} \mid \partial G_u = \partial G_{u_0} \text{ on } \mathcal{D}^2(\mathbb{R}^3 \times \hat{\mathbb{R}}^3), \nu = \nu_0 \text{ on } \partial\mathcal{B}_\nu \right\}.$$

Like the constraints on the structure of the energy, even the choice of the functional space $\mathcal{W}_{r,s}$ has a constitutive nature. Other possible choices of the functional space can be made.

For example, one can imagine another lower bound for the energy density. Any choice of lower bounds has a constitutive nature. The second one adopted here is indicated by (H3).

(H3): The energy density satisfies the growth condition

$$e(x,u,\nu,F,N) \geq C_2(|F|^{d-1} + |\text{adj } F|^{d/d-1} + |N|^s) + \vartheta(\det F) \tag{3.3}$$

for any (x,u,ν,F,N) with $\det F > 0$, $C_2 > 0$ a constant and $\vartheta : (0,+\infty) \to \mathbb{R}^+$ as above.

Consider the energy as defined over the functional class

$$\begin{aligned} \mathcal{W}_{d,\frac{d}{d-1},s} \ : \ &= \Big\{ (u,\nu) \, | u \in W^{1,d-1}(\mathcal{B}, \hat{\mathbb{R}}^d), \text{adj}(Du) \in L^{d/d-1}, \\ &\quad \text{(iv.) in Def. 1 holds}, \nu \in W^{1,s}(\mathcal{B}, \mathcal{M}) \Big\}. \end{aligned}$$

By taking into account the $L \log L$ estimate in [**40**] (see also [**42**]), one can find a relevant existence result.

THEOREM 3 ([**37**]). *Under the assumptions (H1) and (H3) reported above, the functional* $\mathcal{E}$ *achieves its minimum value in the class*

$$\mathcal{W}^d_{d,\frac{d}{d-1},s} := \left\{ (u,\nu) \in \ \mathcal{W}_{d,\frac{d}{d-1},s} \mid u = u_0 \text{ on } \partial\mathcal{B}_u, \nu = \nu_0 \text{ on } \partial\mathcal{B}_\nu \right\},$$

provided that there exists a pair $(u_0, \nu_0) \in \mathcal{W}_{d,\frac{d}{d-1},s}$ *such that* $\mathcal{E}(u_0, \nu_0, \mathcal{B}) < +\infty$.

The presence of the function ϑ in the lower bounds selected in (H2) and (H3) is justified by the need of avoiding physically undesired behaviors such as the extreme deformations obtained by letting $\det F$ go to zero on a set of positive measure.

Different special structural choices of the energy $\mathcal{E}(u,\nu,\mathcal{B})$ can be made. A couple of examples are reported here.

(1) Neglect macroscopic deformation and consider ν to be a scalar coinciding, for example, with the volume fraction of a given phase in a two-phase material. The density

$$\zeta \left(\nu^2 - 1\right)^2 + \varsigma \left|N\right|^2,$$

with ζ and ς two constants, can be selected. With ν in $\tilde{\mathbb{R}}^m$, the previous density becomes $\left(|\nu|^2 - 1\right)^2 + \mu |N|^2$. Both densities are of Ginzburg-Landau type.

(2) Assume the existence of an internal constraint of the type

$$\nu = \nu(F).$$

In this case the microstructure is called *latent* (see [**10**]). The energy density becomes that of a second-grade Cauchy body, that is,

$$e\left(x, u, Du, D^2u\right).$$

The special choice

$$e\left(x, u, Du, D^2u\right) = \left(|Du|^2 - 1\right)^2 + \varsigma^2 \left|D^2u\right|^2$$

is of Aviles-Giga type. It can be obtained even when (a) the macroscopic deformation is neglected and (b) the morphological descriptor ν coincides with the spatial derivative of some field, namely $\nu = D\phi$, with $x \longmapsto \phi(x)$ a differentiable map.

The existence theorems reported above apply to wide classes of complex bodies. An extended list of prominent physical examples can be found in [**37**].

The first variation of the energy $\mathcal{E}(u_0, \nu_0, \mathcal{B})$ can be obtained in different ways.

In the presence of regular minimizers admitting tangential derivatives, since the energy density presented above is in essence a 3-form over the first jet bundle of a bundle $\mathcal{Y}$ over $\mathcal{B}$, namely $\pi : \mathcal{Y} \to \mathcal{B}$, with π the canonical projection and the typical fiber $\pi^{-1}(x) = \hat{\mathbb{R}} \times \mathcal{M}$, one can use the standard *vertical lift* of the first jet bundle (canonical injection) to determine common Euler-Lagrange equations. Elements of them are representations of standard and microstructural interactions arising within the body. In particular, the derivative of the energy with respect to F represents the standard Piola-Kirchhoff stress, the derivative with respect to N the so-called microstress measuring contact interactions due to inhomogeneous microstructural changes, and the derivative of $\hat{e}$ with respect to ν indicates a self-action within every material element. The derivatives of $\hat{w}$ measure the external bulk actions over the body in its whole (gravitational action) and the microstructure (e.g., electric fields determining the polarization in a body).

The tangential derivative of maps in $\mathcal{W}_{r,s}$ or $\mathcal{W}_{d,\frac{d}{d-1},s}$ does not always exist. Moreover, regularity theorems seem not to be available. However, *horizontal variations* can be computed. They are determined by diffeomorphisms of $\mathcal{B}$ into itself such that their restriction over the boundary $\partial\mathcal{B}$ coincides with the identity. Horizontal variations lead to balance equations called the balance of configurational forces, the ones displaying the balance of actions on defects in solids, at least in the conservative setting when dissipative actions are not present. When smooth minimizers are at our disposal, the balance of configurational forces coincides also with the pullback in the reference place $\mathcal{B}$ of the balances of standard and microstructural actions. For nonsmooth minimizers, the two classes of balance equations, namely the ones obtained by vertical variations and the one following from horizontal variations, have a different nature. Examples stressing this difference can be found in [**21**], vol. I, pp. 152-153.

4. Another tool: varifolds

Consider the dimension d of the ambient space greater than or equal to 2. For a positive integer k, $1 \leq k \leq d$, the Grassmann manifold of k-planes through the origin in $\mathbb{R}^d$ is indicated by $\mathcal{G}_{k,d}$ and is also identified with the set of the projectors $\Pi : \mathbb{R}^d \to \mathbb{R}^d$ onto k-planes. They have the well-known characterizing properties $\Pi^2 = \Pi$, $\Pi^2 = \Pi$, $\operatorname{Rank}\Pi = k$. Every projector is an element of a compact subset of $\mathbb{R}^d \otimes \mathbb{R}^d$.

It is then possible to construct a bundle $\mathcal{G}_k(\mathcal{B})$ with a natural projection $\pi : \mathcal{G}_k(\mathcal{B}) \to \mathcal{B}$ and typical fiber coinciding with the Grassmann manifold $\mathcal{G}_{k,d}$.

DEFINITION 2. A k-**varifold** over $\mathcal{B}$ is a nonnegative Radon measure V over the bundle $\mathcal{G}_k(\mathcal{B})$.

In short-hand notation one writes $V \in \mathsf{M}(\mathcal{G}_k(\mathcal{B}))$. Let $\pi_\#$ be the projection of measures over $\mathcal{B}$ associated with the natural projection $\pi : \mathcal{G}_k(\mathcal{B}) \to \mathcal{B}$. The projection $\pi_\#$ allows one to define the *weighed measure* of the varifold V, which is the Radon measure over $\mathcal{B}$ defined by $\mu_V := \pi_\# V$. Such a measure defines the *mass* $\mathbf{M}(V)$ of the varifold through the relation

$$\mathbf{M}(V) := V(\mathcal{G}_k(\mathcal{B})) = \mu_V(\mathcal{B}).$$

It is rather immediate to construct varifolds over a subset $\mathfrak{b}$ of $\mathcal{B}$. For the purpose of the analysis developed here, some assumptions on the structure of $\mathfrak{b}$ are necessary. In fact, $\mathfrak{b}$ is individuated by a measure which is absolutely continuous with integer density θ with respect to the k-dimensional Hausdorff measure $\mathcal{H}^k$ in $\mathbb{R}^d$. It is then assumed that (i) $\mathfrak{b}$ is an $\mathcal{H}^k$-measurable, k-rectifiable subset of $\mathcal{B}$ and (ii) the density θ belongs to $L^1(\mathfrak{b}, \mathcal{H}^k)$ and takes integer values. All these assumptions avoid the selection of too many exotic subsets $\mathfrak{b}$. For example, they assure that for almost every $x \in \mathfrak{b}$, there exists the approximated (in the sense of geometric measure theory, see [**17**]) tangent space $T_x\mathfrak{b}$ to $\mathfrak{b}$ at x.

Under these conditions a varifold associated with the triple $(\mathfrak{b}, \theta, \mathcal{H}^k)$ can be defined through its action over the space of compactly supported C^0 functions over the fiber bundle $\mathcal{G}_k(\mathcal{B})$. Such a measure is indicated by $V_{\mathfrak{b},\theta}(\varphi)$ and is defined by

$$V_{\mathfrak{b},\theta}(\varphi) := \int_{\mathcal{G}_k(\mathcal{B})} \varphi(x, \Pi)\ dV_{\mathfrak{b},\theta}(x, \Pi) := \int_{\mathfrak{b}} \theta(x)\, \varphi(x, \Pi)\ d\mathcal{H}^k,$$

for any $\varphi \in C^0(\mathcal{G}_k(\mathcal{B}))$. The second-rank tensor $\Pi(x)$ is the projection onto $T_x\mathfrak{b}$ at all x's where $T_x\mathfrak{b}$ is defined. This circumstance justifies the assumption of having at disposal a set $\mathfrak{b}$ admitting (an at least approximated) tangent space at almost every x. $V_{\mathfrak{b},\theta}$ is called *integer rectifiable k-varifold associated with* $(\mathfrak{b}, \theta, \mathcal{H}^k)$.

When one selects vector-valued measures in the space of Radon measures over $\mathcal{G}_k(\mathcal{B})$, it is done with the aim of defining a special class of varifolds: the *varifolds with curvature* [**32, 34**]. The generalized curvature associated with $\mathfrak{b}$ is a third-rank tensor A which is at every $x \in \mathfrak{b}$ the value of a tensor field defined over $\mathcal{G}_k(\mathcal{B})$ and taking values over $\mathbb{R}^{d*} \otimes \mathbb{R}^d \otimes \mathbb{R}^{d*}$.

DEFINITION 3. A varifold V is called **curvature k-varifold with boundary** if

(1) *V is an integer, rectifiable k-varifold $V_{\mathfrak{b},\theta}$ associated with the triple* $(\mathfrak{b}, \theta, \mathcal{H}^k)$,

(2) *there exists a function* $A \in L^1\left(\mathcal{G}_k(\mathcal{B}), \mathbb{R}^{d*} \otimes \mathbb{R}^d \otimes \mathbb{R}^{d*}\right)$, *in components* $A_j^{\ell i}$, *and a vector Radon measure* $\partial V \in \mathsf{M}\left(\mathcal{G}_k(\mathcal{B}), \mathbb{R}^d\right)$, *the so-called* ***varifold boundary measure****, such that, for every* $\varphi \in C_c^\infty\left(\mathcal{G}_k(\mathcal{B})\right)$, *one gets*

$$\int_{\mathcal{G}_k(\mathcal{B})} \left(\Pi D_x \varphi + A^t D_\Pi \varphi + AI\varphi\right) \, dV\left(x, \Pi\right) = -\int_{\mathcal{G}_k(\mathcal{B})} \varphi \, d\partial V\left(x, \Pi\right).$$

In the previous formula, I is the second-rank unit tensor. Indices are saturated in such a way that $\left(\Pi D_x \varphi + A^t D_\Pi \varphi + AI\varphi\right)$ is a vector.

As a matter of notation, the subclass of curvature varifolds with curvature field $(x, \Pi) \longmapsto A\left(x, \Pi\right)$ in $L^p\left(\mathcal{G}_k(\mathcal{B}), \mathbb{R}^{d*} \otimes \mathbb{R}^d \otimes \mathbb{R}^{d*}\right)$, with $p \geq 1$, is indicated in what follows by $CV_k^p\left(\mathcal{B}\right)$. Moreover, $\nabla^{\mathfrak{b}}$ indicates the gradient along $\mathfrak{b}$. The symbol $\Pi_\#$ represents the projector acting over vector measures.

Essential properties of curvature varifolds are discussed in [**32, 34**]. Some of them are summarized in the remarks below where V is a k-varifold with boundary ∂V and curvature $A \in L^1\left(\mathcal{G}_k(\mathcal{B})\right)$.

(1) The curvature tensor satisfies the following relations:

$$A_j^{\ell i} = A_\ell^{ji}, \quad A_j^{ji} = 0, \quad A_j^{\ell i} = \Pi_h^\ell A_j^{hi} + \Pi_j^h A_h^{\ell i}, \quad A_j^{\ell h} \Pi_h^i = A_j^{\ell i}, \quad V - a.e.$$

(2) The vector $H^i\left(x\right) := A_j^{\ell j}\left(x, \Pi\left(x\right)\right)$ has the meaning of generalized mean curvature for b, and is $\mu_V - a.e.\ x$ perpendicular to $T_x b$.

(3) The projection map $x \longmapsto \Pi\left(x\right)$ is $\mu_V - a.e.$ approximately differentiable and

$$\left(\nabla^{\mathfrak{b}} \Pi_j^\ell\left(x\right)\right)^i = A_j^{\ell i}\left(x, \Pi\left(x\right)\right) \quad \mu_V - a.e.\ x.$$

(4) The support of $|\partial V|$ is contained in the support of V, also $|\partial V| \perp V$, and ∂V is tangential to b in the sense that $\Pi_{\# j}^i \left(\partial V\right)^j = \left(\partial V\right)^i$ as measures.

(5) V is a varifold with locally bounded first variation and generalized mean curvature in the sense of Allard with generalized mean curvature vector $H\left(x\right)$ and generalized boundary $\pi_\# \partial V$. So, Allard's regularity and compactness theorems apply. In particular, it has been shown in [**32**] that if $V = V_{\mathfrak{b},\theta} \in CV_k^T\left(\mathcal{B}\right)$, with $p > k$, then it is locally the graph of a multivalued function of class $C^{1,\alpha}$, $\alpha = 1 - \frac{p}{k}$, far from ∂V.

THEOREM 4 (Compactness [**34**]). *For* $1 < p < \infty$, *consider a sequence* $\{V_r\} \subset CV_k^p\left(\mathcal{B}\right)$ *of curvature varifolds with boundary and the corresponding sequences of curvatures* $\{A_r\}$ *and boundary measures* $\{\partial V_r\}$. *For every open set* $\Omega \Subset \mathcal{B}$ *and for every* r, *assume the existence of a constant* $c\left(\Omega\right)$, *depending on* Ω, *such that*

$$\mu_{V_r}\left(\Omega\right) + \left|\partial V_r\right|\left(\mathcal{G}_k(\mathcal{B})\right) + \int_{\mathcal{G}_k(\mathcal{B})} \left|A^{(r)}\right|^p \, dV_r \leq c\left(\Omega\right).$$

Under these conditions, there exists a subsequence $\left\{V^{(r_s)}\right\}$ *of* $\left\{V^{(r)}\right\}$ *and a* k*-varifold* $V \in CV_k^p\left(\mathcal{B}\right)$, *with curvature* A *and boundary* ∂V, *such that*

$$V_{r_s} \rightharpoonup V, \quad A_{r_s} dV_{r_s} \rightharpoonup AdV, \quad \partial V_{r_s} \rightharpoonup \partial V,$$

in the sense of measures, as $r_s \to \infty$. *Moreover, for any convex and lower semicontinuous function* $f : \mathbb{R}^{d*} \otimes \mathbb{R}^d \otimes \mathbb{R}^{d*} \to [0, +\infty]$, *one gets*

$$\int_{\mathcal{G}_k(\mathcal{B})} f\left(A\right) \, dV \leq \liminf_{r_s \to \infty} \int_{\mathcal{G}_k(\mathcal{B})} f\left(A_{r_s}\right) \, dV_{r_s}.$$

5. An extended class of weak diffeomorphisms

A new class of weak diffeomorphims with boundary controlled by a varifold can be defined. They are useful to describe cracks or dislocations in continuum mechanics. An essential example of their physical meaning will be presented in the ensuing section.

The basic idea is to have first at disposal a class of curvature varifolds with boundary $\{V_k\}_{k=1}^{d-1}$ such that (i) each element of the class is in $CV_k^{p_k}(\mathcal{B})$ with $p_k > 1$ and (ii) for $k = 2, ..., d-1$,

$$\pi_{\#} |\partial V_k| \leq \mu_{V_{k-1}}.$$

This last relation is not exotic: it is the weak version of the standard relation occurring between a manifold and its boundary. Such a relation is here expressed in terms of varifolds stratified over different dimensions and supported by rectifiable sets. Once this family of varifolds has been selected, one aims to choose a special class of maps which admit jump sets contained in the supports of the selected varifolds and describe standard deformations outside these sets. Such maps are called *extended weak diffeomorphisms with boundary controlled by stratified varifolds*. Essentially they are maps satisfying all items in the definition of weak diffeomorphism but the requirement to be without boundary. Boundaries in the graphs of these maps are admitted but they have to satisfy conditions expressed in terms of the selected family of stratified varifolds. These conditions are of two types and are classified as **type 1** and **type 2** below.

DEFINITION 4. Assigned a class $\{V_k\}_{k=1}^{d-1}$ of curvature varifolds with boundary, an **extended weak diffeomorphism with controlled boundary of type** 1 is an a.e. approximately differentiable map which satisfies the conditions (i), (ii), (iv) in Definition 1, and

$$\pi_{\#} |\partial G_u| \leq \sum_{j=1}^{d-1} \mu_{V_k}.$$

The condition above means that Green formulas hold true outside the support of the stratified varifolds involved in the previous definition. Moreover, it indicates also that the boundary current ∂G_u has finite mass, and that u belongs to the class $SBV_0(\mathcal{B}, \hat{\mathbb{R}}^d)$ (see [**2**]).

DEFINITION 5. Assigned a class $\{V_k\}_{k=1}^{d-1}$ of curvature varifolds with boundary, an **extended weak diffeomorphism with controlled boundary of type 2** is an a.e. approximately differentiable map which satisfies the conditions (i), (ii), (iv) in Definition 1, and

$$\pi_{\#} |\partial G_u| \leq \sum_{j=1}^{d-1} \mu_{V(j)} + \pi_{\#} |\partial V_1| .$$

Comments on the different physical situations described by the two classes are reported in the ensuing section. Here the attention is mainly focused on the class of extended diffeomorphisms with controlled boundary of type 1. To affirm that a map $u : \mathcal{B} \to \hat{\mathbb{R}}^d$ belongs to this class, one writes just $u \in dif^{1,1}(\mathcal{B}, V, \hat{\mathbb{R}}^d)$. The structural properties of $dif^{1,1}(\mathcal{B}, V, \hat{\mathbb{R}}^d)$ are collected in the ensuing theorem.

THEOREM 5. *Consider a sequence of varifolds $\{V_k\}$ on $\mathcal{B}$, chosen in $CV_1^p(\mathcal{B})$, $p > 1$, and with equibounded variation, i.e.* $\sup_k \mu_{V_k}(\mathcal{B}) < \infty$. *Take a sequence*

$\{u_k\}$ *such that* $u_k \in dif^{1,1}(\mathcal{B}, V_k, \hat{\mathbb{R}}^d)$. *Assume that there exist* $u \in L^1(\mathcal{B}, \hat{\mathbb{R}}^d)$, $v \in L^1(\mathcal{B}, \Lambda_d(\mathbb{R}^d \times \hat{\mathbb{R}}^d))$, *and* $V \in CV_1^p(\mathcal{B})$, $p > 1$, *such that* $u_k \rightharpoonup u$, $M(Du_k) \rightharpoonup v$, *and* $V_k \rightharpoonup V$ *as measures. The identity* $v = M(Du)$ *holds. Moreover, if* $\det Du > 0$ *a.e., one also finds that* $u \in dif^{1,1}(\mathcal{B}, V, \hat{\mathbb{R}}^d)$.

PROOF. The assumptions imply that $\mathbf{M}(G_{u_r}) + \mathbf{M}(\partial G_{u_r}) \leq C$ independently of r. In particular, the sequence $\{u_k\}$ is equibounded in $BV(\mathcal{B}, \hat{\mathbb{R}}^d)$ so that, by passing eventually to subsequences, $\{u_k\}$ converges strongly in L^1 and a.e. to u, and G_{u_k} converges to a current S. Moreover, S is an integer multiplicity rectifiable current by the Federer-Fleming compactness theorem. For a more direct proof, see [**27**]. It then follows that $v = M(Du)$ a.e. and $S = G_u$ (see [**27**]). Properties (ii), (iii), and (iv) in Definition 1 hold true. If u satisfies (i), then $u \in dif^{1,1}(\mathcal{B}, V, \hat{\mathbb{R}}^d)$.

In particular, the subclass

$$dif^{p,1}(\mathcal{B}, V, \hat{\mathbb{R}}^d) := \left\{ u \in dif^{1,1}(\mathcal{B}, V, \hat{\mathbb{R}}^d) \mid |M(Du)| \in L^p(\mathcal{B}) \right\}$$

will be useful in the next section.

6. Describing cracks in term of varifolds

Varifolds are an essential tool for describing low-dimensional defects in solids such as discontinuity surfaces, dislocations, cracks.

Assume also that a crack pattern can occur in a Cauchy body which is elastic-brittle. The basic idea is to describe the crack pattern through a family of varifolds of various co-dimensions. Consider the reference configuration $\mathcal{B}$ to be selected in $\mathbb{R}^3$ for the sake of simplicity. Imagine a smooth single crack in the actual configuration of the body which has as pre-image in $\mathcal{B}$ a piece of a certain surface $\mathcal{C}$ which can be assumed smooth just to visualize the situation. A two-dimensional varifold can be used to describe the surface. The Grassmanian is constructed by using the tangent planes to the surface, and the surface itself is the support $\mathfrak{b}$ of the varifold. The boundary of $\mathcal{C}$ is then the support of the boundary of the varifold. A sketch of the situation is described in Figure 1. It is possible to give a special status to a part of the boundary of $\mathcal{C}$, namely the part inside the body (the dashed line in Figure 1), that, is the *tip* of the crack, with the aim of assigning a line energy to it. In this case a one-dimensional varifold can be assigned. It should have support including the tip. Part of the support of such a varifold could also describe line defects such as dislocations occurring ahead of the crack tip. In evaluating the equilibrium of such a body one has two unknowns: (i) the family of varifolds describing the crack pattern, and (ii) the transplacement field. The latter is selected in $dif^{p,1}(\mathcal{B}, V, \hat{\mathbb{R}}^d)$ to assure that the boundary of its graph can be projected over the support of the varifold only. This way one wants to select a transplacement describing a standard deformation outside the crack pattern.

Of course, a generic crack path is more complicated than the one described above (see also Figure 1). Moreover, it is not necessary that the ambient space be three-dimensional. The treatment proposed here can be set in an ambient space with higher dimension, say d.

Energy is assigned to the lateral margins of the crack and to the tip. As an extension of the classical Griffith scheme, the crack energy depends on the curvature of the crack. Special concrete examples justifying such an extension can be found in [**22**]. The energy presented in the introduction is written for the sake of simplicity

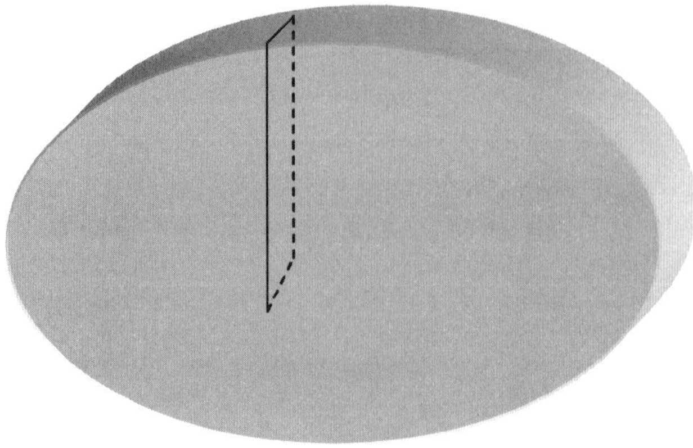

FIGURE 1. Elastic-brittle simple body with a crack which has a planar pre-image in the reference configuration depicted above.

in a three-dimensional ambient space to favour the physical visualization of the meaning of the various terms. Its extension to higher-dimensional spaces is however immediate. In this case, the geometry of the crack pattern is then described by a family $\{V_k\}_{k=1}^{d-1}$ of varifolds stratified over supports at various dimensions. Such a family of varifolds is characterized by the ensuing formal definition.

DEFINITION 6. A family $\{V_k\}_{k=1}^{d-1}$ of curvature varifolds in $CV_k^{p_k}(\mathcal{B})$, with $p_k > 1$, is called **stratified** when

$$\pi_{\#} |\partial V_k| \leq \mu_{V_{k-1}}, \ \forall k = 2, ..., d-1.$$

Stratified cracks describe naturally the geometry of crack patterns in a body placed in $\hat{\mathbb{R}}^d$. The associated energy, written in accordance with the remarks above, reads

$$\begin{aligned} \mathcal{E}(u, \{V_k\}, \mathcal{B}) \ : &= \int_{\mathcal{B}} e(x, Du) \ dx + \sum_{k=1}^{d-1} \alpha_k \int_{\mathcal{G}_k(\mathcal{B})} \left|\mathbf{A}_{(k)}\right|^{p_k} \ dV_k \\ &+ \sum_{k=1}^{d-1} \beta_k \mathbf{M}(V_k) + \gamma \mathbf{M}(\partial V_1). \end{aligned}$$

Physical convenience suggests the introduction of a family of *comparison varifolds* $\left\{\tilde{V}_k\right\}_{k=1}^{d-1}$ such that for any k one gets $\tilde{V}_k \in CV_k^{p_k}(\mathcal{B})$ and $\mu_{\tilde{V}_k} \leq \mu_{V_k}$. The assignment of $\left\{\tilde{V}_k\right\}$ does not mean that one is considering a preexisting crack pattern because the comparison varifold family can be null. However, when an initial crack exists, the condition assures that the competitors in the minimizing procedure can only extend from the initial crack.

The functional setting in which one tries to find minimizers of $\mathcal{E}(u, V, \mathcal{B})$ can then be specified. The space

$$\begin{aligned} \mathcal{A}_{q,p,K,\{\tilde{V}_k\}}(\mathcal{B}) \ : &= \Big\{(u, \{V_k\}) \mid V_k \in CV_k^{p_k}(\mathcal{B}), u \in dif^{q,1}(\mathcal{B}, V_k, \hat{\mathbb{R}}^d), \\ &\{V_k\} \text{ is stratified}, \|u\|_{L^\infty(\mathcal{B})} \leq K, \mu_{\tilde{V}_k} \leq \mu_{V_k}, \forall k = 1, ..., d-1\Big\} \end{aligned}$$

is then the natural ambient in which the existence of minimizers of the energy $\mathcal{E}(u, V, \mathcal{B})$ can be investigated. In particular, the subspace

$$\mathcal{A}^{u_0}_{q,p,K,\{\tilde{V}_k\}}(\mathcal{B}) := \left\{(u, \{V_k\}) \in \mathcal{A}_{q,p,K,\{\tilde{V}_k\}}(\mathcal{B}) \mid u(x) = u_0(x), x \in \partial\mathcal{B}_u\right\},$$

with $\partial\mathcal{B}_u$ the part of the boundary of the body where the transplacement field is prescribed, plays a role.

THEOREM 6. *Assume* $K > 0$, $q, p_k > 1$, *and* $\tilde{V}_k \in CV_k^{p_k}(\mathcal{B})$ *for any* k. *If there exists* $\left(u_0, \left\{V_k^0\right\}\right) \in \mathcal{A}^{u_0}_{q,p,K,\{\tilde{V}_k\}}(\mathcal{B})$ *such that* $\mathcal{E}\left(u_0, \left\{V_k^0\right\}, \mathcal{B}\right) < +\infty$, *then* $\mathcal{E}(u, \{V_k\}, \mathcal{B})$ *attains there the minimum value.*

Further details, proofs and the evaluation of the first variation of $\mathcal{E}(u, V, \mathcal{B})$ can be found in [**22**].

A final remark deserves mention: in fact, in managing an energy such as $\mathcal{E}(u, V, \mathcal{B})$, one is in essence considering *a cracked body such as a complex body*, the difference with the format described in the previous sections resting in the nature of the morphological descriptor which is now a measure. $CV_k^p(\mathcal{B})$ plays here the role of the manifold of substructural shapes.

Acknowledgement. This work has been developed within the programs of the research group in 'Theoretical Mechanics' of the 'Centro di Ricerca Matematica Ennio De Giorgi' of the Scuola Normale Superiore at Pisa.

References

[1] Ambrosio, L., Braides, A. (1995), Energies in SBV and variational models in fracture mechanics, *Homogenization and applications to material sciences* (Nice, 1995), 1–22, *GAKUTO Internat. Ser. Math. Sci. Appl.*, 9, Gakkōtosho, Tokyo, 1995. MR1473974 (98g:73044)

[2] Ambrosio, L., Braides, A., Garroni, A. (1998), Special functions with bounded variation and with weakly differentiable traces on the jump set, *NoDEA Nonlinear Differential Equations Appl.*, **5**, 219–243. MR1618974 (99d:49075)

[3] Aviles, P., Giga, Y. (1991), Variational integrals on mappings of bounded variation and their lower semicontinuity, *Arch. Rational Mech. Anal.*, **115**, 201–255. MR1106293 (92d:49019)

[4] Ball, J. M. (1976/77), Convexity conditions and existence theorems in nonlinear elasticity, *Arch. Rational Mech. Anal.*, **63**, 337-403. MR0475169 (57:14788)

[5] Ball, J. M. (1981), Global invertibility of Sobolev functions and the interpenetration of matter, *Proc. Royal Soc. Edinburgh*, **88A**, 315-328. MR616782 (83f:73017)

[6] Ball, J. M. (2002), Some open problems in elasticity, in *Geometry, Mechanics and Dynamics*, Newton, P., Holmes, P. and Weinstein, A. (eds.), Springer-Verlag, New York, 3-59. MR1919825 (2003g:74001)

[7] Bethuel, F. (1993), On the singular set of stationary harmonic maps, *Manuscripta Math.*, **78**, 417–443. MR1208652 (94a:58047)

[8] Bourdin, B., Francfort, G. A., Marigo, J.-J. (2008), The variational approach to fracture, *J. Elasticity*, **91**, 5–148. MR2390547 (2009b:49013)

[9] Brezis, H., Coron, J.-M., Lieb, E. H. (1986), Harmonic maps with defects. *Comm. Math. Phys.*, **107**, 649–705. MR868739 (88e:58023)

[10] Capriz, G. (1985), Continua with latent microstructure, *Arch. Rational Mech. Anal.*, **90**, 43-56. MR792882 (86m:73002)

[11] Capriz, G., *Continua with microstructure*, Springer-Verlag, Berlin, 1989. MR985585 (90c:73007)

[12] Capriz, G., Virga, E. G. (1990), Interactions in general continua with microstructure, *Arch. Rational Mech. Anal.*, **109**, 323-342. MR1031574 (91c:73001)

[13] Ciarlet, P., Nečas, J. (1987), Injectivity and self-contact in nonlinear elasticity, *Arch. Rational Mech. Anal.*, **97**, 171-188. MR862546 (87j:73028)

[14] Dal Maso, G., Francfort, G. A., Toader, R. (2005), Quasistatic crack growth in nonlinear elasticity, *Arch. Ration. Mech. Anal.*, **176**, 165–225. MR2186036 (2007b:35311)

[15] de Fabritiis, C., Mariano, P. M. (2005), Geometry of interactions in complex bodies, *J. Geom. Phys.*, **54**, 301-323. MR2139085 (2006c:74004)

[16] Eells, J., Jr., Sampson, J. H. (1964), Harmonic mappings of Riemannian manifolds, *Amer. J. Math.*, **86**, 109–160. MR0164306 (29:1603)

[17] Federer, H., *Geometric measure theory*, Springer-Verlag, New York, 1969. MR0257325 (41:1976)

[18] Foss, M., Hrusa, W. J., Mizel, V. J. (2003), The Lavrentiev gap phenomenon in nonlinear elasticity, *Arch. Rational Mech. Anal.*, **167**, 337-365. MR1981861 (2004d:74008)

[19] Francfort, G., Marigo, J.-J. (1998), Revisiting brittle fracture as an energy minimizing problem, *J. Mech. Phys. Solids* **46**, 1319-1342. MR1633984 (99f:73066)

[20] Giaquinta, M., *Multiple integrals in the calculus of variations and nonlinear elliptic systems*, Princeton University Press, Princeton, 1983. MR717034 (86b:49003)

[21] Giaquinta, M., Hildebrandt, S., *Calculus of variations*, vol. I *The Lagrangian formalism*, vol. II *The Hamiltonian formalism*, Springer-Verlag, Berlin, 1996. MR1368401 (98b:49002a); MR1385926 (98b:49002b)

[22] Giaquinta, M., Mariano, P. M., Modica, G., Mucci, D. (2010), Ground states of simple bodies that may undergo brittle fractures, *preprint*.

[23] Giaquinta, M., Modica, G. (2001), On sequences of maps with equibounded energies, *Calc. Var. PDE*, **12**, 213-222. MR1825872 (2002k:49084)

[24] Giaquinta, M., Modica, G., Souček, J. (1989), Cartesian currents, weak diffeomorphisms and existence theorems in nonlinear elasticity, *Arch. Rational Mech. Anal.*, **106**, 97-159. Erratum and addendum, *Arch. Rational Mech. Anal.*, (1990) **109** , 385-392. MR980756 (90c:58044); MR1031576 (91b:58050)

[25] Giaquinta, M., Modica, G., Souček, J. (1989), The Dirichlet energy of mappings with values into the sphere, *Manuscripta Math.*, **65**, 489–507. MR1019705 (90i:49053)

[26] Giaquinta, M., Modica, G., Souček, J. (1994), A weak approach to finite elasticity, *Calc. Var. Partial Diff. Eq.*, **2**, 65–100. MR1384395 (97i:73029)

[27] Giaquinta, M., Modica, G., Souček, J. (1998), *Cartesian currents in the calculus of variations*, vols. I and II, Springer-Verlag, Berlin. MR1645086 (2000b:49001a); MR1645082 (2000b:49001b)

[28] Giaquinta, M., Mucci, D., *Maps into manifolds and currents: Area and $W^{1,2}$, $W^{\frac{1}{2}}$, BV energies*, CRM series, Scuola Normale Superiore, Pisa, 2006. MR2262657 (2008a:49052)

[29] Giaquinta, M., Souček, J. (1985), Harmonic maps into a hemisphere, *Ann. Scuola Norm. Sup. Pisa Cl. Sci.*, **12**, 81-90. MR818802 (87d:58048)

[30] Hardt, R., Lin, F. H. (1986), A remark on H^1 mappings, *Manuscripta Math.*, **56**, 1-10. MR846982 (87k:58068)

[31] Hildebrandt, S., Kaul, H., Widman, K.-O. (1977) An existence theorem for harmonic mappings of Riemannian manifolds, *Acta Math.*, **138**, 1–16. MR0433502 (55:6478)

[32] Hutchinson, J. E. (1986), Second fundamental form for varifolds and the existence of surfaces minimising curvature, *Indiana Univ. Math. J.*, **35**, 45–71. MR825628 (87e:49068)

[33] Lin, F. H. (1999), Gradient estimates and blow-up analysis for stationary harmonic maps, *Ann. of Math.* (2), **149**, 785-829. MR1709303 (2000j:58028)

[34] Mantegazza, C. (1996), Curvature varifolds with boundary, *J. Differential Geom.*, **43**, 807–843. MR1412686 (98e:49104)

[35] Mariano, P. M. (2002), Multifield theories in mechanics of solids, *Adv. Appl. Mech.*, **38**, 1-93.

[36] Mariano, P. M. (2008), Cracks in complex bodies: Covariance of tip balances, *J. Nonlinear Sci.*, **18**, 99-141. MR2386717

[37] Mariano, P. M., Modica, G. (2009), Ground states in complex bodies, *ESAIM Contr. Opt. Calc. Var.*, **15**, 377-402. MR2513091

[38] Mariano, P. M., Stazi, F. L. (2005), Computational aspects of the mechanics of complex bodies, *Arch. Comp. Meth. Eng.*, **12**, 391-478. MR2201460 (2007d:74005)

[39] Morrey, C. B., *Multiple integrals in the calculus of variations*, Springer-Verlag, Berlin, 1966. MR0202511 (34:2380)

[40] Müller, S. (1990), Higher integrability of determinants and weak convergence in L^1, *J. Reine Angew. Math.*, **412**, 20–34. MR1078998 (92b:49026)

[41] Müller, S., Spector, S. (1995), An existence theory for nonlinear elasticity that allows for cavitation, *Arch. Rational Mech. Anal.*, **131**, 1-66. MR1346364 (96h:73017)
[42] Müller, S., Tang, Q. and Yan, B. S. (1994), On a new class of elastic deformations not allowing for cavitation, *Ann. Inst. H. Poincaré Anal. Non Linéaire*, **11**, 217-243. MR1267368 (95a:73025)
[43] Rivière, T. (1992), Applications harmoniques de B^3 dans S^2 partout discontinues, *C. R. Acad. Sci. Paris Sér. I Math.*, **314**, 719–723. MR1163864 (93f:58047)
[44] Schoen, R., Uhlenbeck, K. (1984), Regularity of minimizing harmonic maps into the sphere, *Invent. Math.*, **78**, 89–100. MR762354 (86a:58024)
[45] Segev, R. (1994), A geometrical framework for the statics of materials with microstructure, *Math. Models Methods Appl. Sci.*, **4**, 871-897. MR1305562 (96b:73003)
[46] Sverák, V. (1988), Regularity properties of deformations with finite energy, *Arch. Rational Mech. Anal.*, **100**, 105-127. MR913960 (89g:73013)
[47] Tang, Q. (1988), Almost everywhere injectivity in non-linear elasticity, *Proc. Royal Soc. Edinburgh*, **109A**, 79-85. MR952330 (89g:73014)
[48] Truedell, C. A., Noll, W., *The non-linear field theories of mechanics*, Springer-Verlag, Berlin, third edition 2004, Edited and with a preface by S. S. Antman. MR2056350 (2005a:74002)
[49] Uhlenbeck, K. (1970), Harmonic maps; a direct method in the calculus of variations, *Bull. Amer. Math. Soc.*, **76**, 1082–1087. MR0264714 (41:9305)

Scuola Normale Superiore, piazza dei Cavalieri, Pisa, Italy
E-mail address: m.giaquinta@sns.it

DICeA, University of Florence, via Santa Marta 3, I-50139 Firenze, Italy
E-mail address: paolo.mariano@unifi.it

Dipartimento di Matematica Applicata "G. Sansone", University of Florence, via Santa Marta 3, I 50139 Firenze, Italy
E-mail address: giuseppe.modica@unifi.it

Dipartimento di Matematica, Università di Parma, Parma, Italy
E-mail address: domenico.mucci@unipr.it

Amer. Math. Soc. Transl.
(2) Vol. **229**, 2010

On Classic Solvability of the m-Hessian Evolution Equation

N. M. Ivochkina

Dedicated to N. N. Uraltseva

ABSTRACT. The paper presents the proof of classic solvability of the first initial boundary value problem for nondegenerate m-Hessian evolution equations. In particular, it delivers construction of a priori estimates in $C^{2,1}(\bar{Q})$ of admissible solutions for these equations based on the N. V. Krylov maximum principle.

1. Introduction

Let $\Omega \subset R^n$ be a bounded domain, $Q = \Omega \times (0,T]$, $u \in C^{2,1}(Q)$. The $(m+1)$-Hessian operators E_{m+1} are defined by the following line:

$$E_{m+1}[u] = (-u_t \mathbf{tr}_m u_{xx} + \mathbf{tr}_{m+1} u_{xx})^{\frac{1}{m+1}}, \quad m = 0, \dots, n. \tag{1.1}$$

Here $u_t = \partial u / \partial t$, u_{xx} is the Hessian matrix in spatial variables, $\mathbf{tr}_l u_{xx}$ is the sum of all principal minors of l-th order of u_{xx} if $0 < l < n+1$, $\mathbf{tr}_0 u_{xx} := 1$, $\mathbf{tr}_{n+1} u_{xx} := 0$.

With the operator (1.1) we associate the cone of $(m+1)$-admissible functions $\mathbf{K}_{m+1}(Q) \subset C^{2,1}(Q)$:

$$\mathbf{K}_{m+1}(Q) = \{u : E_{m+1}[u - \sigma t + (\xi, x)^2] > E_{m+1}[u], \quad \sigma + |\xi| > 0\} \tag{1.2}$$

with arbitrary $\sigma \in R^+, \xi \in R^n$, and set up in $\mathbf{K}_{m+1}$ the first initial boundary value problem

$$E_{m+1}[u] = f, \quad u|_{\partial_p Q} = \Phi, \tag{1.3),}$$

where $\partial_p Q$ is the parabolic boundary of Q, $f \geq 0$, Φ are given functions. In (1.4) the surface $\partial\Omega$ is always m-strictly convex (strictly convex if $m = n$), $\partial\Omega \in C^4$.

In [**5**] the following theorem was presented.

THEOREM 1.1. *Assume that $f \in L^{n+1}(Q)$, $f \geq 0$, $\Phi \in C(\partial_p Q)$, $\Phi(x,0) \in \bar{\mathbf{K}}(\Omega) \cap C(\bar{\Omega})$, Φ does not increase in t, when $m = n$. Then there exists a unique $(m+1)$-approximate solution $u \in C(\bar{Q})$ to the problem (1.2).*

2000 *Mathematics Subject Classification.* Primary 35J65.

Key words and phrases. m-Hessian evolution equations, admissible solutions, nondegenerate parabolic equations.

The paper was supported by the grants Sci. Schools RF 227.2008.1 and RFFI 09-01-00729.

The present paper contains the proof of Theorem 1.1 under the assumption of classic solvability of regularized problems associated with (1.2). In order to describe this assumption, we introduce the following notations:

$$\partial_t Q = \partial\Omega \times (0;T], \quad \partial_x Q = \partial\Omega \times \{0\}, \quad \partial_{xt} Q = \partial\Omega \times \{0\},$$

$$\psi(x) = \Phi(x,0), \quad \phi(x,t) = \Phi(x,t), \quad x \in \partial\Omega. \tag{1.3}$$

Then it is sufficient to prove the following theorem.

THEOREM 1.2. *Assume that* $f \in C^{2,1}(\bar{Q})$, $f \geq \nu_0 > 0$, $\psi \in C^4(\bar{\Omega})$, $\phi \in C^{4,2}(\overline{\partial_t Q})$ *and if* $m = n$, *then in addition* $\phi_t \geq \nu_1 > 0$. *Assume also that for* $(x,t) \in \partial_{xt} Q$, *the following equalities hold:*

$$\psi = \phi, \quad E_{m+1}(-\phi_t, \psi_{xx}) = f. \tag{1.4}$$

Then there exists a unique solution $u \in \mathbf{K}_{m+1}(\bar{Q})$ *to the problem*

$$E_{m+1}[u] = f, \quad u|_{\Omega} = \psi, \quad u|_{\partial_t Q} = \phi. \tag{1.5}$$

The existence of admissible solutions of the first initial boundary value problem for general fully nonlinear evolution equations is the subject of the book [**10**, Ch. 15], and at first glance, Theorem 1.2 should be a particular case of Theorem 15.9 in that book. However, it is not easy to overcome the generality and implicitness of assumptions on data in [**10**].

The goal of this paper is to deliver a proof of Theorem 1.2 via the continuity method, which originates in the theory of stationary fully nonlinear equations [**1**]. In the joint papers of the author with O. A. Ladyzhenskaya (see, e.g., [**6**]) this method was adapted to some evolution equations and we follow here such an approach. In particular, to prove the existence of a solution we use the continuity method. Hence, Section 2 contains the description of the homotopy associated with the problem (1.5).

It is well known that the continuity method works in the presence of the a priori estimate of a solution in $C^{2,1}(\bar{Q})$, and the major part of the paper contains a construction of this a priori estimate using methods developed in the theory of stationary fully nonlinear equations.

2. On existence theorems

Let $\tau \in [0,1]$, $u^0 \in \mathbf{K}(\bar{Q})$ be a given smooth function. We relate to u^0 the data

$$f^0 = E_{m+1}[u^0], \quad \psi^0 = u^0(x,0), \quad \phi^0 = u^0(x,t), \quad (x,t) \in \partial_t Q.$$

For f, ψ, ϕ as in (1.5), we denote

$$(f^\tau)^{m+1} = \tau f^{m+1} + (1-\tau)(f^0)^{m+1}, \quad \psi^\tau = \tau\psi + (1-\tau)\psi^0.$$

In order to satisfy the compatibility condition (1.4) we choose

$$\phi^\tau = (\tau\phi + (1-\tau)\phi^0)(x,t) - t(\tau\phi_t + (1-\tau)\phi^0_t + \frac{(f^\tau)^{m+1} - \mathbf{tr}_{m+1}(\psi^\tau_{xx})}{\mathbf{tr}_m \psi^\tau_{xx}})(x,0).$$

Consider the τ-parametric family of problems

$$E_{m+1}[u^\tau] = f^\tau, \quad u^\tau|_{\partial_x Q} = \psi^\tau, \quad u^\tau|_{\partial_t Q} = \phi^\tau. \tag{2.1}$$

LEMMA 2.1. *Let* $m < n$, $u^0 = |x|^2/2$ *and let the functions* f, ψ, ϕ *be as in Theorem 1.2. Then the data of the problems (2.1) satisfy the assumptions of Theorem 1.2 for all* $\tau \in [0,1]$.

Indeed, as is shown in [**5**], the inequalities

$$E_{m+1}[u] > 0, \quad \mathbf{tr}_l u_{xx} > 0, \quad l = 1, \dots, m, \quad (x,t) \in \bar{Q} \tag{2.2}$$

suffice for the inclusion $u \in \mathbf{K}_{m+1}(\bar{Q})$ and hence our $u^0 \in \mathbf{K}_{m+1}(\bar{Q})$ if $m < n$. The compatibility condition (1.4) is fulfilled for all $\phi^\tau, \tau \in [0,1]$, by construction.

If $m = n$, it is necessary to ensure that $\phi^\tau > 0$ on $\partial_t Q$. To proceed we consider the set $\{\lambda_i[\psi]\}$ of the eigenvalues of the matrix ψ_{xx} and let

$$\lambda = \inf_{i,\Omega} \lambda_i[\psi], \quad \mu^n = \sup_{\Omega} \det \psi_{xx}. \tag{2.3}$$

LEMMA 2.2. *Let $m = n$, f, ψ, ϕ be from Theorem 1.2,*

$$u^0 = -t + \frac{1}{2}\lambda x^2. \tag{2.4}$$

Then

$$-\phi_t^\tau \geq \tau\nu_1 + (1-\tau)(\frac{\lambda}{\mu})^n, \quad \tau \in [0,1], \tag{2.5}$$

and the data of the problem (2.1) satisfy the assumptions of Theorem 1.2.

In order to prove inequality (2.5) we notice that, due to the choice (2.3), (2.4), we have

$$\psi_{xx}^\tau = \psi_{xx} + (1-\tau)(\lambda I - \psi_{xx}) \leq \psi_{xx}.$$

Now, the monotonicity of the function $\det S := \mathbf{tr}_n S$ on the set of positive definite matrices and assumption (1.4) yield the inequality

$$\frac{\tau f^{n+1}(x,0) + (1-\tau)\lambda^n}{\mathbf{tr}_n \psi_{xx}^\tau} \geq \tau\nu_1 + (1-\tau)(\frac{\lambda}{\mu})^n.$$

The latter inequality ensures the validity of (2.5).

Function (2.4) would certainly service $m < n$ with, say, $\lambda = 1$. We have separated two cases to emphasize the simplicity of the case $m < n$ compared to $m = n$.

The continuity method provides existence theorems in the presence of a uniform in τ a priori estimate of sufficiently strong norm of u^τ. In our case it is the norm in $C^{2+\alpha,1+\alpha/2}(\bar{Q})$. Due to the famous results of N. V. Krylov on Hölder continuity of u_{xx}, u_t for solutions of uniformly parabolic Bellman equations [**9**, Ch. 5] it is sufficient to bound the norm of u in $C^{2,1}(\bar{Q})$ and this is the subject of Sections 3–6. Namely, the goal of consideration there is the following statement.

THEOREM 2.3. *Let $u \in \mathbf{K}_{m+1}(\bar{Q}) \cap C^{4,2}(Q)$ be a solution to the problem (1.3), (1.5). Suppose that theassumptions of Theorem 1.2 are fulfilled. Then*

$$\|u\|_{C^{2,1}(\bar{Q})} \leq C(m, T, \partial\Omega, \psi, \phi, f). \tag{2.6}$$

In the book [**10**], there is a different and perhaps more natural approach to the existence problem for evolution equations. Namely, due to (1.1), (1.2), equation (1.3) is nondegenerate parabolic in $\mathbf{K}_m(\bar{Q})$ and hence enjoys solvability for small time intervals. The a priori estimate (2.6) and the N. V. Krylov results then allow us to prolong this process up to an arbitrary finite T.

3. The minimum principle. Reduction to the estimates on the boundary

Consider $u \in \mathbf{K}_{m+1}(Q)$ and introduce the following notation:

$$E^0_{m+1} = -\frac{\partial E_{m+1}}{\partial u_t}, \quad E^{ij}_{m+1} = \frac{\partial E_{m+1}}{\partial u_{ij}}, \quad i, j = 1, \dots, n.$$

With u we associate the parabolic linear operator

$$L[m; v] = -v_t E^0_{m+1} + E^{ij}_{m+1} v_{ij}.$$

It was noticed in [**5**] that the structure of such linear operators ensures the following version of the Alexandrov–Krylov maximum principle [**8**] (see also [**12**], [**13**]).

THEOREM 3.1. *Let $v \in C^{2,1}(Q)$. Then*

$$v \geq \inf_{\partial_p Q} v - c(n, \Omega)\|L^+[m; v]\|_{n+1,Q^+}, \tag{3.1}$$

where $Q^+ = \{(x,t) \in Q : -v_t \geq 0, v_{xx} \geq \mathbf{0}\}$. Moreover, if $v \geq 0$ on $\partial_p Q$, $v(x_0, t_0) = 0$ and $L^+[v] = 0$ in Q^+, then $v_n(x_0, t_0) \geq 0$, where n is the interior normal to $\partial\Omega$ at x_0.

It turns out that Theorem 3.1 services all estimates of this paper. We start with the following proposition.

LEMMA 3.2. *Let $u \in \mathbf{K}_{m+1}(Q)$ be a solution to equation (1.3), $\xi \in R^n$. Then*

$$|u| \leq \sup_{\partial_p Q} |u| + c(n, \Omega)\|f\|_{n+1,Q}, \tag{3.2}$$

$$|u_x| \leq \sup_{\partial_p \Omega} |u_x| + c(n, \Omega)\|f_x\|_{n+1,Q}, \tag{3.3}$$

$$|u_t| \leq \sup_{\partial_p \Omega} |u_t| + c(n, \Omega)\|f_t\|_{n+1,Q} \tag{3.4}$$

and eventually

$$u_{\xi\xi} = (u_{xx}\xi, \xi) \leq \sup_{\partial_p Q} u_{\xi\xi} + c(n, \Omega)\|f^-_{\xi\xi}\|_{n+1,Q}. \tag{3.5}$$

Indeed, since due to 1-homogeneity, $E_{m+1}[u] \equiv L[m; u] = f$ and by differentiation, $L[m; u_\xi] = f_\xi$, $L[m; u_t] = f_t$, the inequalities (3.2)–(3.4) directly follow from (3.1). Due to concavity of the function $E_{m+1}(-u_t, u_{xx})$ we also have $L^+[m; -u_{\xi\xi}] \leq f^-_{\xi\xi}$ and Theorem 3.1 applied to $v = -u_{\xi\xi}$ yields (3.5).

The estimate (3.5) bounds $u_{\xi\xi}$ only from above. But it is well known that $E_1[u] > 0$ for all $u \in \mathbf{K}_{m+1}(Q)$, which provides the estimate of $u_{\xi\xi}$ also from below and the following reduction to the estimates on the boundary is true.

THEOREM 3.3. *Let $u \in \mathbf{K}_{m+1}(Q)$ be a solution to the problem (1.5) and let the assumptions of Theorem 1.2 be fulfilled. Then the following inequalities,*

$$|u| \leq \sup_{\partial_p Q} |\Phi| + c\|f\|_{n+1}, \quad |u_t| \leq \sup_{\partial_p Q} |\phi_t| + c\|f_t\|_{n+1},$$

$$|u_{xx}| \leq (n-1) \sup_{\eta, \partial_p Q} u^+_{\eta\eta} + \sup_Q (f + u^-_t) + c \sup_\eta \|f^-_{\eta\eta}\|_{n+1}, \quad |\eta| = 1,$$

together with (3.3) hold.

If $m = n$, then the inequality $-u_t > 0$ is necessary for the admissibility of solution u. Hence we need to provide the a priori estimate from below for $-u_t$ in this case,m and the following proposition gives it.

THEOREM 3.4. *Let $m = n$, $u \in \mathbf{K}_{n+1}$ be a solution to the problem (1.5). Then*

$$-u_t \geq \exp(-\alpha T)\max\{\nu_1; (\frac{\nu_0}{\mu})^n\}, \quad \alpha = \sup_Q \frac{f_t^-}{f}, \tag{3.6}$$

where ν_0, ν_1 are the constants from Theorem 1.2, μ is from (2.3).

Indeed, due to the choice of α,

$$L[n; v] = -\exp(\alpha t)(L[n; u_t] + \alpha E_{n+1}[u]) = -\exp(\alpha t)(f_t + \alpha f) \leq 0$$

for $v = -u_t \exp(\alpha t)$. Hence $L^+[n; v] = 0$, and Theorem 3.1 guarantees the validity of (3.6).

4. Geometric equipment

Consider a hypersurface $\Gamma \subset R^n$, its local parametrization $\{\theta^k, k = 1, \ldots, n-1\}$ and let $X = X(\theta)$, N be its position vector and interior unit normal respectively. We represent its metric tensor $g = (g_{kl})$, $g_{kl} = (X_k, X_l)$, $X_k = \partial X / \partial \theta^k$ in the form $g = \eta^T \eta$. Let $\tau = \eta^{-1}$ and $X_{(l)} := X_p \tau_l^p$, $X_{(kl)} := X_{pq} \tau_k^p \tau_l^q$. Then the collection of vectors $\{X_{(l)}, N, l = 1, \ldots, n-1\}$ is the moving frame generated by Γ, i.e., $(X_{(k)}, X_{(l)}) = \delta_{kl}$, $(X_{(k)}, N) = 0$.

We introduce the curvature matrix $\mathbf{K}[\Gamma]$ as

$$\mathbf{K}[\Gamma] = (X_{(\theta\theta)}, N) = \sum_1^n X^i_{(\theta\theta)} N^i. \tag{4.1}$$

The eigenvalues of the symmetric matrix (4.1) coincide with the principal curvatures of Γ, and we say that $\mathbf{k}_m := \mathbf{tr}_m \mathbf{K}$ is the m-th curvature of Γ. The hypersurface Γ is called strictly m-convex at $M \in \Gamma$ if $\mathbf{k}_i[\Gamma](M) > 0$, $i = 1, \ldots, m$.

Now we assign to $0 \in \partial\Omega$ the Cartesian basis $\{e_k(0), N(0)\}$ and denote by $\tilde{x}$ the tangential variables. So, we have our parametrization fixed by $\theta = \tilde{x}$. Let the function $\omega = \omega(\tilde{x})$, $\omega(0) = 0$, $\omega_x(0) = 0$ be generated by $\Gamma_r = \partial\Omega \cap \{|\tilde{x}| < r\}$ (r is sufficiently small) in the sense that

$$\Gamma_r = \{|\tilde{x}| < r, x^n = \omega(\tilde{x})\}. \tag{4.2}$$

Let $\varkappa > 0$ and

$$y = x^n - \hat{\omega}(\tilde{x}), \quad \hat{\omega} = \omega - \frac{\varkappa}{2}\tilde{x}^2. \tag{4.3}$$

Similarly to (4.2) we associate with $\hat{\omega}$ the surface $\hat{\Gamma}_r$ and also

$$\Omega_r = \{x : |\tilde{x}| < r, \frac{\varkappa}{2}\tilde{x}^2 < y < \frac{\varkappa}{2}r^2\}. \tag{4.4}$$

All proceeding calculations from here will be performed in the domain (4.4). Let $x_0 \in \Omega_r$,

$$e_k^0 = \hat{X}_{(k)} = (\hat{\tau}_k^1, \ldots, \hat{\tau}^{n-1}, \hat{N})(\tilde{x}_0), \quad e_n^0 = \frac{(-\hat{\omega}_{\tilde{x}}, 1)}{\sqrt{1 + \hat{\omega}_{\tilde{x}}^2}}(\tilde{x}_0). \tag{4.5}$$

Denote by $\hat{x}$ the variables in the basis (4.5). The validity of the following formulas at x_0 is the matter of direct computations:

$$\frac{\partial^2 y}{\partial \hat{x}^k \partial \hat{x}^l} = (y_{xx} e_k^0, e_l^0) = -\hat{\omega}_{(kl)} = -\sqrt{1+\hat{\omega}_{\tilde{x}}^2}\mathbf{K}_{kl}[\hat{\Gamma}] \quad k,l = 1,\dots,n-1, \tag{4.6}$$

and actually (4.6) is the main goal of the above consideration. Also the formulas

$$y_x = \sqrt{1+\hat{\omega}_{\tilde{x}}^2}N, \quad \frac{\partial y}{\partial \hat{x}^k} = 0,$$

$$\frac{\partial^2 y}{\partial \hat{x}^k \partial \hat{x}^n} = -\frac{(\hat{\omega}_{\tilde{x}(k)}, \hat{\omega}_{\tilde{x}})}{\sqrt{1+\hat{\omega}_{\tilde{x}}^2}}, \quad \frac{\partial^2 y}{\partial \hat{x}^n \partial \hat{x}^n} = -\frac{(\hat{\omega}_{\tilde{x}\tilde{x}}\hat{\omega}_{\tilde{x}}, \hat{\omega}_{\tilde{x}})}{\sqrt{1+\hat{\omega}_{\tilde{x}}^2}} \tag{4.7}$$

are true at x_0 and will be used in further arguments.

The above technique allows us to explicitly construct barriers in terms of the curvature matrix (4.1) of $\partial\Omega$. Actually, this approach has been applied in [**3**] in order to treat stationary curvature equations and after that was consistently used in the subsequent papers by the author (see, e.g., [**4**]). Its refined version is presented here.

5. The estimates of gradient and second mixed derivatives on the boundary

In this section the main role belongs to

$$W = \frac{y}{\varkappa r^2}\left(\frac{y}{2\varkappa r^2} - 1\right) \tag{5.1}$$

considered in the domain (4.4) with y from (4.3). It is obvious that W satisfies the following relations :

$$W|_{\partial\Omega_r \cap \partial\Omega} \le -\frac{3}{8}\frac{\tilde{x}^2}{r^2}, \quad W(0) = 0, \quad W|_{\partial\Omega_r \cap \Omega} \le -\frac{3}{8}. \tag{5.2}$$

To describe the principal property of function (5.1), we reduce the cone (1.2), (2.2) to the cone $\mathbf{K}_{m+1}(\Omega_r)$ letting $u_t = 0$.

LEMMA 5.1. *Assume that Γ_r is strictly m-convex and $\mathbf{k}_m[\Gamma_r] \ge \mathbf{k}_m^0 > 0$. Then there exist $\varkappa = \varkappa(n, \|\omega\|_{C^3}, \mathbf{k}_m^0)$, $c_1 = c_1(n, \|\omega\|_{C^2}, \varkappa)$ such that $W \in \mathbf{K}_{m+1}[\bar{\Omega}_r]$ and*

$$\mathbf{tr}_{m+1} W_{xx} \ge \left(\frac{1}{2\varkappa r^2}\right)^{m+2}\mathbf{k}_m^0, \quad \mathbf{tr}_m W_{xx} \le c_1\left(\frac{1}{\varkappa r^2}\right)^{m+1}. \tag{5.3}$$

Indeed, formulas (4.5)–(4.7) and direct computations yield the equality

$$\mathbf{tr}_l W_{xx} = \left(1 - \frac{y}{\varkappa r^2}\right)^{l-1}\left(\frac{\sqrt{1+\tilde{\omega}_{\tilde{x}}^2}}{\varkappa r^2}\right)^l\left(\frac{\sqrt{1+\tilde{\omega}_{\tilde{x}}^2}}{\varkappa r^2}\mathbf{k}_{l-1}[\hat{\Gamma}_r] + \left(1 - \frac{y}{\varkappa r^2}\right)(\mathbf{k}_l[\hat{\Gamma}_r] + O(r^2))\right).$$

The latter guarantees the existence of $\varkappa$, c_1 and from now on we fix $\varkappa$, $r < 1$.

In further considerations, we associate with an arbitrary $x \in \partial\Omega$ the cylinder $Q_r = \Omega_r \times (0;T]$, where Ω_r is the domain (4.4). We also extend the function ϕ from (1.5) to Q_r by $\phi(\tilde{x}, x^n) := \phi(\tilde{x}, \omega(\tilde{x}))$.

LEMMA 5.2. *Let $u \in \mathbf{K}_{m+1}(Q_r) \cap C^{2,1}(\bar{Q}_r)$ be a solution to equation (1.5),*

$$u(x,0) = \psi, \quad u|_{\partial_t Q \cap \partial_t Q_r} = \phi. \tag{5.4}$$

Then

$$u_n(0,t) \geq -c_2(c_1, \|\psi\|_{C^1}, \|\phi\|_{C^{2,1}}, \sup_{Q_r}(f+|u|)). \tag{5.5}$$

In the case $m = n$, c_2 also depends on $\nu_1 > 0$, $-\phi_t \geq \nu_1$.

To prove (5.5) we introduce

$$v = u - w_1, \quad w_1 = MW + \phi \tag{5.6}$$

with W from (5.1) and choose M so large that

$$v|_{\partial_p Q_r} \geq 0, \quad v(0,t) = 0, \quad L^+[m; v] = 0 \tag{5.7}$$

and again apply Theorem 3.1 to conclude (5.6).

Indeed, due to (5.2) we have

$$v|_{\partial_t Q_r \cap \partial_t Q} \geq 0, \quad v(0,t) = 0, \quad v|_{\partial_t Q_r \cap Q} \geq \frac{3}{8}M - \sup_{Q_r}(|u| - \phi)$$

and also

$$v(x,0) \geq \psi(\tilde{x}, x^n) - \psi(\tilde{x}, \omega(\tilde{x})) + \frac{M}{2}\frac{y}{\varkappa r^2} \geq (x^n - \omega(\tilde{x}))(-\sup|\psi_x| + \frac{M}{2\varkappa r^2}).$$

On the other hand, due to (5.3), $w_1 \in \mathbf{K}_{m+1}(Q_r)$ for $M \gg 1$, due to which 1-homogeneity with concavity of E_{m+1} in $\mathbf{K}_{m+1}$ validates the inequality $L[m; w_1] \geq E_{m+1}[w_1]$. Hence,

$$L[m; v] - E_{m+1}[u] - L[m; w_1] \leq f - E_{m+1}[w_1],$$

and since M is large we have $L^+[m; v] = 0$.

The basis of a priori estimates of mixed derivatives is a remarkable consequence of orthogonal invariance of functions defined on the space of symmetric matrices [**2**], [**1**], [**7**]. In our case it reads as follows.

LEMMA 5.3. *Let H be a constant matrix, $H = -H^T$, $\gamma \in R$, $x = (x^1, \ldots, x^n)(\gamma)$ be a solution of the system of ordinary differential equations*

$$\frac{d}{d\gamma}x(\gamma) = H(x - x_1), \quad x_1 \in R^n.$$

Then

$$\frac{d}{d\gamma}E_{m+1}[u] = L[m; u_\gamma] = f_\gamma, \quad u_\gamma = u_i \frac{dx^i(\gamma)}{d\gamma}. \tag{5.8}$$

To describe H related to the problem (5.4) we first direct the tangential axes in such way that $\omega_{ql}(0) = k_l \delta_{ql}, l = 1, \ldots, n-1$, where $\{k_l\}$ is the collection of principal curvatures of $\Gamma_r(0)$. To every k_l we assign its own H, x_1. Let, for instance, $l = 1$. If $k_1 = 0$ we choose $H = \mathbf{0}$, $x_1 = 0$. In the more interesting case, $k_1 \neq 0$, we take

$$H_i^1 = k_1 \delta_i^n, \quad x_1^i = \frac{1}{k_1}\delta_n^1, \quad i = 1, \ldots, n. \tag{5.9}$$

Then $u_\gamma = u_1 + k_1(u_n x^1 - u_1 x^n)$ and this is exactly the function involved in the relevant piece of argument in the paper [**1**]. The following relations explain the choice (5.9):

$$u_\gamma|_{\partial_t Q_r \cap \partial_t Q} = \phi_\gamma + O(\tilde{x}^2), \quad \psi_\gamma = \phi_\gamma + O(y), \quad x \in \Omega_r, \quad t = 0.$$

The arguments similar to the proof of Lemma 5.2 applied instead of (5.6) to the function

$$v = u_\gamma + M_1 u - w_2, \quad w_2 = \phi_\gamma + M_1\phi + M_2 W$$

with $M_1 \gg 1$, $M_2 \gg M_1$ under control lead to the a priori bound of $u_{1n}(0,t)$. Varying the choice of l and taking (5.8) with $\hat\gamma = -\gamma$ we deduce the following result.

LEMMA 5.4. *Let $u \in \mathbf{K}(Q_r) \cap C^{2,1}(Q_r)$ be a solution to the problem (1.5), (5.4). Then*

$$|u_{n\tilde x}(0,t)| \le c_4(n, \mathbf{k}_m^0, \|f,u\|_{C^{1,0}(Q_r)}, \|\phi\|_{C^{2,1}(Q_r)}, \|\psi,\omega_x\|_{C^2(\Omega_r)}, \delta_m^n\nu_1). \tag{5.10}$$

6. The estimate of the normal second derivative

Following the idea from [**1**] for stationary equations we plan to estimate u_{nn}, $(x,t) \in \partial_t Q$, directly from the equation (1.5). In the coordinates of Section 4 it may be written at $(0,t_0)$ in the form

$$u_{nn}^0 G_m(s^0, S^0(u_n^0)) + h = f^{m+1},$$

$$G_m(s,S) := s \times \mathbf{tr}_{m-1}(S) + \mathbf{tr}_m S, \quad s \in R, \quad S \in Sym(n-1), \tag{6.1}$$

where

$$s^0 = -\phi_t^0, \quad S^0(b) = \phi_{\tilde x\tilde x}^0 - b\omega_{\tilde x\tilde x}^0, \tag{6.2}$$

and h depends on $u_{n\tilde x}^0$ and the given values.

Let $P_m(b) = G_m(s^0, S^0(b))$. Denote by $b_0 = b^0(\omega_{\tilde x\tilde x}^0, \phi_{\tilde x\tilde x}^0)$ the number such that $P_m(b_0) = 0$, $P_m(b) > 0$ for all $b < b_0$. In particular, if $u \in \mathbf{K}_{m+1}(\bar Q)$, then $u_n^0 < b_0$. To make use of (6.1) we ought to construct $\underline{b} < b_0$ as an upper bound for u_n^0.

To proceed, we apply Theorem 3.1 to $v = w - u$ and without loss of generality substitute in (3.1) u for w; i.e., this time,

$$L[m;v] = -v_t E_{m+1}^0[w] + E_{m+1}^{ij}[w]v_{ij}.$$

Notice that under the conditions of Theorem 1.2, the problem (1.5) is solvable in a cylinder $Q(\delta) = \Omega \times (0,\delta]$ with sufficiently small δ under control, and we will apply Theorem 3.1 in the cylinders $Q_r = \Omega_r \times (t_0 - r; t_0]$ with $t_0 \ge \delta$, $r \le t_0$.

LEMMA 6.1. *Let $u \in \mathbf{K}_{m+1}(\bar Q_r)$, $v = w - u$. Then either Q_r^+ is empty or $w \in \mathbf{K}_{m+1}(Q_r^+)$ and*

$$L[m;v] \le (E_1[w])^{\frac{1}{m+1}}\left(G_m(-w_t, w_{\tilde x\tilde x})^{\frac{1}{m+1}} - E_{m+1}[u]\right), \quad (x,t) \in Q_r^+. \tag{6.3}$$

Indeed, 1-homogeneity and concavity of E_{m+1} in $\mathbf{K}_{m+1}(Q_r^+)$ give $L[m;w] \le E_{m+1}[w] - E_{m+1}[u]$. On the other hand, as a consequence of the monotonicity of the quotients established in [**11**],

$$\frac{G_{m+1}[w]}{G_1[w]} := \frac{(E_{m+1}[w])^{m+1}}{E_1[w]} \le G_m(-w_t, w_{\tilde x\tilde x}).$$

Denote by $\mathbf{S}$ a symmetric $n \times n$ matrix such that $s_{00} = s$, $s_{0i} = 0, i = 1,\dots,n-1$, $S = (s_{ij}) \in Sym(n-1)$ and relate to $G_m = G_m(\mathbf{S})$ the following reduction of the cone $\mathbf{K}_m$ in the form (2.2) to the space of symmetric matrices

$$K_m = \{\mathbf{S} : G_m(\mathbf{S}) > 0, \quad \mathbf{tr}_i S > 0, \quad i = 1,\dots,m-1\}.$$

Due to the concavity of $G_m^{1/m}$ in K_m, the inequality

$$G_m^{\frac{m-1}{m}}(\mathbf{S}^0)G_m^{\frac{1}{m}}(\mathbf{S}) \leq m(G_m^0(\mathbf{S}^0)s + G_m^{ij}(\mathbf{S}^0)s_{ij}) \tag{6.4}$$

is true for $\mathbf{S}, \mathbf{S}^0 \in K_m$.

Let $(0,t) \in \partial_t Q$, $b < b_0$, $\mathbf{S}^0 = \mathbf{S}^0(b)$ be defined by (6.2) and let $G_m(\mathbf{S}^0) = \beta^m$, $\beta = \beta(b) > 0$. In further developments we direct axes tangential at 0 in such a way that $G_m^{ij}(\mathbf{S}^0) = \delta^{ij}G_m^{ii}(\mathbf{S}^0)$, $i = 1,\dots,n-1$. Finally, let

$$w = \phi + (b + (\tilde{a},\tilde{x}))(x^n - \omega) + \frac{y}{\beta^{m-1}}(M_1(t_0 - t) + \frac{1}{2}M_2 y), \tag{6.5}$$

where

$$\tilde{a} = (a^1,\dots,a^{n-1}), \quad a^i = \frac{-G^0{}_m(\mathbf{S}^0)\phi^0_{it} + G^{kk}{}_m(\mathbf{S}^0)(\phi^0{}_{kki} - b\omega^0_{kki})}{G_m^{kk}(\mathbf{S}^0)\omega^0_{kk} + 2\varkappa G_m^{ii}(\mathbf{S}^0)}, \tag{6.6}$$

where M_1, M_2 are constants to be chosen and y is from (4.3).

To the function (6.5) and a point $(x,t) \in Q_r$, we associate the matrix $\mathbf{S} = \mathbf{S}(x,t) : s = -w_t$, $S = (w_{xx}e_i, e_j)$, $i,j = 1,\dots,n-1$, where $\{e_1,\dots,e_n\}$ is the basis (4.5) associated to (x,t). In order to cover the cases $m < n$, $m = n$ simultaneously we introduce in further statements the character $\bar{\mathbf{k}}_m$ equal to $\mathbf{k}_m$ for $m < n$ and to $-\phi^0_t\mathbf{k}_{n-1}$ otherwise.

Denote by $\boldsymbol{\mu}$ the following collection of data:

$$\boldsymbol{\mu} = \{m, \varkappa, \|\phi_{\tilde{x}\tilde{x}}, \omega_{\tilde{x}\tilde{x}}\|_{2,1}, \|\phi_t\|_{2,1}, \sup_{Q_r}|u_x|, \bar{\mathbf{k}}_m[\Gamma_r], -\delta_n^m\phi_t^0\}. \tag{6.7}$$

Denote also by d the parabolic distance, i.e., $d^2 = \tilde{x}^2 + t_0 - t$.

LEMMA 6.2. *Let Γ_r be strictly $(m-1)$-convex, $\bar{\mathbf{k}}_m[\Gamma_r] \geq k_0^m > 0$. Then there are constants M_1, M_2 depending on $\boldsymbol{\mu}$ such that $G_m(\mathbf{S}[w]) \leq m\beta^m$, $(x,t) \in Q_r^+$.*

PROOF. We assign to a point $(x,t) \in Q_r^+$ the matrix $\mathbf{S}^1$ by the formulas

$$s = s^0 - s^1 + \phi^0_{it}x^i,$$

$$s_{ij} = s^0_{ij} - s^1_{ij} + (\phi^0_{ij\tilde{x}} - b\omega^0_{ij\tilde{x}} - \tilde{a}\hat{\omega}^0_{ij}) - \varkappa(a^i\tilde{x}^j + a^j\tilde{x}^i), \quad i,j = 1,\dots,n-1. \tag{6.8}$$

In our assumptions, $\mathbf{S}, \mathbf{S}^0 \in K_m$. Therefore (6.4) is valid, which together with (6.6) brings out the inequality

$$G_m(\mathbf{S}) \leq m\beta^m - (G_m^0(\mathbf{S}^0)s^1 + G_m^{ij}(\mathbf{S}^0)s^1_{ij}). \tag{6.9}$$

Notice also that $|\tilde{a}| \leq c(\boldsymbol{\mu})/\beta^{m-1}$ and computations by means of (4.5)–(4.7) and the Taylor formulae applied to (6.8) carry out the inequalities

$$\begin{aligned} s^1 &\geq -(\frac{M_1}{\beta^{m-1}}y + c(\boldsymbol{\mu})d^2), \\ S^1 &\geq \frac{1}{\beta^{m-1}}(M_1(t_0 - t) + M_2 y)\hat{\omega}^0_{\tilde{x}\tilde{x}} - c(\boldsymbol{\mu})(\frac{\tilde{x}^2}{\beta^{m-1}} + d^2)I. \end{aligned} \tag{6.10}$$

If $m < n$, then strict m-convexity of $\hat{\Gamma}_r$ and the definition of K^m together with (6.10) provide independent of β quantities $M_1, M_2 > 1$, $M_2 \geq c(\boldsymbol{\mu})M_1$ such that $\mathbf{S}^1 \in K_m$ and the inequality (6.9) ensures $G_m(\mathbf{S}) \leq m\beta^m$.

Let $m = n$, i.e., Γ_r is strictly convex, $-\phi_t^0 > 0$. It follows from (6.10) that $S^1 \in K^{n-1}$ and hence

$$G_n^{ij}(\mathbf{S}^0)s_{ij}^1 \geq n(\beta^{\frac{n(n-2)}{n-1}})(-\phi_t^0 \mathbf{tr}_{n-1} S^1)^{\frac{1}{n-1}}$$
$$\geq \frac{n}{2}\beta^{\frac{n(n-2)}{n-1}} \frac{M_1(t^0-t)+M_2 y}{\beta^{n-1}}(\bar{\mathbf{k}}_{n-1}[\hat{\Gamma}_r])^{\frac{1}{n-1}}$$

if $M_1, M_2 \gg 1$. On the other hand,

$$G_n^0(\mathbf{S}^0) = \frac{\beta^n}{-\phi_t^0}, \quad G_n^0(\mathbf{S}^0)s^1 \geq \frac{\beta^n}{\phi_t^0}(\frac{M_1}{\beta^{n-1}}y + c(\boldsymbol{\mu})M_1)d^2)$$

and again the choice $M_2 \geq c(\boldsymbol{\mu})M_1$ and (6.9) yield the validity of Lemma 6.2 □

Lemma 6.2 prescribes the lower bound for M_1, M_2 to be sufficiently large depending on parameters (6.7). In the following proposition, we fix this upper bound and choose $\beta = \underline{\beta}(\boldsymbol{\mu}, r, \inf_{Q_r} f(x,t))$ after that.

LEMMA 6.3. *Assume that $u \in \mathbf{K}_{m+1}(\bar{Q}_r)$ is a solution to equation (1.5) equal to ϕ on $\partial_t Q_r \cap \partial_t Q$. Then there are $b = \underline{b}(\boldsymbol{\mu}, r) < b_0$, $\beta = \underline{\beta}(\underline{b}, \inf_{Q_r} f) > 0$ such that*

$$u_n(0,t_0) \leq \underline{b}, \quad u_{nn}(0,t_0) \leq c(\underline{\beta}, \boldsymbol{\mu}, \sup_{Q_r} f, |u_{n\tilde{x}}(0,t_0)|). \tag{6.11}$$

PROOF. First, since M_1, M_2 are large, we have

$$v = w - u|_{\partial_p Q_r} \geq 0, \quad v(0,t_0) = 0. \tag{6.12}$$

It is obvious that (6.12) is true for $(x,t) \in \partial_t Q_r \cap \partial_t Q$. Since

$$|\phi - u| = |u(\tilde{x}, \omega(\tilde{x}), t) - u(\tilde{x}, x^n, t)| \leq y \sup_{Q_r} |u_x|, \quad y \leq \frac{\varkappa}{2} r^2,$$

the inequalities

$$v|_{\Omega_r \times \{t_0 - r\}} \geq -|\phi - u| - \frac{y}{\beta^{m-1}}(c(\boldsymbol{\mu}) - M_1 r) \geq 0,$$
$$v|_{\partial_t Q_r \cap Q} \geq |\phi - u| - \frac{r^2}{8\beta^{m-1}}(c(\boldsymbol{\mu}) - \varkappa^2 r^2)$$

hold if $M_1 \geq c(\boldsymbol{\mu})/r$, $M_2 \geq c(\boldsymbol{\mu})/r^2$ are large and (6.12) has been proven.

Now increase if necessary M_1, M_2 to keep Lemma 6.2 valid. From now on, $M_1 = c(\boldsymbol{\mu}, r)$, $M_2 = c(\boldsymbol{\mu}, r)$ are given numbers and

$$E_1[w] = -w_t + \mathbf{tr}\, u_{xx} \leq \frac{c(\boldsymbol{\mu}, r)}{\beta^{m-1}} \tag{6.13}$$

for all $\beta \in (0,1]$. Appoint $\underline{\beta}$ by the equality

$$\underline{\beta} m c(\boldsymbol{\mu}, r) = \inf_{Q_r} f(x,t),$$

where $c(\boldsymbol{\mu}, r)$ is the constant from (6.13). If $Q_r^+(\underline{\beta})$ is empty, then Theorem 3.1 in the presence of (6.12) brings out the first inequality in (6.11). The second relation in (6.11) is then a sequel to (6.1), (6.2). Otherwise $L^+[m; v] = 0$, which leads to the same conclusion. □

Theorem 3.3, the inequality (3.6) and the estimates (5.5), (5.10), (6.11) extended to the whole $\partial_t Q$ prove Theorem 2.3 and hence Theorem 1.2.

References

[1] L. Caffarelli, L. Nirenberg, and J. Spruck, *The Dirichlet problem for nonlinear second order elliptic equations III. Functions of the Hessian.* Acta Math. **155** (1985), 261–301. MR806416 (87f:35098)

[2] N. M. Ivochkina, *The integral method of barrier functions and the Dirichlet problem for equations with operators of the Monge–Ampère type.* Mat. Sb. **112(156)** (1980), 193–206; English transl. in Math. USSR Sb. **40** (1981). MR585774 (82d:35049)

[3] ———, *The Dirichlet problem for m-th order curvature equations.* Algebra i Analiz **2** (1990), 192–217; English transl. in Leningrad Math. J. **2** (1991). MR1073214 (92e:35072)

[4] ———, *Geometric evolution equations preserving convexity.* Amer. Math. Soc. Transl. (2), **220** (2007), 91–121. MR2343608 (2009b:35187)

[5] ———, *On approximate solutions to the first initial boundary value problem for the m-Hessian evolution equations.* J. Fixed Point Th. Appl. **4** (2008), no. 1, 47–56. MR2447961 (2009i:35156)

[6] N. M. Ivochkina and O. A. Ladyzhenskaya, *On parabolic problems generated by some symmetric functions of the Hessian.* Top. Meth. Nonlinear Anal. **4** (1994), 19–29. MR1321807 (96e:35088)

[7] N. M. Ivochkina, N. Trudinger, and X.-J.Wang, *The Dirichlet problem for degenerate Hessian equations.* Commun. Partial Differ. Equations **29** (2004), 219–235. MR2038151 (2005d:35076)

[8] N. V. Krylov, *Sequences of convex functions and estimates of the maximum of the solutions of a parabolic equation.* Sibirsk. Mat. Zh. **17** (1976), 226–236; English transl., Siberian Math. J. **17** (1976), 226–236. MR0420016 (54:8033)

[9] ———, *Nonlinear elliptic and parabolic equations of second order.* Nauka, Moscow (1985). English transl.: Reidel, Dordrecht, 1987. MR815513 (87h:35002)

[10] G. M. Lieberman, *Second order parabolic differential equations.* World Sci. Singapore, 2006. MR1465184 (98k:35003)

[11] M. Lin and N. Trudinger, *On some inequalities for elementary symmetric functions.* Bull. Austr. Math. Soc. **50** (1994), 317–326. MR1296759 (95i:26036)

[12] A. I. Nazarov and N. N. Uraltseva, *Convex-monotone hulls and an estimate of the maximum of the solution of a parabolic equation.* Zap. Nauchn. Sem. LOMI **147** (1985), 95–109; English transl., J. Soviet Math. **37** (1987), 851–859. MR821477 (87h:35039)

[13] K. Tso, *On an Aleksandrov–Bakel'man type maximum principle for second-order parabolic equations.* Commun. Partial Differ. Equations **10** (1985), 543–553. MR790223 (87f:35031)

St.-Petersburg State University of Architecture and Civil Engineering, 2 Krasnoarmeiskaya, 4 198005 St.-Petersburg, Russia

E-mail address: ninaiv@NI1570.spb.edu

Amer. Math. Soc. Transl.
(2) Vol. **229**, 2010

About an Example of N. N. Ural'tseva and Weak Uniqueness for Elliptic Operators

N. V. Krylov

To N. N. Ural'tseva with admiration of her talent

ABSTRACT. We discuss an example of Ural'tseva which shows the impossibility of W_p^2, $p \neq 2$, estimates for elliptic operators. It turns out that in her example and, actually, for more general operators, weak uniqueness still holds and W_p^2 estimates are available if the operator is fixed and p is close to 2, the closeness depending on the operator. We also prove estimates for parabolic operators with part of the coefficients measurable in (t, x) and the remaining being constant.

1. Introduction

First, we will be dealing with elliptic operators of the form

$$Lu = L_0u - \mu u = a^{ij}u_{x^ix^j} + b^iu_{x^i} - cu - \mu u \tag{1.1}$$

acting on functions u given in $\mathbb{R}^d$, where, generally, $a^{ij} = a^{ji}$, b^i, and c are Borel-measurable real-valued functions on $\mathbb{R}^d$, $\mu \geq 0$ is a constant and, for a constant $\delta > 0$ and the vector $b = (b^1, ..., b^d)$, we have

$$\delta|\xi|^2 \leq a^{ij}\xi^i\xi^j \leq \delta^{-1}|\xi|^2, \quad |b| \leq \delta^{-1}, \quad 0 \leq c \leq \delta^{-1} \tag{1.2}$$

for all $\xi \in \mathbb{R}^d$ on $\mathbb{R}^d$. Denote by Du and D^2u the gradient and the Hessian matrix, respectively, of a function u. For $p \in (1, \infty)$ and smooth domain $D \subset \mathbb{R}^d$ we denote by $\overset{0}{W}{}_p^2(D)$ the Sobolev space of functions u such that $u, Du, D^2u \in \mathcal{L}_p(D)$ and u vanishes on ∂D. We also use the space $W_p^2 = W_p^2(\mathbb{R}^d)$. Recently it was proved in [**1**] that under some quite mild assumptions the equation

$$Lu = f \tag{1.3}$$

has a unique solution in W_p^2 for any $f \in \mathcal{L}_p = \mathcal{L}_p(\mathbb{R}^d)$ if p is sufficiently close to 2 and μ is large enough. Two questions immediately arose:

1) Can one prove the same result for larger p, say $p \geq d$?

2000 *Mathematics Subject Classification.* Primary 35K10, 35J15.

Key words and phrases. Linear elliptic and parabolic equations, weak uniqueness, Sobolev spaces.

The work was partially supported by NSF grant DMS-0653121.

2) Does weak uniqueness hold for the operators from [**1**]?

In Section 4 we present a situation somewhat similar to the one in [**1**], including the parabolic equations and addressing the issue of weak uniqueness in Remarks 4.2 and 4.3.

If $d \geq 3$ or $p > d$, the answer to the first question is negative due to an example of N.N. Ural'tseva since the class of operators in [**1**] (as well as in Section 4) includes the operators in her example. However, in her example, weak uniqueness still holds (see Remarks 3.3 and 4.2) along with the unique solvability in W^2_p for p close to 2 (see [**1**] or Remark 4.3). In general we do not know the answer to the second question for the operators from [**1**] (also see Remark 4.3). Of course, Nadirashvili's example shows that weak uniqueness may fail if the coefficients are merely measurable and $d \geq 3$. But what happens if we have the unique solvability in W^2_p, albeit with not large p, remains an extraordinarily intriguing question.

The usual way to prove the solvability of equation (1.3) in a smooth bounded domain $D \subset \mathbb{R}^d$ in $\overset{0}{W}{}^2_p(D)$ for any $f \in \mathcal{L}_p(D)$ is to show that

$$\begin{aligned} &\|u\|_{W^2_p(D)} \leq N\|(\lambda L_0 + (1-\lambda)\Delta)u - \mu u\|_{\mathcal{L}_p(D)}, \\ &\qquad \forall u \in \overset{0}{W}{}^2_p(D), \quad \lambda \in [0,1], \end{aligned} \tag{1.4}$$

with a constant N independent of λ and u.

In 1967, N.N. Ural'tseva, see [**5**] or [**4**], gave an example showing that for any given $p \neq 2$ there exists an operator L_0 such that the a priori estimate (1.4) is violated for some $\lambda \in [0,1]$ and $u \in \overset{0}{W}{}^2_p(D)$ no matter how big N is. The emphasis is of course on large p because the sharpness of the Alexandrov maximum principle shows that without additional restrictions on the coefficients such estimates do not exist for $p < d$.

The present article revolves about this example and since neither of [**4**] and [**5**] was translated into English we present the example in Section 2 in a somewhat modified form. Actually, the example shows the impossibility of an even weaker estimate (see (2.2)); however, this difference disappears if $p \geq d$ (see Remark 2.2). Anyway, the example ruins the hope to prove the unique $\overset{0}{W}{}^2_p(D)$ solvability by the above approach based on the method of continuity even for a rather restricted class of operators if p is unrelated to the operator and $p \neq 2$.

Then one can think about the unique solvability in a weaker sense. Since long ago it has been known that for each operator L (satisfying the conditions in the beginning) there is a Borel measurable function $G_\mu(x,y) \geq 0$, $x, y \in D$, such that for any

$$u \in C^2_0(D) = C^2(\bar{D}) \cap \{u : u|_{\partial D} = 0\}$$

and any $x \in \bar{D}$, we have

$$u(x) = \int_D G_\mu(x,y)(\mu - L_0)u(y)\, dy.$$

Any such G_μ is called a Green's function of $L_0 - \mu$ in D. One knows (Alexandrov's estimate) that $G_\mu(x,\cdot) \in \mathcal{L}_{d/(d-1)}(B)$ for any x and the $\mathcal{L}_{d/(d-1)}(D)$-norms of $G_\mu(x,\cdot)$ are bounded in D. These Green's functions are of interest because the function

$$G_\mu f(x) := \int_D G_\mu(x,y) f(y)\, dy,$$

if it is well defined (say, if $f \in \mathcal{L}_d(D)$), could be called a generalized solution of the equation $(L_0 - \mu)u = -f$ in D with zero boundary condition. But then the issue of uniqueness arises: Does there exist only one Green's function for a particular operator $L_0 - \mu$? This is the so-called weak uniqueness problem. It is well known that weak uniqueness holds for an operator $L_0 - \mu$ if and only if weak uniqueness holds for the diffusion processes corresponding to the operator L_0, and thus the issue is independent of μ. Weak uniqueness almost trivially holds if $LC_0^2(D)$ is dense in $\mathcal{L}_d(D)$. Indeed, in this case one can approximate each $f \in C(\bar{D})$ in $\mathcal{L}_d(D)$ by functions Lu_n, with $u_n \in C_0^2(D)$ and then

$$G_\mu f(x) = \lim_{n\to\infty} G_\mu L u_n(x) = -\lim_{n\to\infty} u_n(x),$$

no matter which Green's function we take. This denseness property is somewhat weaker than the unique solvability in $W_p^2(D)$ or estimate (1.4), the latter leading to the unique solvability in $W_p^2(D)$. For instance, the equation $\Delta u = f$ is not solvable in W_p^2 for many $f \in \mathcal{L}_p$, but $\Delta C_0^\infty(\mathbb{R}^d)$ is dense in $\mathcal{L}_p$ for any $p \in (1, \infty)$. By the way, what was said above regarding weak uniqueness also applies if we replace D with $\mathbb{R}^d$ and assume that $\mu > 0$.

In Section 3 we show that for any given $p \neq 2$ a slight change in the operators in Ural'tseva's example leads to an operator L such that the set $LC_0^\infty(\mathbb{R}^d)$ is *not* dense in $\mathcal{L}_p$ and its co-dimension can be as large as we like (see Remark 3.1).

The author believes that for the same operator L the set $LC_0^2(D)$ is not dense in $\mathcal{L}_p(D)$ either, thus ruining the hope for proving weak uniqueness by the above method even for particular operators, say, the ones from Ural'tseva's example. It turns out that regardless of the bad properties of the operators from Sections 2 and 3 weak uniqueness still holds for them and it is proved by probabilistic means (see Remarks 3.2 and 3.3). At present, there are no known analytic tools for doing that, even though, as our results and the results of [**1**] show, the a priori estimate (1.4) with $\mathbb{R}^d$ in place of D and $\mu > 0$ holds for p close to 2 (with the closeness depending on the operator) in Ural'tseva's example and the examples in Section 2 and, if μ is large enough, for the operators in Section 3 (see Remarks 4.1 and 4.3). Here the point is that in the above argument concerning denseness and weak uniqueness we needed $p \geq d$ and it does not work if $d \geq 3$ and p is close to 2.

Coming back to the operators from Section 4, which are more general than the ones in Sections 2 and 3, we reiterate question 2) from above: For an operator L, could it be that the a priori estimate (1.4) holds with $p = 2$ and weak uniqueness does not? Nadirashvili's example shows that there can be no weak uniqueness if the coefficients of L are only measurable. What happens if they are such that (1.4) holds?

2. An example of N.N. Ural'tseva

We represent points in $\mathbb{R}^d$ as couples $z = (x, y)$, where $x \in \mathbb{R}^2$ and y consists of the remaining coordinates of z. The reader understands that if $d = 2$, then there is no y component in z.

For a parameter $\beta > -1$ set

$$L^\beta u = \Delta u + \beta \frac{x^i x^j}{|x|^2} u_{x^i x^j}, \tag{2.1}$$

where, naturally, Δ is the Laplacian with respect to $x^1, x^2, y^1, ..., y^{d-2}$ and the summation is conducted only over $i, j = 1, 2$. Observe that L^β is a uniformly elliptic operator with bounded coefficients which are infinitely differentiable outside the set $\{z : x = 0\}$.

Let B be the ball in $\mathbb{R}^d$ of radius 1 centered at the origin. Following N.N Ural'tseva we are going to show that for any

$$p \neq 2 \quad (p \in (1, \infty))$$

there exists a $\beta_0 > -1$ such that the estimate

$$\|D^2 u\|_{\mathcal{L}_p} \leq N(\|L^\beta u\|_{\mathcal{L}_p} + \|u\|_{\mathcal{L}_p}) \tag{2.2}$$

cannot hold with the same constant N for all β in a small neighborhood of β_0 and $u \in W^2_p$ such that $u = 0$ outside B. This will imply that, no matter which $\mu \geq 0$ we take, (1.4) is impossible if we take $L_0 = L^\theta$ with a $\theta > \beta_0$ since $\lambda L^\theta + (1-\lambda)\Delta = L^{\lambda\theta}$ and for λ running through $(0, 1)$ the value $\lambda\theta$ will run through $(0, \theta) \ni \beta_0$.

Take and fix an integer $n \geq 0$ and set

$$\gamma = \frac{\beta + \sqrt{\beta^2 + 4n^2(1+\beta)}}{2(1+\beta)}, \quad v^{(\beta)}(z) = |x|^\gamma \cos n\phi,$$

where ϕ is the polar angle on the (x^1, x^2)-plane changing in $(-\pi, \pi]$. Observe that $\gamma \geq 0$ is a root of the equation

$$(1+\beta)\gamma^2 - \beta\gamma - n^2 = 0. \tag{2.3}$$

The function v depends only on x; hence while computing Lv we may disregard differentiations with respect to y and restrict ourselves to $\mathbb{R}^2$. In polar coordinates in $\mathbb{R}^2$ on functions $u(x)$, the operator L is written as

$$L^\beta u = (1+\beta)u_{rr} + \frac{1}{r}u_r + \frac{1}{r^2}u_{\phi\phi}.$$

This allows us to obtain

$$L^\beta v^{(\beta)} = \left[(1+\beta)\gamma(\gamma-1) + \gamma - n^2\right]r^{-2}v^{(\beta)} = 0. \tag{2.4}$$

Also observe that v is a positive-homogeneous function of degree γ which is infinitely differentiable outside the origin. Its first-order and second-order derivatives are also homogeneous and, as is easy to see,

$$\|D^2 v^{(\beta)}\|_{\mathcal{L}_p(B)} < \infty \quad \text{if} \quad (\gamma - 2)p + 1 > -1.$$

One can rewrite the latter condition in three equivalent ways:

$$\gamma > \frac{2}{q} =: \gamma_0, \quad q = \frac{p}{p-1},$$

or (cf. (2.3))

$$(1+\beta)\gamma_0^2 - \beta\gamma_0 - n^2 < 0, \tag{2.5}$$

or else ($p \neq 2$)

$$(\beta - \beta_0)(2 - q) < 0, \tag{2.6}$$

where

$$\beta_0 = \frac{n^2 q^2 - 4}{2(2-q)}$$

is the value of β which makes γ_0 to be a root of (2.3). We also have that

$$(\gamma - 2)p + 1 \to -1 \quad \text{iff} \quad \gamma \to \gamma_0 \quad (\text{that is,} \quad \beta \to \beta_0). \tag{2.7}$$

Furthermore,

$$\Delta v^{(\beta)} = -\beta\gamma(\gamma - 1)r^{-2}v^{(\beta)},$$

which along with (2.7) easily implies that, if $\beta_0 \neq 0$ and $\beta_0 > -1$, then

$$\|D^2 v^{(\beta)}\|_{\mathcal{L}_p(B)} \to \infty \quad \text{as} \quad \beta \to \beta_0. \tag{2.8}$$

Impossibility of (2.2) *if* $p \in (1, 2)$. Take $n = 0$ and observe that $q > 2$ and $\beta_0 = 2/(q - 2) > 0$. Hence for any $\beta > \beta_0$ we have that $\beta > -1$ and (2.6) holds implying that $\gamma > 2/q$. In that case one sees readily that

$$\|Dv^{(\beta)}\|_{\mathcal{L}_p(B)} + \sup_B |v^{(\beta)}| \leq N, \tag{2.9}$$

where the constant N independent of β as long as $\beta > \beta_0$ is close to β_0.

Now take a $\zeta \in C_0^\infty(B)$ such that $\zeta(z) = 1$ for $|z| \leq 1/2$ and introduce $u^{(\beta)} = \zeta v^{(\beta)}$. Then owing to (2.4) and (2.9) we have

$$\|L^\beta u^{(\beta)}\|_{\mathcal{L}_p} \leq N$$

with N independent of β as long as $\beta > \beta_0$ is close to β_0. Therefore, if (2.2) were true with N independent of $u \in \overset{0}{W}{}_p^2$, then we would have

$$\|D^2 u^{(\beta)}\|_{\mathcal{L}_p} \leq N$$

with N is independent of β. However, this does not hold owing to (2.8) and the fact that $\beta_0 \neq 0$.

Impossibility of (2.2) *if* $p \in (2, \infty)$. We take $n \geq 2$, observe that in this case $\beta_0 > -1$ since $1 \leq q < 2$, and restrict β to the range $(-1, \beta_0)$. Then (2.6) is again satisfied and $\gamma > 2/q > 1$. It follows that

$$\sup_B |Dv^{(\beta)}| + \sup_B |v^{(\beta)}| \leq N,$$

where the constant N is independent of β as long as β is close to β_0. This leads to the impossibility of (2.2) in the same way as above.

REMARK 2.1. The above arguments are not applicable if $p = 2$ because in that case (2.5) becomes $1 < n^2$, in which case, for any $\beta > -1$, we have $\gamma > 2/q = 1$ and

$$\|D^2 v^{(\beta)}\|_{\mathcal{L}_2(D)}$$

stays bounded on any closed finite interval of values of β in $(-1, \infty)$ since $\gamma \downarrow 1$ only if $\beta \to \infty$. The latter easily follows from the fact that

$$\gamma^2 = \frac{n^2 + \beta\gamma}{1 + \beta},$$

which shows that if $\gamma \downarrow 1$, then β cannot stay bounded.

The fact that we do not have a counterexample for $p = 2$ is not a flaw of the example. We will see later in Remark 4.1 that in the above example for any $d \geq 2$ and $\beta_0 > -1$, estimate (2.2) does hold for p close to 2 and $u^{(\beta)}$ in place of u as long as β is close to β_0. In addition, if $p = 2$, the constant N can be chosen to be the same as long as $\beta \in [r, s]$ with fixed r and s such that $-1 < r < s < \infty$.

REMARK 2.2. If for an operator L_0 and $p \geq d$ there is a constant N such that for all $u \in W_p^2$ we have

$$\|D^2 u\|_{\mathcal{L}_p} \leq N(\|L_0 u\|_{\mathcal{L}_p} + \|u\|_{\mathcal{L}_p}),$$

then for any $\mu > 0$ and any $u \in W_p^2$,

$$\|u\|_{W_p^2} \leq N\|(L_0 - \mu)u\|_{\mathcal{L}_p}$$

with perhaps a different N. This follows from the well-known fact that $\mu\|u\|_{\mathcal{L}_p} \leq N\|(L_0 - \mu)u\|_{\mathcal{L}_p}$ with N depending only on d, p, δ.

3. An example in which LC_0^∞ is not dense in $\mathcal{L}_p$

The absence of an a priori estimate (1.4) is a big obstacle if one wants to show weak uniqueness for an operator L by showing that equation (1.3) is uniquely solvable in W_p^2 for $p \geq d$. On the other hand, as we pointed out in the Introduction, it is sufficient for weak uniqueness that LC_0^∞ be dense in $\mathcal{L}_p$ with $p \geq d$ (p can be even slightly less than d). In this section we show that this denseness does not happen if $p > 2$ by constructing an operator depending on p. The existence of such operators with merely measurable coefficients, of course, follows from Nadirashvili's example in which there is even no weak uniqueness. But the operators we construct possess the weak uniqueness property (see Remark 3.3) and the corresponding equations are uniquely solvable in W_2^2 (see Remark 4.3).

Let $v = v(y)$ be a fixed smooth function of $y \in \mathbb{R}^d$ exponentially decreasing as $|y| \to \infty$ and such that $|v''| \leq v$, say $v(y) = 1/\cosh(\varepsilon y)$ with ε small enough.

Fix some parameters $\beta > -1$ and $\mu > 0$ and in the 3-dimensional space of points (x, y), $x \in \mathbb{R}^2$, $y \in \mathbb{R}$, consider the operator

$$Lu = \Delta u + \beta \frac{x^i x^j}{|x|^2} u_{x^i x^j} - cu - \mu u, \tag{3.1}$$

where (the summation is conducted over $i, j = 1, 2$ and) $c = c(y) = 1 + v_{yy}(y)/v(y)$ (≥ 0).

LEMMA 3.1. *Take some numbers $B \geq 0, C > 0$. Then there exists a unique bounded continuous function $f(r)$, $r \geq 0$, such that $f(0) = 1$, f is twice continuously differentiable on $(0, \infty)$ and satisfies*

$$f_{rr} - \frac{B}{r} f_r - C^2 f = 0. \tag{3.2}$$

Furthermore, $f(r) \leq \exp(-Cr/2)$ for $r \geq 2B/(3C)$, f_r is bounded and $f(r) \geq \exp(-Cr)$.

PROOF. *Uniqueness.* The coefficients of (3.2) are bounded on each interval (ε, ∞) if $\varepsilon > 0$. Furthermore, $C^2 > 0$. Therefore, the maximum principle implies that, if we have two bounded continuous classical solutions $f(r)$ and $g(r)$, $r \geq 0$, of (3.2), then $|f(r) - g(r)| \leq |f(\varepsilon) - g(\varepsilon)|$. Hence, if $f(1) = g(1)$, then $f \equiv g$, which proves uniqueness.

Existence. Take a point $r_0 > 0$ and, in the class of bounded functions, solve the equation on $[r_0, \infty)$ with boundary condition 1 at r_0. Denote this solution by f^{r_0}. Observe that the function $g(r) = \exp(-Cr)$ satisfies

$$g_{rr} - \frac{B}{r} g_r - C^2 g \geq 0, \quad r > 0.$$

This and the maximum principle imply that $1 \geq f^{r_0}(r) \geq f^{r_1}(r) \geq g(r)$ if $0 < r_1 \leq r_0 \leq r$, which in turn implies that

$$\lim_{r_0 \downarrow 0} f^{r_0}(r) =: f(r)$$

exists for any $r > 0$, and if we set $f(0) = 1$, then f is a bounded continuous function on $[0, \infty)$. The elliptic regularity theory or the theory of ODE allows us to conclude that f is a classical solution of (3.2).

Next, the function $h(r) = \exp(-Cr/2)$ satisfies

$$h_{rr} - \frac{B}{r} h_r - C^2 h \leq 0$$

for $r \geq 2B/(3C) =: \bar{r}$. By the maximum principle, $f \leq f^{r_0} \leq h/h(\bar{r})$ on $[\bar{r}, \infty)$ if $r_0 \leq \bar{r}$. It follows that

$$(r^{-B} f_r)_r = C^2 r^{-B} f, \quad r^{-B} f_r(r) = -C^2 \int_r^\infty s^{-B} f(s)\, ds,$$

$$|f_r(r)| \leq N r^B \int_r^\infty s^{-B} h(s)\, ds. \tag{3.3}$$

If $B = 0$, the boundedness of the right-hand side of (3.3) is obvious. For $B > 0$ the boundedness at infinity is obtained either from the elliptic regularity theory or by using l'Hôpital's rule. Near the origin one can also use this rule if $B \geq 1$. Finally, if $B \in (0,1)$, then the right-hand side of (3.3) goes to zero as $r \downarrow 0$. This proves the boundedness of f_r and finishes the proof of the lemma.

Now take an integer $n \geq 2$ and set

$$\alpha = n^2 - 1, \quad \gamma = \frac{-(2+\beta) + \sqrt{(2+\beta)^2 + 4\alpha(1+\beta)}}{2(1+\beta)},$$

$$B = \frac{1 + 2(1+\beta)\gamma}{1+\beta}, \quad C^2 = \frac{1+\mu}{1+\beta}.$$

Observe that $\gamma > 0$ and

$$w(\gamma) := (1+\beta)\gamma(\gamma+1) + \gamma - \alpha = (1+\beta)\gamma^2 + (2+\beta)\gamma - \alpha = 0. \tag{3.4}$$

Since $B > 0$ and $C > 0$, we can take f from the lemma. Also notice that

$$\gamma < 1 \Longleftrightarrow w(1) > 0 \Longleftrightarrow 3 + 2\beta > \alpha.$$

One more useful observation is that $\gamma = 1$ if $3 + 2\beta = \alpha$ and $\gamma \to 0$ as $\beta \to \infty$ and $\alpha = \text{const}$.

Next, set

$$h(x) = |x|^{-\gamma-1} f(|x|) \cos n\phi, \quad g(x,y) = h(x) v(y),$$

where ϕ is the polar angle on the (x^1, x^2)-plane changing in $(-\pi, \pi]$.

THEOREM 3.2. *Take β such that $3 + 2\beta > \alpha$. Then $\beta > 0$, $\gamma < 1$ and $h \in \mathcal{L}_q(\mathbb{R}^2)$, $g \in \mathcal{L}_q(\mathbb{R}^3)$ for any $q \in [1, 2/(\gamma+1))$. Furthermore, for any $u \in C_0^\infty(\mathbb{R}^3)$ we have*

$$\int_{\mathbb{R}^3} gLu\, dxdy = 0. \tag{3.5}$$

PROOF. That $\beta > 0$ follows from the fact that $\alpha \geq 3$. The asserted summability of h and g is almost obvious. To prove (3.5), we first observe that g and L are symmetric with respect to the change of variables $x \to -x$. Therefore, we may concentrate only on u such that $u(x,y) = u(-x,y)$. Then we integrate by parts in (3.5) with respect to y noticing that $v_{yy} - cv = -v$ and

$$\int_{\mathbb{R}^3} g(u_{yy} - cu)\,dxdy = \int_{\mathbb{R}^2} h(x)\Big(\int_{\mathbb{R}} (v_{yy} - cv)(y)u(x,y)\,dy\Big)\,dx$$
$$= -\int_{\mathbb{R}^3} gu\,dxdy.$$

Hence the integral in (3.5) equals

$$\int_{\mathbb{R}} v(y)\Big(\int_{\mathbb{R}^2} h(x)\big(\Delta_x u + \beta\frac{x^i x^j}{|x|^2}u_{x^i x^j} - \nu u\big)(x,y)\,dx\Big)\,dy,$$

where $\nu = \mu + 1$ and Δ_x is the Laplacian with respect to x^1, x^2. It is seen that to prove the lemma it suffices to show that for any even function $u \in C_0^\infty(\mathbb{R}^2)$ we have

$$\int_{\mathbb{R}^2} h\big(\Delta_x u + \beta u_{rr} - \nu u\big)\,dx = 0, \tag{3.6}$$

where u_{rr} is the second-order derivative of u in the radial direction. Equivalently, it suffices to show that $I_\varepsilon \to 0$ as $\varepsilon \downarrow 0$, where

$$I_\varepsilon = \int_{|x|\geq\varepsilon} h\big(\Delta_x u + \beta u_{rr} - \nu u\big)\,dx.$$

Passing to the polar coordinates yields

$$I_\varepsilon = \int_{r\geq\varepsilon} r^{-\gamma}f(r)\Big(\int_{-\pi}^{\pi}[(1+\beta)u_{rr} + r^{-1}u_r + r^{-2}u_{\phi\phi} - \nu u]\cos n\phi\,d\phi\Big)\,dr.$$

Integrating by parts with respect to ϕ leads to

$$I_\varepsilon = \int_{-\pi}^{\pi} \cos n\phi\Big(\int_{r\geq\varepsilon} r^{-\gamma}f(r)\big((1+\beta)u_{rr} + r^{-1}u_r - n^2r^{-2}u - \nu u\big)\,dr\Big)\,d\phi.$$

Now we integrate by parts with respect to r noticing that

$$\begin{aligned}
&r^{\gamma+2}[(1+\beta)(r^{-\gamma}f(r))_{rr} - (r^{-\gamma-1}f(r))_r]\\
&\quad= (1+\beta)[\gamma(\gamma+1)f - 2\gamma r f_r + r^2 f_{rr}] + (\gamma+1)f - rf_r\\
&\quad= f[(1+\beta)\gamma(\gamma+1) + \gamma + 1] - rf_r(2(1+\beta)\gamma + 1) + r^2(1+\beta)f_{rr}\\
&\quad= n^2 f + (1+\beta)[r^2 f_{rr} - rBf_r - r^2C^2 f] + \nu r^2 f = n^2 f + \nu r^2 f.
\end{aligned}$$

It is seen that after integrating by parts we obtain

$$I_\varepsilon = -[\gamma(1+\beta)+1]f(\varepsilon)I_\varepsilon^1 + (1+\beta)f_r(\varepsilon)I_\varepsilon^2 - (1+\beta)f(\varepsilon)I_\varepsilon^3,$$

where

$$I_\varepsilon^1 = \varepsilon^{-\gamma-1}\int_{-\pi}^{\pi} u(\varepsilon,\phi)\cos n\phi\,d\phi,$$
$$I_\varepsilon^2 = \varepsilon^{-\gamma}\int_{-\pi}^{\pi} u(\varepsilon,\phi)\cos n\phi\,d\phi,$$
$$I_\varepsilon^3 = \varepsilon^{-\gamma}\int_{-\pi}^{\pi} u_r(\varepsilon,\phi)\cos n\phi\,d\phi.$$

Since we assumed that u is an even function, $Du(0) = 0$ and $|Du(x)| \le N|x|$, and since $\gamma < 1$, we have $|I_\varepsilon^3| \le N\varepsilon^{1-\gamma} \to 0$. The integral defining I_ε^2 will not change if we replace the integrand with $u - u(0)$. Then, as above, we get $|I_\varepsilon^2| \le N\varepsilon^{1-\gamma} \to 0$. To deal with I_ε^1 we use Taylor's formula, showing that

$$|u(x) - u(0) - x^i u_{x^i}(0)| \le N|x|^2.$$

We also use the fact that, since $n \ge 2$, $\cos n\phi$ is orthogonal to $\cos\phi$ and $\sin\phi$. Then for $u_1 = u_{x^1}(0), u_2 = u_{x^2}(0)$ we obtain

$$I_\varepsilon^1 = \varepsilon^{-\gamma-1} \int_{-\pi}^{\pi} [u(\varepsilon,\phi) - u(0) - u_1\varepsilon\cos\phi - u_2\varepsilon\sin\phi] \cos n\phi \, d\phi,$$

implying that $|I_\varepsilon^1| \le N\varepsilon^{1-\gamma} \to 0$. This proves the theorem.

REMARK 3.1. Theorem 3.2 implies that the set $LC_0^\infty(\mathbb{R}^3)$ is not dense in $\mathcal{L}_p$ if $p = q/(q-1)$ and $q \in [1, 2/(\gamma+1))$. In terms of p the latter means that $p > 2/(1-\gamma)$ or $\gamma < 1 - 2/p$ or else

$$p > 2 \quad \text{and} \quad w(1 - 2/p) > 0, \tag{3.7}$$

where w is introduced in (3.4). The latter in terms of n is equivalent to

$$n^2 < 1 + (1+\beta)(p-2)^2 p^{-2} + (2+\beta)(p-2)p^{-1}. \tag{3.8}$$

By the way, (3.7) obviously implies that $w(1) > 0$ and $3+2\beta > \alpha$, which is required in Theorem 3.2.

Now we see that for any given $p > 2$ and integer $k \ge 2$ one can find a large β such that (3.8) is satisfied for $n = 2, ..., k$, in which case there exist $k-1$ linearly independent functions $g_1, ..., g_{k-1} \in \mathcal{L}_q(\mathbb{R}^3)$ for which (3.5) is valid for all $u \in C_0^\infty(\mathbb{R}^3)$ with g_i in place of g.

REMARK 3.2. The proof of Theorem 3.2 implies that $LC_0^\infty(\mathbb{R}^2)$, where $Lu = \Delta u + \beta u_{rr} - \mu u$, is not dense in $\mathcal{L}_p(\mathbb{R}^2)$ if $p > 2$ and β are chosen so that (3.8) is valid for an $n \ge 2$. One knows that for $d = 2$ for any operator L with the properties listed before equation (1.3) and with $\mu > 0$, equation (1.3) is uniquely solvable in $W_2^2(\mathbb{R}^2)$ for any $f \in \mathcal{L}_2(\mathbb{R}^2)$ and estimate (1.4) holds (with $\mathbb{R}^2$ in place of D). In particular, this fact implies that $LC_0^\infty(\mathbb{R}^2)$ is dense in $\mathcal{L}_2(\mathbb{R}^2)$ and weak uniqueness holds in $\mathbb{R}^2$ $(p = d)$.

In this connection it would be interesting to have an example showing that for $d \ge 3$ and $p \in (1, 2]$ there are uniformly nondegenerate operators such that $LC_0^\infty(\mathbb{R}^d)$ is not dense in $\mathcal{L}_p(\mathbb{R}^d)$.

We finish the section with a discussion of weak uniqueness for the operators L^β from (2.1) and L from (3.1). The following result is proved in [**3**] in a more general form admitting degenerate operators. It is stated there in probabilistic terms but the way to translate it into purely analytic language is also provided.

THEOREM 3.3. *Take an operator L as in* (1.1) *with $\mu > 0$ satisfying the conditions listed after* (1.1). *Let d' and d'' be integers such that $1 \le d' < d$, $d'' = d - d'$. For $x \in \mathbb{R}^d$ we write $x = (x', x'')$, where $x' \in \mathbb{R}^{d'}$, $x'' \in \mathbb{R}^{d''}$. Assume that $a^{ij}(x) = a^{ij}(x')$ if $i, j = 1, ..., d'$ and assume that weak uniqueness holds for the operator*

$$\sum_{i,j=1}^{d'} a^{ij}(x') u_{x^i x^j}(x') - \mu u(x')$$

as an operator acting on functions given on $\mathbb{R}^{d'}$. *Finally, assume that for* $i,j = 1,...,d$ *and* $x' \in \mathbb{R}^{d'}$, $y'', z'' \in \mathbb{R}^{d''}$ *we have*

$$|a^{ij}(x', y'') - a^{ij}(x', z'')| \le K|y'' - z''|, \tag{3.9}$$

where K *is a constant.*

Then weak uniqueness holds for the operator L *in* $\mathbb{R}^d$.

REMARK 3.3. Obviously Theorem 3.3 and Remark 3.2 yield weak uniqueness for L from (3.1) and $L^\beta - \mu$ if $\mu > 0$ and we consider these operators in the whole space. By the way, one of the results in [**3**] is that weak uniqueness in the whole space implies weak uniqueness in any domain.

4. Some positive results

Here we present a result somewhat similar to one of the results of [**1**], which however does not follow from [**1**] because we only assume that a^{ij}, $i,j = 1,2$, are measurable in $(t, x^1, ..., x^d)$.

The arguments in this section are not any shorter for elliptic equations than for parabolic ones. Therefore, for $d \ge 3$ and a constant $K > 0$, we consider the operator L of the form

$$Lu = \sum_{i,j=1}^{2} a^{ij} u_{x^i x^j} + K \sum_{j=3}^{d} u_{x^j x^j} \tag{4.1}$$

with the coefficients depending on $(t, x) \in \mathbb{R}^{d+1}$, satisfying conditions (1.2) on $\mathbb{R}^{d+1}$. We also assume that

$$\operatorname{tr}_2 a := a^{11} + a^{22} \equiv 1. \tag{4.2}$$

We discuss this assumption in Remark 4.1. Denote

$$W^{1,2}_p = W^{1,2}_p(\mathbb{R}^{d+1}).$$

THEOREM 4.1. *There is a small constant* $\varepsilon > 0$, *depending only on* δ, *such that for any* $p \in (2-\varepsilon, 2+\varepsilon)$ *and any* $u \in W^{1,2}_p$ *and* $\mu \ge 0$ *we have*

$$\mu\|u\|_{\mathcal{L}_p(\mathbb{R}^{d+1})} + \sqrt{\mu}\|Du\|_{\mathcal{L}_p(\mathbb{R}^{d+1})} + \|D^2u\|_{\mathcal{L}_p(\mathbb{R}^{d+1})} \le N\|u_t + Lu - \mu u\|_{\mathcal{L}_p(\mathbb{R}^{d+1})},$$

where N *depends only on* K, d, *and* δ.

The proof of this theorem closely follows some arguments from [**2**]. Set

$$\Delta' u = \sum_{j=1}^{2} u_{x^j x^j}, \quad \Delta'' u = \sum_{j=3}^{d} u_{x^j x^j}.$$

LEMMA 4.2. *Under the assumptions of Theorem 4.1 additionally assume that* $2a^{ij} = \delta^{ij}$ *for* $i,j \le 2$. *Then there is a continuous function* $N = N(p)$, $p \in (1,\infty)$ *such that* $N(2) \le 2$ *and for any* $u \in W^{1,2}_p$ *and* $\mu > 0$ *we have*

$$\|U\|_{\mathcal{L}_p(\mathbb{R}^{d+1})} \le N(p)\|u_t + Lu - \mu u\|_{\mathcal{L}_p(\mathbb{R}^{d+1})}, \tag{4.3}$$

where $U \ge 0$ *is defined by*

$$U^2 = |u_t + u_{x^1x^1} + K\Delta'' u - \mu u|^2 + 2|u_{x^1x^2}|^2$$
$$+ |u_t + u_{x^2x^2} + K\Delta'' u - \mu u|^2.$$

PROOF. By using scalings in the x'' coordinate we reduce the general case to the one with $K = 1/2$. Obviously, we may assume that $u \in C_0^\infty(\mathbb{R}^{d+1})$.

Denote $f = u_t + Lu - \mu u$. Then

$$4\|f\|_{\mathcal{L}_2}^2 = \|2u_t + \Delta u - 2\mu u\|_{\mathcal{L}_2}^2.$$

Since

$$2u_t + \Delta u - 2\mu u = (u_t + u_{x^1x^1} + \tfrac{1}{2}\Delta'' u - \mu u) + (u_t + u_{x^2x^2} + \tfrac{1}{2}\Delta'' u - \mu u),$$

we have

$$4\|f\|_{\mathcal{L}_2}^2 = \|U\|_{\mathcal{L}_2}^2 + 2\int_{\mathbb{R}^{d+1}} [u_t^2 + u_t(\Delta u - 2\mu u) + \phi_1\phi_2 - u_{x^1x^2}^2]\,dxdt,$$

where

$$\phi_i = u_{x^ix^i} + \tfrac{1}{2}\Delta'' u - \mu u.$$

An easy integration by parts shows that the integral of $u_t(\Delta u - 2\mu u)$ vanishes. Therefore,

$$4\|f\|_{\mathcal{L}_2}^2 \geq \|U\|_{\mathcal{L}_2}^2 + 2\int_{\mathbb{R}^{d+1}} [\phi_1\phi_2 - u_{x^1x^2}^2]\,dxdt.$$

We evaluate the last integral by Parseval's identity, showing that it is equal to the integral with respect to $d\xi dt$ of

$$[(|\xi^1|^2 + \tfrac{1}{2}|\xi''|^2 + \mu)(|\xi^2|^2 + \tfrac{1}{2}|\xi''|^2 + \mu) - |\xi^1|^2|\xi^2|^2]|\tilde{u}|^2,$$

where $\tilde{u}$ is the Fourier transform of u with respect to x. The last expression is nonnegative and this proves (4.3) for $p = 2$ with $N(2) = 2$.

Estimate (4.3) is, of course, well known for any $p \in (1, \infty)$ since, actually, by the classical theory,

$$\mu\|u\|_{\mathcal{L}_p} + \sqrt{\mu}\|Du\|_{\mathcal{L}_p} + \|D^2u\|_{\mathcal{L}_p} \leq N(p,d)\|f\|_{\mathcal{L}_p}. \tag{4.4}$$

By using the Riesz-Torin theorem one proves that, for the best possible constant $N(p)$ we have that $\frac{1}{p}\ln N(p)$ is a convex function of $1/p$. For a reference the reader is sent, for instance, to Lemma 5 in [**2**], where $N(p)$ is, actually, defined as the best constant for which (4.3) is valid for all complex-valued u. The above derivation that $N(2) \leq 2$ is applicable for real and imaginary parts of such functions and yields (4.3) again with $N(2) \leq 2$.

Since finite convex functions are continuous, the lemma is proved.

Next, we need the following, which is Lemma 7 of [**2**].

LEMMA 4.3. *Let $\bar{a} = (\bar{a}^{ij})$ be a 2×2 symmetric matrix with $\operatorname{tr}_2 \bar{a} = 1$ and such that*

$$\delta \leq \bar{a}^{ij}\xi^i\xi^j \leq \delta^{-1}$$

for all units $\xi \in \mathbb{R}^2$. Then for any symmetric 2×2 matrix σ and $p > 1$ we have

$$|\operatorname{tr}_2 \sigma|^p \leq \kappa(p)(\operatorname{tr}_2 \sigma\sigma^*)^{p/2} + \chi(p)|\operatorname{tr}_2 \bar{a}\sigma|^p,$$

where $\kappa(p) = (1 - \delta^2/2)^p$, $\chi(p) = (2/\delta)^p$.

PROOF OF THEOREM 4.1. As usual it suffices to concentrate on $u \in C_0^\infty(\mathbb{R}^{d+1})$. Set $f = u_t + Lu - \mu u$ and

$$\bar{a}^{ij} = a^{ij}, \quad i,j = 1,2, \quad \sigma^{kr} = u_{x^kx^r}, \quad k,r = 1,2, k \neq r,$$

$$\sigma^{ii} = u_t + u_{x^ix^i} + K\Delta'' u - \mu u, \quad i = 1,2.$$

Then

$$f = \operatorname{tr}_2 \bar{a}\sigma, \quad \operatorname{tr}_2 \sigma = 2u_t + \Delta' u + 2K\Delta'' u - 2\mu u, \quad \operatorname{tr}_2 \sigma\sigma^* = U^2,$$

where U is taken from Lemma 4.2. Hence by Lemmas 4.3 and 4.2 we obtain

$$\begin{aligned} &2^p \|u_t + \tfrac{1}{2}\Delta' u + K\Delta'' u - \mu u\|^p_{\mathcal{L}_p(\mathbb{R}^{d+1})} \\ &\quad \le \kappa(p)\|U\|^p_{\mathcal{L}_p(\mathbb{R}^{d+1})} + \chi(p)\|f\|^p_{\mathcal{L}_p(\mathbb{R}^{d+1})} \\ &\quad \le \chi(p)\|f\|^p_{\mathcal{L}_p(\mathbb{R}^{d+1})} + \kappa(p)N^p(p)\|u_t + \tfrac{1}{2}\Delta' u + K\Delta'' u - \mu u\|^p_{\mathcal{L}_p(\mathbb{R}^{d+1})}. \end{aligned}$$

Thus,

$$(2^p - \kappa(p)N^p(p))\|u_t + \tfrac{1}{2}\Delta' u + K\Delta'' u - \mu u\|^p_{\mathcal{L}_p(\mathbb{R}^{d+1})} \le \chi(p)\|f\|^p_{\mathcal{L}_p(\mathbb{R}^{d+1})}$$

and using the continuity of $N(p)$ and the facts that $N(2) \le 2$ and $\kappa(2) < 1$, we find an appropriate $\varepsilon > 0$ such that for $p \in (2-\varepsilon, 2+\varepsilon)$ we have

$$\|u_t + \tfrac{1}{2}\Delta' u + K\Delta'' u - \mu u\|^p_{\mathcal{L}_p(\mathbb{R}^{d+1})} \le N\|f\|^p_{\mathcal{L}_p(\mathbb{R}^{d+1})}.$$

Now it only remains to use (4.4) again. The theorem is proved.

Here is an elliptic counterpart of Theorem 4.1.

THEOREM 4.4. *Assume that a^{ij} are independent of t. Then for the constant $\varepsilon > 0$ from Theorem* 4.1, *any $p \in (2-\varepsilon, 2+\varepsilon)$, any $u \in W^2_p$, and any $\mu \ge 0$ we have*

$$\mu\|u\|_{\mathcal{L}_p} + \sqrt{\mu}\|Du\|_{\mathcal{L}_p} + \|D^2 u\|_{\mathcal{L}_p} \le N\|Lu - \mu u\|_{\mathcal{L}_p}, \tag{4.5}$$

where N is the constant from Theorem 4.1.

PROOF. Take a $\zeta \in C_0^\infty(\mathbb{R})$ such that $\zeta \not\equiv 0$ and set $\zeta^n(t) = \zeta(t/n)$ and $u^n(t,x) = \zeta^n(t)u(x)$. By Theorem 4.1,

$$\begin{aligned} &\mu\|u^n\|_{\mathcal{L}_p(\mathbb{R}^{d+1})} + \sqrt{\mu}\|u^n_x\|_{\mathcal{L}_p(\mathbb{R}^{d+1})} + \|u^n_{xx}\|_{\mathcal{L}_p(\mathbb{R}^{d+1})} \\ &\quad \le N\|u^n_t\|_{\mathcal{L}_p(\mathbb{R}^{d+1})} + N\|Lu^n - \mu u^n\|_{\mathcal{L}_p(\mathbb{R}^{d+1})}, \end{aligned}$$

which easily implies that (4.5) holds if we add on the right the term

$$Nn^{-1}\|\zeta'\|_{\mathcal{L}_p(\mathbb{R})}\|\zeta\|^{-1}_{\mathcal{L}_p(\mathbb{R})}\|u\|_{\mathcal{L}_p}.$$

By letting $n \to \infty$ we come to (4.5) as is. The theorem is proved.

REMARK 4.1. Observe that a simple scaling of time shows that condition (4.2) can be replaced with the condition that $\operatorname{tr}_2 a$ is a constant. If one uses a time change, then one sees that in Theorem 4.1 it suffices to assume that $\operatorname{tr}_2 a$ is a function only of t. Anyway, in the example in Section 2, we have $\operatorname{tr}_2 a \equiv 1 + \beta$ so that Theorem 4.4 is applicable.

This theorem with $\mu = 0$ shows that for any $\beta_0 > -1$ there exists a neighborhood of 2 such that for any p from that neighborhood and $u \in W^2_p$, estimate (2.2) holds if $\beta = \beta_0$ with N independent of p and u. The perturbation technique shows that the same is true for β close to β_0. The fact that N in Theorem 4.1 depends only on d, K, δ shows that N in (2.2) depends only on d and β. In particular, (2.2) holds for $p = 2$ and N independent of β as long as $\beta \in [r, s]$ with fixed r and s such that $-\infty < r < s < \infty$. We have proved Theorems 4.1 and 4.4 only for $d \ge 3$ just for convenience of notation. That the corresponding results are also true for $d = 2$ is known from [**2**].

REMARK 4.2. By using probabilistic means it is not hard to show that weak uniqueness holds for operators of type (4.1) satisfying the conditions in the beginning of the section. The author does not have any idea how to do this analytically, and this is one of the reasons for an open question in the next remark.

REMARK 4.3. An immediate corollary of the above theorems and the method of continuity is the fact that the corresponding equations are uniquely solvable if $\mu > 0$. For μ large enough, by using perturbations, linear transformations in x'' coordinates, and partitions of unity, one can get further results on a priori estimates and solvability in Sobolev spaces for p close to 2 of equations with $a^{ik} = 0$ for $i = 1, 2$ and $k \geq 3$ and a^{kr}, $k, r \geq 3$, uniformly continuous in x (uniformly with respect to t for the parabolic case). One can also add lower-order terms. We leave doing all of this to the interested reader, but unlike the situation in Remark 4.2, we have no idea whether, while going to the more general operators described above, weak uniqueness is preserved.

We finish the paper with the one-dimensional counterpart of operators of type (4.1).

THEOREM 4.5. *The assertion of Theorem* 4.1 *holds if we take* $d \geq 2$, *drop condition* (4.2), *and take*

$$Lu = a^{11} u_{x^1 x^1} + K \sum_{j=2}^{d} u_{x^j x^j}.$$

The proof of this theorem is obtained by following the same lines as in the proof of Theorem 4.4, but first we find a function $a^{d+1,d+1} \geq \delta$ such that $a^{11} + a^{d+1,d+1}$ is a constant, then apply Theorem 4.4 to the operator

$$a^{11} u_{x^1 x^1} + a^{d+1,d+1} u_{x^{d+1} x^{d+1}} + K \sum_{j=2}^{d} u_{x^j x^j},$$

and finally take $u(x, x^{d+1}) = u(x)\zeta(x^{d+1}/n)$ and let $n \to \infty$ in the estimate obtained from Theorem 4.4.

All of what was said after the proof of Theorem 4.1 applies to Theorem 4.5, including the fact that in case $d \geq 3$ we do not know if weak uniqueness holds after we generalize this theorem for variable coefficients a^{ij}, $i, j \geq 2$.

References

[1] Hongjie Dong and N. V. Krylov, *Second-order elliptic and parabolic equations with* $B(\mathbb{R}^2, VMO)$ *coefficients*, submitted to Trans. Amer. Math. Soc. (To appear) http://arxiv.org/abs/0810.2739

[2] N. V. Krylov, *On equations of minimax type in the theory of elliptic and parabolic equations in the plane*, Matem. Sbornik **81**(1970), no. 1, 3–22; English transl., Math. USSR Sbornik, **10**(1970), 1–20. MR0255954 (41:614)

[3] ———, *On weak uniqueness for some diffusions with discontinuous coefficients.* Stoch. Proc. and Appl. **113**(2004), no. 1, 37–64. MR2078536 (2005e:60119)

[4] O. A. Ladyzhenskaya and N. N. Ural'tseva, *Linear and quasilinear equations of elliptic type.* Second edition, "Nauka", Moscow, 1973; English transl. of the first edition, Academic Press, New York and London, 1968. MR0509265 (58:23009)

[5] N. N. Ural'tseva, *On impossibility of* W_p^2 *estimates for multidimensional elliptic equations with discontinuous coefficients.* Zap. nauchn. seminarov LOMI im. Steklova, Vol. 5, 1967, "Nauka", Leningrad. (Russian)

127 Vincent Hall, University of Minnesota, Minneapolis, Minnesota, 55455, USA
E-mail address: krylov@math.umn.edu

Amer. Math. Soc. Transl.
(2) Vol. **229**, 2010

On the Fundamental Solution of an Elliptic Equation in Nondivergence Form

Vladimir Maz'ya and Robert McOwen

Dedicated to Nina Ural'tseva with admiration

ABSTRACT. We consider the existence and asymptotics for the fundamental solution of an elliptic operator in nondivergence form, $\mathcal{L}(x,\partial_x) = a_{ij}(x)\partial_i\partial_j$, for $n \geq 3$. We assume that the coefficients have modulus of continuity satisfying the square Dini condition. For fixed y, we construct a solution of $\mathcal{L}Z_y(x) = 0$ for $0 < |x-y| < \varepsilon$ with an explicit leading order term which is $O(|x-y|^{2-n}e^{I(x,y)})$ as $x \to y$, where $I(x,y)$ is given by an integral and plays an important role for the fundamental solution: if $I(x,y)$ approaches a finite limit as $x \to y$, then we can solve $\mathcal{L}(x,\partial_x)F(x,y) = \delta(x-y)$, and $F(x,y)$ is asymptotic as $x \to y$ to the fundamental solution for the constant coefficient operator $\mathcal{L}(y,\partial_x)$. On the other hand, if $I(x,y) \to -\infty$ as $x \to y$, then the solution $Z_y(x)$ violates the "extended maximum principle" of Gilbarg and Serrin and is a distributional solution of $\mathcal{L}(x,\partial_x)Z_y(x) = 0$ for $|x-y| < \varepsilon$ although $Z_y(x) \to \infty$ as $x \to y$.

0. Introduction

Background. Consider an elliptic operator in nondivergence form,

$$\mathcal{L}(x,\partial_x)\,u(x) = a_{ij}(x)\,\partial_i\partial_j u(x), \tag{1}$$

where $\partial_i = \partial/\partial x_i$ and we have used the summation convention for repeated indices. The coefficients $a_{ij} = a_{ji}$ are real-valued functions defined on $\mathbb{R}^n$ for $n \geq 3$, and we denote the symmetric and positive definite matrix $(a_{ij}(x))$ by $\mathbf{A}_x$. (The case $n = 2$ can be treated with a similar analysis, but additional complications arise, which we have chosen to avoid here.) A *fundamental solution* for $\mathcal{L}$ in an open set U is a function $F(x,y)$ satisfying $F(x,\cdot) \in L^1_{\ell oc}(U)$ and

$$-\mathcal{L}(x,\partial_x)F(x,y) = \delta(x-y) \quad \text{for } x,y \in U \tag{2}$$

in a distributional sense that needs to be made clear; for this, some regularity of the coefficients will be required. If $F(x,y)$ satisfying (2) exists, then the operator

Key words and phrases. Fundamental solution, maximum principle, nondivergence form, elliptic operator, modulus of continuity, asymptotics, Dini condition, square-Dini condition.

The support of the first author by the UK Engineering and Physical Sciences Research Council via the grant EP/F005563/1 is gratefully acknowledged.

Q defined by

$$Q\phi(x) = -\int_U F(x,y)\,\phi(y)\,dy \tag{3}$$

provides a right-inverse for $\mathcal{L}$ on $C_0(U)$, the space of continuous functions with compact support in U.

In the "classical" case that the coefficient functions are λ-Hölder continuous in a bounded domain U for some $\lambda \in (0,1)$, it is well known (cf. [**21**]) that a fundamental solution exists in U and is asymptotic (as $x \to y$) to the fundamental solution for the constant coefficient operator obtained by freezing the coefficients a_{ij} at y: for $n \geq 3$ this means

$$F(x,y) = \tilde{F}_y(x-y)(1 + H(x,y)), \tag{4}$$

where, letting $\langle , \rangle$ denote the inner product in $\mathbb{R}^n$,

$$\tilde{F}_y(x) = \frac{\langle \mathbf{A}_y^{-1} x, x\rangle^{\frac{2-n}{2}}}{(n-2)\,|S^{n-1}|\,\sqrt{\det \mathbf{A}_y}} \tag{5}$$

is the fundamental solution for the constant coefficient operator $\mathcal{L}(y, \partial_x) = a_{ij}(y)\partial_i\partial_j$, and the remainder term $H(x,y)$ in (4) satisfies

$$|H(x,y)| + r|D_x H(x,y)| + r^2|D_x^2 H(x,y)| \leq c\, r^\lambda \quad \text{as} \quad r = |x-y| \to 0, \tag{6}$$

for all y in a compact subset of U. This fundamental solution may be used (cf. [**21**]) to show the classical regularity result: if $u \in C^2(U)$ and $\mathcal{L}u$ is λ-Hölder continuous in U, then $\partial_i\partial_j u$ is λ-Hölder continuous in U.

The Hölder continuity may be generalized by assuming the coefficients have a weaker modulus of continuity, i.e. $a_{ij} \in C^\omega(U)$, where $\omega(r)$ is a continuous, nondecreasing function for $0 \leq r < 1$ satisfying $\omega(0) = 0$, and

$$C^\omega(U) = \{f \in C(U) : |f(x) - f(y)| \leq c\,\omega(|x-y|) \quad \text{for } x, y \in U\}. \tag{7}$$

If ω satisfies the Dini condition at zero, i.e. $\int_0^1 \omega(t)t^{-1}\,dt < \infty$, then we say that the coefficients are *Dini continuous*. In this case, there are regularity results analogous to the case of Hölder continuity (cf. Proposition 1.14 in Chapter 3 of [**24**]); however, we could not find in the literature an asymptotic description of the fundamental solution such as (4) with estimates on the second-order derivatives $D_x^2 H(x,y)$.

Dini continuity is also essential for the "extended maximum principle" of Gilbarg and Serrin [**8**] to hold: a C^2-solution of

$$\mathcal{L}u \geq 0 \text{ for } 0 < |x| \leq r_0$$

with

$$u(x) = o(|x|^{2-n}) \quad \text{as} |x| \to 0$$

must satisfy

$$u(x) < M := \max\{u(y) : |y| = r_0\} \quad \text{for } 0 < |x| < r_0,$$

and $\limsup_{|x|\to 0} u(x) < M$. In fact, they give an example (which we will discuss in Section 2) in which the coefficients are not Dini continuous and the extended maximum principle fails.

The above regularity assumptions (Hölder or Dini continuity) on the coefficients are required to study the behavior of the fundamental solution as a function of x (for fixed y). If we instead fix x and consider the behavior in y, then regularity of the coefficients a_{ij} is not required; however, we cannot expect to achieve as precise an asymptotic description as (4). This is most conveniently described in terms of the *Green's function* for (1) on a smooth, bounded domain U, which may be defined (as in [**4**]) to be $G(x,\cdot) \in L^1_{\ell oc}(U)$ satisfying

$$\phi(x) = -\int_U G(x,y)\,\mathcal{L}(y,\partial_y)\,\phi(y)\,dy \text{ for } \phi \in C^2(\overline{U}) \text{ with } \phi = 0 \text{ on } \partial U. \tag{8}$$

Notice that (8) can be expressed formally as $-\mathcal{L}^*(y,\partial_y)G(x,y) = \delta(x-y)$ and implies that $Q\phi(x) = -\int_U G(x,y)\,\phi(y)\,dy$ defines a left-inverse for $\mathcal{L}$ on $C_0^2(U)$. When the a_{ij} are measurable, bounded, and uniformly elliptic in U, then the Green's function is known to exist; Fabes and Strook[**6**] showed that $G(x,\cdot) \in L^q(U)$ for some $q > n/(n-1)$, while Bauman [**1**], [**2**], [**3**], and Escauriaza [**4**] obtained pointwise estimates on $G(x,y)$ as $y \to x$ in terms of a nonnegative "adjoint solution" $W(y)$ which satisfies $\mathcal{L}^*(y,\partial_y)W(y) = 0$ in U. However, our paper is not concerned with such general coefficients, and for us a Green's function will also be a fundamental solution in the sense of (2).

Our results. In this paper, we allow the coefficients a_{ij} to be less regular than Dini continuous, and we want to study the solutions of $\mathcal{L}u(x) = 0$ with an isolated singularity at $x = y$, as well as the existence and asymptotics of a fundamental solution $F(x,y)$ satisfying (2) in an appropriate distributional sense. We assume that the coefficients have modulus of continuity ω satisfying the "square-Dini condition"

$$\int_0^1 \omega^2(t)\,\frac{dt}{t} < \infty. \tag{9}$$

Condition (9) has been encountered by other authors in different contexts: cf. [**5**], [**7**], [**11**], [**23**].

To construct our solution of (2), we first fix y and seek a solution of

$$\mathcal{L}(x,\partial_x)Z_y(x) = 0 \quad \text{for } x \in B_\varepsilon(y)\backslash\{y\}, \tag{10}$$

where $B_\varepsilon(y) = \{x : |x-y| < \varepsilon\}$ for ε sufficiently small, and $Z_y(x)$ has the appropriate singularity as $x \to y$. Assuming that the modulus of continuity at y satisfies (9), we shall construct a solution of (10) with the asymptotic description

$$Z_y(x) \sim \langle \mathbf{A}_y^{-1}(x-y),(x-y)\rangle^{\frac{2-n}{2}}\, e^{I(x,y)} \quad \text{as } x \to y, \tag{11}$$

where the factor $e^{I(x,y)}$ adjusts for lack of regularity in the coefficients: if the a_{ij} are Hölder continuous, then we can take $I(x,y) \equiv 0$ and $c_y\,Z_y(x)$ is asymptotic to $\tilde{F}_y(x-y)$ as $x \to y$. In general, however, we find that

$$I(x,y) = I_y\left(\sqrt{\langle \mathbf{A}_y^{-1}(x-y),(x-y)\rangle}\right), \tag{12}$$

where $I_y(r)$ is given by

(13)
$$\frac{1}{|S^{n-1}|}\int_{r<|z-y|<\varepsilon}\left(\operatorname{tr}(\mathbf{A}_z\mathbf{A}_y^{-1}) - n\,\frac{\langle \mathbf{A}_z\mathbf{A}_y^{-1/2}(z-y),\mathbf{A}_y^{-1/2}(z-y)\rangle}{|z-y|^2}\right)\frac{dz}{|z-y|^n}$$

with tr denoting the matrix trace. As $r \to 0$, $I_y(r)$ need not even be bounded, so the singularity of $Z_y(x)$ need not be $O(|x-y|^{2-n})$ as it was in the Hölder case. These formulas simplify significantly if we use an affine change of variables in which y corresponds to $x = 0$ and $a_{ij}(0) = \delta_{ij}$:

$$Z(x) \sim |x|^{2-n}e^{I(|x|)} \quad \text{as } |x| \to 0, \tag{14}$$

where

$$I(r) = \frac{1}{|S^{n-1}|} \int_{r<|z|<\varepsilon} \left(\mathrm{tr}(\mathbf{A}_z) - n \, \frac{\langle \mathbf{A}_z z, z \rangle}{|z|^2} \right) \frac{dz}{|z|^n}. \tag{15}$$

We can verify that the absolute value of the parenthetical term in (15) is bounded by $\omega(|z|)$, so the coefficients being Dini continuous implies that the improper integral defining $I(0)$ converges absolutely. Even if $I(0)$ does not converge, we shall see that for any $\lambda > 0$, there exists $C_\lambda > 0$ such that

$$|I(r)| \leq \lambda |\log r| + C_\lambda \quad \text{for } 0 < r < \varepsilon, \tag{16}$$

so the singularity of Z at $x = 0$ is never very far from $|x|^{2-n}$. Nevertheless, the behavior of $I(r)$ as $r \to 0$ plays an important role in the extended maximum principle and the existence of a fundamental solution. There are three principal cases to consider.

1. $I(0) = \lim_{r \to 0} I(r)$ exists and is finite.
In this case, we can scale $Z(x)$ by a constant multiple to obtain a solution that is asymptotic to the fundamental solution for the Laplacian. In fact, we can solve the distributional equation

$$-\mathcal{L}(x, \partial_x) Z(x) = C_0 \, \delta(x), \tag{17}$$

and find

$$C_0 = (n-2) \, |S^{n-1}| \, e^{I(0)}. \tag{18}$$

Note that the improper integral defining $I(0)$ may converge even if the modulus of continuity does *not* satisfy the Dini condition.

2. $I(r) \to -\infty$ as $r \to 0$.
We see that $Z(x) = o(|x|^{2-n})$ as $|x| \to 0$; we still have $Z(x) \to +\infty$ as $x \to 0$, so this violates the extended maximum principle of [**8**]. Nevertheless, we can solve (17) to find $C_0 = 0$, and this yields the surprising result that there is a distributional solution of

$$\mathcal{L} u = 0 \quad \text{in } B_\varepsilon \tag{19}$$

that is quite singular at $x = 0$.

3. $I(r) \to \infty$ as $r \to 0$.
Now we find $Z(x)|x|^{n-2} \to \infty$ as $|x| \to 0$, so this solution grows more rapidly than the fundamental solution for the Laplacian. However, we cannot solve (17) for C_0 in this case.

We next allow y to vary over U. Provided that we are in Case 1 at each $y \in U$, we can use $Z_y(x)$ to construct the fundamental solution $F(x, y)$ in U. Our main result (Theorem 3) states that, provided $a_{ij} \in C^\omega(U)$, where ω satisfies (9) and $I_y(0) = \lim_{r \to 0} I_y(r)$ exists and is finite for each $y \in U$, a fundamental solution $F(x, y)$ exists in the form (4), where the remainder term $H(x, y)$ may be estimated in L^p for any $p \in (1, \infty)$ in terms of ω and the rate of convergence $I_y(r) \to I_y(0)$. More specifically, let us assume that $\omega(r) \, r^{-1+\kappa}$ is nonincreasing for $0 < r < 1$, where $\kappa \in (0, 1)$, and introduce

$$\sigma(r) = \int_0^r \omega^2(t) \, \frac{dt}{t}. \tag{20}$$

Further, let us assume that

$$|I_y(r) - I_y(0)| \leq \theta(r) \quad \text{for all } y \in U, \tag{21}$$

where $\theta(r)$ is a nondecreasing function of r satisfying $\theta(0) = 0$. Then for y in a compact subset of U, we have

$$r^2 M_p(D^2 H(\cdot, y), r; y) \leq c \max(\omega(r), \sigma(r), \theta(r)) \quad \text{as } r \to 0, \tag{22}$$

where $M_p(f(\cdot, y), r; y)$ denotes the L^p-mean of $f(x, y)$ as a function of x (for fixed y) over the annulus $A_r(y) = \{x : r < |x - y| < 2r\}$. (In fact, when ω satisfies the Dini condition, the right-hand side of (22) reduces to just $c\,\omega(r)$.) Taking $p > n$, we obtain pointwise bounds on $|H(x, y)| + r|D_x H(x, y)|$, but we no longer have pointwise bounds on $D_x^2 H(x, y)$ as we did in (6) when the coefficients a_{ij} were Hölder continuous.

Note. In (22) and throughout this paper, c is used to denote a constant whose value may change line-by-line. It may depend upon n and the a_{ij}, but not on r.

Organization of this paper and comparison with other works. The organization of this paper is as follows. In Section 1 we discuss some preliminary estimates for solutions of the Poisson equation $\Delta u = f$. In Section 2 we construct the solution $Z_y(x)$ of (10) by first considering the case $a_{ij}(0) = \delta_{ij}$ and then performing a change of coordinates. In Section 3 we analyze the equation $-\mathcal{L}(x, \partial_x) Z_y(x) = C_y \delta(x - y)$ and calculate C_y when $I_y(0)$ is finite or $-\infty$. Finally, in Section 4 we apply the previous results to construct our fundamental solution $F(x, y)$ in the form (4) and obtain estimates on the remainder term $H(x, y)$. The analysis in each section makes use of L_p-means on annuli $A_r = \{x : r < |x| < 2r\}$ to measure the growth and decay of functions as $r \to 0$; these are defined in Section 1.

The results obtained and the techniques used in this paper are closely related to those in our previous paper [**17**]. In [**17**] we studied the asymptotics for solutions of the adjoint equation for $\mathcal{L}$ and discussed their relationship to the estimates obtained in [**4**] as $y \to x$. In fact, the condition that $I_y(r)$ is bounded (above and below) as $r \to 0$ not only allows one to conclude that the solution Z_y of (10) satisfies $c_y|x - y|^{2-n} \leq Z_y(x) \leq c'_y|x - y|^{2-n}$ as $x \to y$, but that $\mathcal{L}^*(x, \partial_x)u(x) = 0$ admits solutions in $B_\varepsilon(y)$ whose L_p-mean is bounded between positive constants as $r = |x - y| \to 0$; if $I_y(0)$ exists and is finite for every y, then this bounded solution of $\mathcal{L}^*(x, \partial_x)u(x) = 0$ is continuous.

Both [**17**] and the present work are independent of, but related to, the asymptotic theory developed in [**14**]. In particular, L_p-means were extensively used in [**12**] and [**13**]. The asymptotic formulas that we obtain are analogous to those of [**15**], where an asymptotic representation near the boundary was obtained for solutions to the Dirichlet problem for elliptic equations in divergence form with discontinuous coefficients. In particular, note that we do not make use of the maximum principle in this paper; in fact, most results of this paper hold for complex-valued coefficients, although this introduces some technical difficulties, which we have chosen to avoid.

1. Preliminaries

Throughout this paper, to measure the growth of functions as $x \to y$, it will be convenient to use the L^p-mean for some $p \in (1,\infty)$:

$$M_p(w,r;y) = \left(\fint_{A_r(y)} |w(x)|^p \, dx \right)^{1/p}, \tag{23}$$

where $A_r(y) = \{x : r < |x-y| < 2r\}$; here (and elsewhere in this paper) the slashed integral denotes mean value. It is natural to also introduce

$$M_\infty(w,r;y) = \sup_{A_r(y)} |w(x)|. \tag{24}$$

We may apply (23) and (24) to vector-valued or matrix-valued functions w with $|w|$ denoting the norm. We also need to estimate derivatives, so we define

$$M_{1,\infty}(w,r;y) = rM_\infty(Dw,r;y) + M_\infty(w,r;y), \tag{25}$$

where Dw represents the gradient of w, and for $p \in (1,\infty)$,

$$M_{2,p}(w,r;y) = r^2 M_p(D^2 w,r;y) + rM_p(Dw,r;y) + M_p(w,r;y), \tag{26}$$

where $D^2 w$ represents the Hessian matrix of w. Sobolev estimates show that

$$p > n \quad \Rightarrow \quad M_{1,\infty}(w,r;y) \le c\, M_{2,p}(w,r;y). \tag{27}$$

When $y = 0$, we shall abbreviate $M_p(w,r;0)$ as $M_p(w,r)$ (and similarly for $M_{1,\infty}$ and $M_{2,p}$).

For $x \in \mathbb{R}^n \backslash \{0\}$, let $\theta = x/|x| \in S^{n-1}$ and let $d\theta$ denote the standard surface measure on S^{n-1}. We will use the spherical mean of a function w:

$$\overline{w}(r) = \fint_{S^{n-1}} w(r\theta)\, d\theta. \tag{28}$$

In particular, in this section we consider the equation

$$\Delta v = f \qquad \text{in } \mathbb{R}^n \backslash \{0\} \tag{29}$$

when $\overline{f} = 0$ and investigate the behavior of the L^p-mean of the solution as $x \to 0$; our results are quite analogous to those of [**12**] and [**14**]. We shall let $\Gamma(|x|) = c_n |x|^{2-n}$ denote the fundamental solution for the Laplacian in $\mathbb{R}^n$.

PROPOSITION 1. *Suppose $n \ge 2$, $p \in (1,\infty)$, and $f \in L^p_{\ell oc}(\mathbb{R}^n \backslash \{0\})$ satisfies $\overline{f} = 0$,*

$$\int_{|x|<1} |x|\,|f(x)|\, dx < \infty, \quad \textit{and} \quad \int_{|x|>1} \frac{|f(x)|}{|x|^{n-1}}\, dx < \infty. \tag{30}$$

Then $v = Kf = \Gamma \star f$ defines a distribution solution of (29) that satisfies

$$M_{2,p}(Kf,r) \le c \left(r^2 \tilde{M}_p(f,r) + r^{1-n} \int_{|x|<r} |x|\,|f(x)|\, dx + r \int_{|x|>r} \frac{|f(x)|}{|x|^{n-1}}\, dx \right), \tag{31}$$

where we have introduced

$$\tilde{M}_p(w,r) := \left(\fint_{r/2<|x|<4r} |w(x)|^p \, dx \right)^{1/p}.$$

Elementary estimates can be used to show that

$$\int_{|x|<r} |g(x)|\,dx \le c\int_0^r M_p(g,\rho)\rho^{n-1}\,d\rho \tag{32}$$

with an analogous estimate for $|x|>r$. These estimates enable us to obtain the following corollary from Proposition 1.

COROLLARY 1. *Under the conditions of Proposition* 1,

$$M_{2,p}(Kf,r) \le c\left(r^{1-n}\int_0^r M_p(f,\rho)\rho^n\,d\rho + r\int_r^\infty M_p(f,\rho)\,d\rho\right). \tag{33}$$

Proof of Proposition 1. Let $v=Kf$ and let χ denote the characteristic function for the annulus $\tilde{A}_r = B_{4r}\backslash B_{r/2}$. Using $\overline{f}=0$, we see that

$$\begin{aligned} v(x) = \int_{\mathbb{R}^n} \Gamma(|x-y|)\chi(y)f(y)\,dy &+ \int_{|y|<r/2} (\Gamma(|x-y|)-\Gamma(|x|))f(y)\,dy \\ &+ \int_{|y|>4r} (\Gamma(|x-y|)-\Gamma(|y|))f(y)\,dy. \end{aligned}$$

We want to estimate $M_p(v,r)$, $rM_p(Dv,r)$, and $r^2M_p(D^2v,r)$, and show that each is bounded by the right-hand side of (31).

Using Stein's inequality [**22**], for $0<a<n/p$ and $0<b<n/p'$ with $a+b=2$ we have

$$\begin{aligned} \left\| \int_{\mathbb{R}^n} \Gamma(|x-y|)\chi(y)f(y)\,dy \right\|_{L^p(A_r)} &\le c\,r^a \|\chi(y)|y|^b f(y)\|_{L^p(\mathbb{R}^n)} \\ &= c\,r^a \||y|^b f(y)\|_{L^p(\tilde{A}_r)} \le c\,r^2\|f\|_{L^p(\tilde{A}_r)}. \end{aligned}$$

It is elementary to show that for $|y|<r/2$ and $r<|x|<2r$ we have $|\Gamma(|x-y|)-\Gamma(|x|)| \le c\,|x|^{1-n}|y|$, where c is independent of r, so for $x\in A_r$ we have

$$\left| \int_{|y|<r/2} (\Gamma(|x-y|)-\Gamma(|x|))f(y)\,dy \right| \le c\,|x|^{1-n}\int_{|y|<r/2} |f(y)||y|\,dy.$$

Similarly, we can show that for $|y|>4r$ and $r<|x|<2r$ we have $|\Gamma(|x-y|)-\Gamma(|y|)| \le c|x||y|^{1-n}$, so $x\in A_r$ implies that

$$\left| \int_{B_{4r}^c} (\Gamma(|x-y|)-\Gamma(|y|))f(y)\,dy \right| \le c\,|x|\int_{|y|>4r} \frac{|f(y)|}{|y|^{n-1}}\,dy.$$

Using these estimates, we easily conclude that $M_p(v,r)$ is bounded by the right-hand side of (31).

Next we consider

$$\partial_i v(x) = \int_{\mathbb{R}^n} \Gamma_i(x-y)\chi(y)f(y)\,dy + \int_{B_{r/2}} \Gamma_i(x-y)f(y)\,dy + \int_{B_{4r}^c} \Gamma_i(x-y)f(y)\,dy,$$

where

$$\Gamma_i(x) = \Gamma'(|x|)\frac{x_i}{|x|}.$$

Applying Stein's inequality as above but with $a+b=1$, we conclude that

$$\left\| \int_{\mathbb{R}^n} \Gamma_i(x-y)\chi(y)f(y)\,dy \right\|_{L^p(A_r)} \le c\,r\,\|f\|_{L^p(\tilde{A}_r)}.$$

Elementary estimates for $r < |x| < 2r$ show that

$$\left| \int_{B_{r/2}} \Gamma_i(x-y) f(y)\, dy \right| = \left| \int_{B_{r/2}} (\Gamma_i(x-y) - \Gamma_i(x)) f(y)\, dy \right|$$
$$\leq c\,|x|^{-n} \int_{|y|<r/2} |f(y)||y|\, dy$$

and

$$\left| \int_{B^c_{4r}} \Gamma_i(x-y) f(y)\, dy \right| = \left| \int_{B^c_{4r}} (\Gamma_i(x-y) - \Gamma_i(y)) f(y)\, dy \right| \leq c \int_{|y|>4r} \frac{|f(y)|}{|y|^{n-1}}\, dy.$$

From these estimates we easily conclude that $rM_p(Dv, r)$ is bounded by the right-hand side of (31).

Finally, we consider

$$\partial_i \partial_j v(x) = \int_{\mathbb{R}^n} \Gamma_{ij}(x-y)\chi(y) f(y)\, dy + \int_{B_{r/2}} \Gamma_{ij}(x-y) f(y)\, dy + \int_{B^c_{4r}} \Gamma_{ij}(x-y) f(y)\, dy,$$

where Γ_{ij} is the singular kernel given by

$$\Gamma_{ij}(x) = \Gamma''(|x|) \frac{x_i x_j}{|x|^2} + \Gamma'(|x|) \frac{\delta_{ij}|x|^2 - x_i x_j}{|x|^3}.$$

Using the L^p-boundedness of singular integral operators, we conclude that

$$\left\| \int_{\mathbb{R}^n} \Gamma_{ij}(x-y)\chi(y) f(y)\, dy \right\|_{L^p(A_r)} \leq c\,\|f\|_{L^p(\tilde{A}_r)}.$$

Elementary estimates for $r < |x| < 2r$ show that

$$\left| \int_{B_{r/2}} \Gamma_{ij}(x-y) f(y)\, dy \right| = \left| \int_{B_{r/2}} (\Gamma_{ij}(x-y) - \Gamma_{ij}(x)) f(y)\, dy \right|$$
$$\leq c\,|x|^{-n-1} \int_{|y|<r/2} |f(y)||y|\, dy$$

and

$$\left| \int_{B^c_{4r}} \Gamma_{ij}(x-y) f(y)\, dy \right| = \left| \int_{B^c_{4r}} (\Gamma_{ij}(x-y) - \Gamma_{ij}(|y|)) f(y)\, dy \right|$$
$$\leq c\,|x|^{-1} \int_{|y|>4r} \frac{|f(y)|}{|y|^{n-1}}\, dy.$$

These estimates show that $r^2 M_p(D^2 v, r)$ is bounded by the right-hand side of (31). This completes the proof. $\square$

2. Constructing the Singular Solution $Z_y(x)$ in $B_\varepsilon(y)\backslash\{y\}$

In this section, we fix y and construct a solution of (10) for ε sufficiently small. Since the result is localized near y, the ellipticity and continuity of the coefficients of $\mathcal{L}$ need only be assumed at y. In fact, we first consider the case when $y = 0$ and $a_{ij}(0) = \delta_{ij}$:

(34) $$\sup_{|x|=r} \|\mathbf{A}_x - \mathbf{I}\| \leq \omega(r) \quad \text{for } 0 < r < 1,$$

where $\mathbf{I}$ is the identity matrix and ω is a continuous, nondecreasing function for $0 \leq r < 1$ satisfying the square-Dini condition (9). For convenience, we shall also assume the monotonicity condition that for some $\kappa \in (0,1)$ we have

$$\omega(r)\, r^{-1+\kappa} \quad \text{is nonincreasing for } 0 < r < 1. \tag{35}$$

The significance of (35) is that it requires $\omega(r)$ to vanish more slowly than r as $r \to 0$, which is a natural constraint. (As long as $\omega(r)\, r^{-1+\kappa}$ is nonincreasing for $0 < r < \varepsilon$ with some $\varepsilon > 0$, then ω may be modified for $\varepsilon < r < 1$ to satisfy (35).) We seek a solution of

$$\mathcal{L}(x,\partial_x)\, Z(x) = 0 \quad \text{for } x \in B_\varepsilon \backslash \{0\}, \tag{36}$$

where $B_\varepsilon = \{x : |x| < \varepsilon\}$ with ε sufficiently small. Although we generally assume that $\mathcal{L}$ has real-valued coefficients, the theorem below holds when the a_{ij} are complex-valued.

THEOREM 1. *For $n \geq 3$ and $p \in (1,\infty)$, suppose the coefficients $a_{ij}(x)$ are bounded, measurable functions satisfying (34). For $\varepsilon > 0$ sufficiently small, there exists a solution of (36) in the form*

$$Z(x) = h(|x|) + v(x), \tag{37}$$

where h is of the form

$$h(r) = \int_r^\varepsilon s^{1-n}\, e^{I(s)}\, ds\, (1 + \zeta(r)), \tag{38}$$

with $I(r)$ given by (15) and

$$M_{2,p}(\zeta, r) \leq c \max(\omega(r), \sigma(r)), \tag{39}$$

where σ is given in (20), and v in (37) satisfies

$$M_{2,p}(v,r) \leq c\, r^{2-n}\, e^{I(r)}\, \omega(r). \tag{40}$$

Moreover, for any $u \in W^{2,p}_{\ell oc}(\overline{B_\varepsilon} \backslash \{0\})$ that is a strong solution of $\mathcal{L}(x,\partial_x)u = 0$ in $\overline{B_\varepsilon} \backslash \{0\}$ subject to the growth condition

$$M_{2,p}(u,r) \leq c\, r^{1-n+\varepsilon_0}, \quad \textit{where } \varepsilon_0 > 0, \tag{41}$$

there exist constants $C, C_0, C_1, \ldots, C_n$ (depending on u) such that

$$u(x) = C Z(x) + C_0 + \sum_{j=1}^{n} C_j x_j + w(x), \tag{42}$$

where w satisfies

$$M_{2,p}(w,r) \leq c\, r^{2-\varepsilon_1} \quad \textit{for any } \varepsilon_1 > 0. \tag{43}$$

We shall prove this theorem below, but first let us make some observations. In general, we do not know whether $I(r)$ is bounded as $r \to 0$, but we can verify that $|I'(r)| \leq c\, r^{-1} \omega(r)$, so integration by parts in (38) shows that

$$h(r) = \frac{r^{2-n}}{n-2}\, e^{I(r)} + h_1(r), \tag{44}$$

where $h_1(r)$ satisfies $M_{1,\infty}(h_1, r) \leq c\, r^{2-n} e^{I(r)} \max(\omega(r), \sigma(r))$. If we take $p > n$ and apply (27) to v, we conclude that

$$(45) \qquad Z(x) = \frac{|x|^{2-n} e^{I(|x|)}}{n-2} (1 + \xi(x)) \quad \text{as } |x| \to 0,$$

where $M_{1,\infty}(\xi, r) \leq c \max(\omega(r), \sigma(r))$. Obviously, we can multiply the Z of (45) by $n-2$ to obtain the Z of (14).

Even when $I(r)$ is not bounded as $r \to 0$, we can derive useful bounds on $Z(x)$ as $|x| \to 0$. It is not difficult to verify that a symmetric matrix $\mathbf{A}$ satisfies

$$(46) \quad -2(n-1)\|\mathbf{A} - \mathbf{I}\| \leq \operatorname{tr}(\mathbf{A}) - n\langle \mathbf{A}y, y\rangle\, |y|^{-2} \leq 2(n-1)\|\mathbf{A} - \mathbf{I}\| \quad \text{for } |y| = 1,$$

so there exist constants $c, C > 0$ so that Z satisfies (assuming $n \geq 3$)

$$(47) \quad c\,|x|^{2-n} \exp\left(-c_n \int_{|x|}^{\varepsilon} \omega(t) \frac{dt}{t}\right) \leq |Z(x)| \leq C\,|x|^{2-n} \exp\left(c_n \int_{|x|}^{\varepsilon} \omega(t) \frac{dt}{t}\right)$$

as $x \to 0$, where $c_n = 2(n-1)/|S^{n-1}|$. Using (47) and the fact that $\omega(r) \to 0$ as $r \to 0$, we obtain (16), which shows that the singularity of Z at $x = 0$ is very close to the classical case.

An interesting class of examples is obtained by letting

$$(48) \qquad a_{ij}(x) = \delta_{ij} + g(|x|) x_i x_j |x|^{-2},$$

where $g(0) = 0$ but vanishes slowly as $r \to 0$. Gilbarg and Serrin [**8**] used (48) with certain specific functions g to show that Dini continuity is essential for their extended maximum principle to hold. In our formulation,

$$(49) \qquad \operatorname{tr}(\mathbf{A}_z) - n \frac{\langle \mathbf{A}_z z, z\rangle}{|z|^2} = (1-n) g(|z|),$$

so

$$(50) \qquad I(r) = (1-n) \int_r^{\varepsilon} g(\rho) \frac{d\rho}{\rho}.$$

Thus any $g(r) > 0$ which does *not* satisfy the Dini condition at $r = 0$ (but does satisfy the square-Dini condition) will yield $I(r) \to -\infty$ as $r \to 0$, so the $Z(x)$ of Theorem 1 gives an example of a solution of (36) with singularity at $x = 0$ even though $Z(x) = o(|x|^{2-n})$ as $|x| \to 0$; i.e., the extended maximum principle fails. The specific function in [**8**] is $g(r) = -(1 + (n-1)\log r)^{-1}$.

Proof of Theorem 1. Instead of showing the existence of Z in a very small ball B_ε, we shall replace the condition that ω satisfies the square-Dini condition by

$$(51) \qquad \sigma(1) = \int_0^1 \omega^2(t) \frac{dt}{t} < \delta,$$

where δ is sufficiently small, and show existence in the unit ball B_1. In fact, using (51) and (35), we see that

$$(52) \qquad \omega(r) < c_{\kappa,n} \sqrt{\delta} \quad \text{for } 0 < r < 1,$$

where $c_{\kappa,n}$ depends only on κ and n:

$$\delta > \int_{r/2}^{r} \omega^2(t) \frac{dt}{t} \geq \omega^2(r)\, r^{-2+2\kappa} \int_{r/2}^{r} t^{1-2\kappa}\, dt = \omega^2(r)\, c'_{\kappa,n}.$$

Moreover, it will be useful to consider $\mathcal{L}$ as defined on all of $\mathbb{R}^n$ with $\mathcal{L} = \Delta$ outside of B_1. Therefore, we shall assume that

$$a_{ij}(x) = \delta_{ij} \qquad \text{for } |x| > 1, \tag{53}$$

and investigate a solution of $\mathcal{L}Z = 0$ in $\mathbb{R}^n \backslash \{0\}$.

To construct $Z(x)$, we let $h(r) = \overline{Z}(r)$ denote the spherical mean as in (28), and let $v(x) = Z(x) - h(|x|)$, so $\overline{v}(r) = 0$. We shall reduce the problem to solving an operator equation of the form $(I + S + T)v = f$, where S and T have small operator norm on a Banach space X defined as follows: for fixed $p \in (1, \infty)$, let us consider the functions v in $W^{2,p}_{\ell oc}(\mathbb{R}^n \backslash \{0\})$ for which the norm

$$\|v\|_X = \sup_{0<r<1} \frac{M_{2,p}(v,r) r^{n-2}}{\omega(r) e^{I(r)}} + \sup_{r>1} \frac{M_{2,p}(v,r)\, r^{n-1}}{\sqrt{\delta}} \tag{54}$$

is finite, and take the closure to form a Banach space X. The reduction to the operator equation will take several steps.

If we take the spherical mean of $\mathcal{L}(h+v) = 0$, we obtain an ordinary differential equation for h:

$$\alpha(r) h'' + \frac{\alpha_n(r) - \alpha(r)}{r} h' + \overline{a_{ij}\partial_i\partial_j v}(r) = 0,$$

where

$$\alpha_n(r) = \fint_{S^{n-1}} a_{ii}(r\theta)\, d\theta \quad \text{and} \quad \alpha(r) = \fint_{S^{n-1}} a_{ij}(r\theta)\theta_i\theta_j\, d\theta. \tag{55}$$

From (34) it is clear that these functions satisfy

$$|\alpha_n(r) - n| \le \omega(r) \text{ and } |\alpha(r) - 1| \le \omega(r) \text{ for } 0 < r < 1. \tag{56}$$

Using the fact that $\overline{\Delta v} = \Delta \overline{v} = 0$, the equation for h becomes

$$\alpha(r) h'' + \frac{\alpha_n(r) - \alpha(r)}{r} h' + \overline{\beta_{ij}\partial_i\partial_j v}(r) = 0, \tag{57}$$

where $\beta_{ij}(x) = a_{ij}(x) - \delta_{ij}$ satisfies $|\beta_{ij}(x)| \le \omega(|x|)$. Notice that v satisfies the partial differential equation

$$-\Delta v = \beta_{ij}\partial_i\partial_j h - \overline{\beta_{ij}\partial_i\partial_j h} + \beta_{ij}\partial_i\partial_j v - \overline{\beta_{ij}\partial_i\partial_j v}. \tag{58}$$

We shall first solve (57) and use that to eliminate h from (58); then we will be able to apply the results of Section 1.

Let us introduce $g = h'$ and rewrite (57) as

$$g' + \frac{n - 1 + R(r)}{r} g = B[D^2 v](r), \tag{59}$$

where

$$R(r) = \frac{\alpha_n(r)}{\alpha(r)} - n \tag{60}$$

and

$$B[D^2 v](r) = -\alpha^{-1}(r)\overline{\beta_{ij}\partial_i\partial_j v}(r).$$

Notice that

$$|R(r)| \le c\,\omega(r) \text{ for } 0 < r < 1 \quad \text{and} \quad R(r) = 0 \text{ for } r > 1, \tag{61}$$

and

$$|B[D^2 v](r)| \le c\,\omega(r) |\overline{D^2 v}(r)| \text{ for } 0 < r < 1 \quad \text{and} \quad B[D^2 v](r) = 0 \text{ for } r > 1.$$

Moreover, the monotonicity of $\omega(r)$ together with (35) implies that

$$\max_{r\leq\rho\leq 2r} \omega(\rho) \leq c\,\omega(r), \tag{62}$$

so we consequently obtain

$$M_p(B[D^2v],r) \leq c\,\omega(r) M_p(D^2v,r) \quad \text{for } 0<r<1. \tag{63}$$

To solve (59), let us introduce

$$E_\pm(r) = \exp\left[\pm\int_r^\infty R(t)\,\frac{dt}{t}\right] = \exp\left[\pm\int_r^1 R(t)\,\frac{dt}{t}\right] = \frac{1}{E_\mp(r)} \tag{64}$$

and observe that $E_\pm(r)\equiv 1$ for $r>1$. It is useful to observe that

$$E_-(r)E_+(\rho) = \exp\left(\int_\rho^r R(t)\,\frac{dt}{t}\right),$$

so as a consequence of (61) and (52), we obtain

$$\left(\frac{\rho}{r}\right)^{c\sqrt{\delta}} \leq \exp\left(\pm\int_\rho^r R(t)\,\frac{dt}{t}\right) \leq \left(\frac{r}{\rho}\right)^{c\sqrt{\delta}} \qquad \text{for } 0<\rho\leq r\leq 1. \tag{65}$$

In particular,

$$c_1\,E_\pm(r) \leq E_\pm(\rho) \leq c_2\,E_\pm(r) \quad \text{for } r<\rho<2r, \tag{66}$$

and for any $g\in L^p_{\ell oc}(\mathbb{R}^n\backslash\{0\})$ we can readily verify that

$$M_p(|x|^\nu\,E_\pm(|x|)\,g(x),r) \leq c\,r^\nu\,E_\pm(r)\,M_p(g,r), \tag{67}$$

for any fixed $\nu\in\mathbb{R}$.

It will be more convenient for us to use $E_\pm(r)$ than $e^{\pm I(r)}$, but these functions are equivalent: if we note that (15) can be written as

$$I(r) = \int_r^1 [\alpha_n(\rho) - n\alpha(\rho)]\,\frac{d\rho}{\rho}, \tag{68}$$

then we see that

$$E_+(r) = A\,e^{I(r)}(1+\tau(r)), \tag{69}$$

where $A=\exp[\int_0^1 R(\rho)[1-\alpha(\rho)]\rho^{-1}d\rho]$ is finite and positive, and

$$\tau(r) = \exp\left[-\int_0^r R(\rho)(1-\alpha(\rho))\,\frac{d\rho}{\rho}\right] - 1 \tag{70}$$

satisfies $|\tau(r)|\leq c\,\sigma(r)$. Thus for some constants c_1,c_2 we have

$$c_1E_+(r) \leq e^{I(r)} \leq c_2E_+(r) \quad \text{for } 0<r<1. \tag{71}$$

Now if we introduce $\phi(r)=r^{n-1}E_-(r)g(r)$, then we can rewrite (59) as

$$\phi'(r) = r^{n-1}E_-(r)B[D^2v](r). \tag{72}$$

But (72) may be integrated to obtain

$$\phi(r) = \phi(0) + \int_0^r \rho^{n-1}E_-(\rho)B[D^2v](\rho)\,d\rho, \tag{73}$$

where $\phi(0)$ is an arbitrary constant. Of course, to conclude (73), we must verify that ϕ' is integrable on $(0,1)$. But $v \in X$ implies $M_p(D^2v,r) \leq c\,\omega(r)r^{-n}E_+(r)$, so we can use (62), (63), (66), and Hölder's inequality to obtain

$$\left|\int_r^{2r} \rho^{n-1}E_-(\rho)B[D^2v](\rho)\,d\rho\right| \leq c\,E_-(r)\omega(r)\int_{r<|x|<2r}|D^2v(x)|dx \leq c\,\omega^2(r).$$

Now if we write

$$\int_0^s \rho^{n-1}E_-(\rho)B[D^2v](\rho)\,d\rho = \sum_{j=0}^{\infty}\int_{s/2^{j+1}}^{s/2^j}\rho^{n-1}E_-(\rho)B[D^2v](\rho)\,d\rho,$$

then we obtain the estimate

$$\begin{aligned}\left|\int_0^s \rho^{n-1}E_-(\rho)B[D^2v](\rho)\,d\rho\right| &\leq c\sum_{j=0}^{\infty}\omega^2\left(\frac{s}{2^{j+1}}\right)\\ &\leq c\int_0^s \omega^2(\rho)\,\frac{d\rho}{\rho} = c\,\sigma(s) < c\,\delta.\end{aligned}\tag{74}$$

In particular, we see that ϕ' is integrable and so (73) is valid. This enables us to write

$$h'(r) = g(r) = r^{1-n}E_+(r)\left[\phi(0) + \int_0^r \rho^{n-1}E_-(\rho)B[D^2v](\rho)\,d\rho\right]\tag{75}$$

and

$$h''(r) = \frac{1-n-R(r)}{r^n}E_+(r)\left[\phi(0) + \int_0^r \rho^{n-1}E_-(\rho)B[D^2v](\rho)\,d\rho\right] + B[D^2v](r).\tag{76}$$

We can now use (75) and (76) to express

$$\begin{aligned}\beta_{ij}\partial_i\partial_j h = r^{-n}E_+(r)&\left[\phi(0) + \int_0^r \rho^{n-1}E_-(\rho)B[D^2v](\rho)d\rho\right]\psi(r\theta)\\ &+B[D^2v](r)\,\beta_{ij}\theta_i\theta_j,\end{aligned}$$

where

$$\psi(r\theta) = \beta_{ii}(r\theta) - (n+R(r))\beta_{ij}(r\theta)\theta_i\theta_j\tag{77}$$

also satisfies $|\psi(r\theta)| \leq c\,\omega(r)$. Thus we can apply $K = \Gamma\star$ to (58) to obtain

$$v + Sv + Tv = \phi(0)w,\tag{78}$$

where

$$w(x) = K_{y\to x}\left[|y|^{-n}E_+(|y|)(\psi(y) - \overline{\psi}(|y|)\right],$$

$$Sv(x) = -K_{y\to x}\left[|y|^{-n}E_+(|y|)\int_0^{|y|}\rho^{n-1}E_-(\rho)B[D^2v](\rho)\,d\rho\,(\psi(y)-\overline{\psi}(|y|)\right],$$

and

$$Tv = -K\left[B[D^2v]\,(\beta_{ij}\theta_i\theta_j - \overline{\beta_{ij}\theta_i\theta_j}) + \beta_{ij}\partial_i\partial_j v - \overline{\beta_{ij}\partial_i\partial_j v}\right].$$

For a given value of $\phi(0)$, we can solve (78) provided we show that $w \in X$ and both $S : X \to X$ and $T : X \to X$ have small operator norms.

To show $w \in X$, we must estimate $M_{2,p}(w,r)$ for $0 < r < 1$ and for $r > 1$. For $0 < r < 1$ apply Proposition 2 to $f(x) = |x|^{-n} E_+(|x|)(\psi(x) - \overline{\psi}(|x|))$ (which vanishes for $|x| > 1$) to conclude that

$$M_{2,p}(w,r) \leq c \left(r^{1-n} \int_0^r E_+(\rho)\omega(\rho)\, d\rho + r \int_r^1 E_+(\rho)\omega(\rho)\rho^{-n}\, d\rho \right).$$

Using (65) and the fact that $r^{1-\kappa}\omega(r)$ is nondecreasing (since both $r^{1-\kappa}$ and ω are), we find

$$\begin{aligned}\int_0^r E_+(\rho)\omega(\rho)\, d\rho &\leq E_+(r)\, r^{c\sqrt{\delta}} \int_0^r \rho^{-c\sqrt{\delta}}\omega(\rho)\, d\rho \\ &\leq E_+(r)\,\omega(r)\, r^{c\sqrt{\delta}+1-\kappa} \int_0^r \rho^{-c\sqrt{\delta}-1+\kappa}\, d\rho.\end{aligned}$$

Taking δ small enough that $\kappa - \sqrt{c\delta} > 0$, we obtain

$$\int_0^r E_+(\rho)\omega(\rho)\, d\rho \leq c\, r\, E_+(r)\,\omega(r). \tag{79}$$

Using (65) and the fact that $r^{-1+\kappa}\,\omega(r)$ is nonincreasing (by (35)), we find

$$\begin{aligned}\int_r^1 E_+(\rho)\omega(\rho)\rho^{-n}\, d\rho &\leq E_+(r)\, r^{-c\sqrt{\delta}} \int_r^1 \rho^{c\sqrt{\delta}-n}\omega(\rho)\, d\rho \\ &\leq E_+(r)\,\omega(r)\, r^{-c\sqrt{\delta}-1+\kappa} \int_r^1 \rho^{c\sqrt{\delta}+1-\kappa-n}\, d\rho.\end{aligned}$$

For δ small enough that $n - 2 + \kappa - c\sqrt{\delta} > 0$, we have

$$\int_r^1 E_+(\rho)\omega(\rho)\, \rho^{-n}\, d\rho \leq c\, r^{1-n}\, E_+(r)\,\omega(r). \tag{80}$$

Using (79) and (80), we obtain

$$\frac{M_{2,p}(w,r) r^{n-2}}{\omega(r)E_+(r)} \leq c \quad \text{for all} \quad 0 < r < 1.$$

We can then use (71) to replace $E_+(r)$ by $e^{I(r)}$ as required in the norm for X in (54). Meanwhile, for $r > 1$ we use $\omega(1) \leq c\sqrt{\delta}$ and $E_+(\rho) \leq \rho^{-c\sqrt{\delta}}$ for $0 < \rho < 1$ to find

$$M_{2,p}(w,r) \leq c\, r^{1-n} \int_0^1 E_+(\rho)\omega(\rho)\, d\rho \leq c\sqrt{\delta}\, r^{1-n} \int_0^1 \rho^{-c\sqrt{\delta}}\, d\rho \leq c\sqrt{\delta}\, r^{1-n},$$

provided δ is sufficiently small. Consequently,

$$\frac{M_{2,p}(w,r)\, r^{n-1}}{\sqrt{\delta}} \leq c \quad \text{for all } r > 1,$$

and this confirms that $w \in X$.

Next let us show that S maps X to itself with small operator norm. We suppose that $\|v\|_X \leq 1$ and estimate $M_{2,p}(Sv,r)$ separately for $0 < r < 1$ and for $r > 1$. For $0 < r < 1$, we examine the proof of (74) and observe that the condition $\|v\|_X \leq 1$ enables us to choose the constant c to be independent of v. Thus the function

$$f_1(y) = |y|^{-n} E_+(|y|) \int_0^{|y|} \rho^{n-1} E_-(\rho)\, B[D^2 v](\rho)\, d\rho\, (\psi(y) - \overline{\psi}(|y|))$$

satisfies

$$M_p(f_1, r) \leq c\,\delta\, E_+(r)\,\omega(r)\, r^{-n} \text{ for } 0 < r < 1$$

and $M_p(f_1,r)=0$ for $r>1$. For $0<r<1$, we apply Proposition 2 to $Sv=-Kf_1$ to obtain

$$M_{2,p}(Sv,r)\le c\,\delta\left(r^{1-n}\int_0^r E_+(\rho)\,\omega(\rho)\,d\rho + r\int_r^1 E_+(\rho)\,\omega(\rho)\,\rho^{-n}\,d\rho\right),$$

and then use (79) and (80) to conclude (for δ sufficiently small) that

$$\frac{M_{2,p}(Sv,r)r^{n-2}}{\omega(r)E_+(r)}\le c\,\delta \quad \text{for all } \ 0<r<1.$$

On the other hand, for $r>1$, Proposition 2 implies (for δ sufficiently small) that

$$M_{2,p}(Sv,r)\le c\,\delta\, r^{1-n}\int_0^1 E_+(\rho)\,\omega(\rho)\,d\rho\le c\,\delta^{3/2}\,r^{1-n}\int_0^1 \rho^{-c\sqrt{\delta}}\,d\rho\le c\,\delta^{3/2}\,r^{1-n}.$$

Thus we have

$$\frac{M_{2,p}(Sv,r)r^{n-1}}{\sqrt{\delta}}\le c\,\delta \quad \text{for all } \ r>1.$$

Combining these inequalities, we see that $S:X\to X$ has a small operator norm.

Finally, we show that T maps X to itself with a small operator norm. We suppose that $\|v\|_X\le 1$ and estimate $M_{2,p}(Tv,r)$ separately for $0<r<1$ and for $r>1$. Notice that the function

$$f_2=B[D^2v]\,(\beta_{ij}\theta_i\theta_j-\overline{\beta_{ij}\theta_i\theta_j})$$

satisfies

$$M_p(f_2,r)\le\omega(r)M_p(B[D^2v],r)\le c\,\omega^3(r)\,E_+(r)\,r^{-n} \text{ for } 0<r<1,$$

where c is independent of v, and $M_p(f_2,r)=0$ for $r>1$. Similarly, the function

$$f_3=\beta_{ij}\partial_i\partial_j v-\overline{\beta_{ij}\partial_i\partial_j v}$$

satisfies

$$M_p(f_3,r)\le\omega(r)M_p(D^2v,r)\le\omega^2(r)\,E_+(r)\,r^{-n} \text{ for } 0<r<1,$$

and $M_p(f_3,r)=0$ for $r>1$. For $0<r<1$, we apply Proposition 2 to $Tv=-K(f_2+f_3)$ to obtain

$$M_{2,p}(Tv,r)\le c\left(r^{1-n}\int_0^r\omega^2(\rho)\,E_+(\rho)\,d\rho+r\int_r^1\omega^2(\rho)\,E_+(\rho)\rho^{-n}\,d\rho\right).$$

Using (52), (79), and (80),

$$\frac{M_{2,p}(Tv,r)\,r^{n-2}}{\omega(r)E_+(r)}\le c\sqrt{\delta}\quad\text{for all } 0<r<1.$$

On the other hand, for $r>1$, we use (52) and (65) to estimate

$$M_{2,p}(Tv,r)\le c\,r^{1-n}\int_0^1\omega^2(\rho)\,E_+(\rho)\,d\rho\le c\,\delta\,r^{1-n}\int_0^1\rho^{-c\sqrt{\delta}}\,d\rho\le c\,\delta\,r^{1-n}.$$

Consequently,

$$\frac{M_{2,p}(Tv,r)\,r^{n-1}}{\sqrt{\delta}}\le c\sqrt{\delta}\quad\text{for all } r>1.$$

Combining these estimates, we see that $T:X\to X$ has a small operator norm.

Since both S and T have small operator norms on X, we conclude that (78) has a unique solution v, depending on the choice of the constant $c_1 = \phi(0)$. But once c_1 and v are known, we obtain $g(r)$ from (75), and $h(r)$ by integration of $g(r)$:

$$h(r) = \int_r^\infty s^{1-n} E_+(s) \left[c_1 + \int_0^s \rho^{n-1} E_-(\rho) B[D^2 v](\rho)\, d\rho \right] ds + c_2, \tag{81}$$

where c_2 is an arbitrary constant. To obtain the desired solution of Theorem 1, we choose c_1 to enable us to replace $E_+(r)$ by $e^{I(r)}$ for $0 < r < 1$. Using (69) we see that we should choose $c_1 = A^{-1}$ and write $h(r) = h_0(r) + h_1(r) + c$, where

$$h_0(r) = \int_r^1 s^{1-n} e^{I(s)}\, ds \tag{82}$$

and (recalling τ from (70))

$$h_1(r) = \int_r^1 s^{1-n} e^{I(s)} \tau(s) ds + \int_r^1 s^{1-n} E_+(s) \int_0^s \rho^{n-1} E_-(\rho) B[D^2 v](\rho) d\rho ds. \tag{83}$$

Now integrate by parts to obtain

$$h_0(r) = \frac{r^{2-n}}{n-2} e^{I(r)} + c + \frac{1}{n-2} \int_r^1 s^{2-n} e^{I(s)} I'(s)\, ds.$$

But $|I'(s)| \le c\,\omega(s)/s$ and, similar to (65), we can show that

$$e^{I(s)} e^{-I(r)} \le \left(\frac{s}{r}\right)^{c\sqrt{\delta}} \qquad \text{for } s > r,$$

so we may use (35) to obtain

$$\begin{aligned} \left| \int_r^1 s^{2-n} e^{I(s)} I'(s)\, ds \right| &\le c \int_r^1 s^{1-n} e^{I(s)} \omega(s)\, ds \le c\, r^{-c\sqrt{\delta}} e^{I(r)} \int_r^1 s^{1-n+c\sqrt{\delta}} \omega(s)\, ds \\ &\le c\,\omega(r) e^{I(r)} r^{-c\sqrt{\delta}-1+\kappa} [r^{3-n+c\sqrt{\delta}-\kappa} + 1] \le c\, r^{2-n} \omega(r) e^{I(r)} \end{aligned}$$

provided δ is sufficiently small. Thus we find

$$\left| h_0(r) - \frac{r^{2-n}}{n-2} e^{I(r)} \right| \le c\, r^{2-n} e^{I(r)} \omega(r) \qquad \text{for } 0 < r < 1. \tag{84}$$

To estimate h_1 we use $|\tau(s)| \le c\,\sigma(s)$ together with (74) and a similar analysis to the above to obtain

$$|h_1(r)| \le c\, r^{2-n} e^{I(r)} \max(\omega(r), \sigma(r)) \qquad \text{for } 0 < r < 1. \tag{85}$$

Define $\zeta(r)$ by

$$\zeta(r) = \frac{h_1(r)}{h_0(r)} \quad \text{for } 0 < r < 1. \tag{86}$$

Using (82)–(85) we can estimate $|\zeta(r)|, |r\zeta'(r)| \le c\,\max(\omega(r), \sigma(r))$. To estimate ζ'', we write $h_0 \zeta'' = h_1'' - h_0'' \zeta - 2h_0' \zeta'$, where

$$h_0''(r) = (n-1)\, r^{-n} e^{I(r)} - r^{1-n} e^{I(r)} I'(r) \tag{87}$$

and

$$\begin{aligned} h_1''(r) &= r^{-n} e^{I(r)} [(n-1)\tau(r) - rI'(r)\tau(r) - r\tau'(r)] \\ &\quad + r^{-n} E_+(r)(n-1+R(r)) \int_0^r \rho^{n-1} R(\rho) B[D^2 v](\rho) d\rho - B[D^2 v](r). \end{aligned} \tag{88}$$

The terms $h_0''\zeta$ and $2h_0'\zeta'$ may be estimated pointwise as before, but $h_1''(r)$ involves the term $B[D^2v](r)$, which cannot be estimated pointwise. However, from (63) and $v \in X$ we conclude that $M_p(r^2\zeta'', r) \leq c \max(\omega(r), \sigma(r))$. Putting this together with the lower-order derivatives, we obtain the desired estimate (39). Summarizing so far, we have found a solution Z of (36) in the desired form (37).

Next we need to verify that any strong solution $u \in W^{2,p}_{\ell oc}(\overline{B_1}\backslash\{0\})$ of $\mathcal{L}u = 0$ that satisfies the growth estimate (41) must be of the form (42). To do this, we shall invoke well-known results for weighted Sobolev spaces. To begin with, let us introduce the weighted L^p-norm on $B_\circ = B_1\backslash\{0\}$:

$$\|u\|^p_{L^p_\beta(B_\circ)} = \int_{0<|x|<1} |x|^{\beta p}\,|u(x)|^p\,dx. \tag{89}$$

To relate this to the M_p-norm, notice that

$$\int_{0<|x|<1} |x|^{\beta p}\,|u(x)|^p\,dx = \sum_{k=1}^{\infty} \int_{2^{-k}<|x|<2^{-k+1}} |x|^{\beta p}|u(x)|^p\,dx.$$

Moreover, there exist constants c_1, c_2 (independent of k) such that

$$\begin{aligned} c_1 2^{-k(\beta p+n)}\, M_p(u, 2^{-k})^p &\leq \int_{2^{-k}<|x|<2^{-k+1}} |x|^{\beta p}|u(x)|^p\,dx \\ &\leq c_2 2^{-k(\beta p+n)}\, M_p(u, 2^{-k})^p. \end{aligned}$$

Consequently, $M_p(u,r) \leq c\,r^\alpha$ for $0 < r < 1$ implies that $u \in L^p_\beta(B_\circ)$ if $\alpha + \beta > -n/p$, and conversely, $u \in L^p_\beta(B_\circ)$ implies $M_p(u, 2^{-k}) \leq c\,2^{k\beta+kn/p}$, which implies $M_p(u,r) \leq c_\alpha\, r^\alpha$ for $0 < r < 1$ if we choose $\alpha = -\beta - n/p$. We obtain analogous relationships between the M_p-norm and the L^p_β-norm of the terms $|x|^{|\alpha|}|\partial^\alpha u(x)|$ for $|\alpha| \leq 2$.

Now let us introduce a weighted L^p-norm for functions on $\mathbb{R}^n_\circ = \mathbb{R}^n\backslash\{0\}$ with separate weights at the origin and infinity:

$$\begin{aligned} \|u\|^p_{L^p_{\beta,\gamma}(\mathbb{R}^n_\circ)} &= \|u\|^p_{L^p_\beta(B_\circ)} + \|u\|^p_{L^p_\gamma(B^c)} \\ &= \int_{0<|x|<1} |u(x)|^p\,|x|^{\beta p}\,dx + \int_{|x|>1} |u(x)|^p\,|x|^{\gamma p}\,dx, \end{aligned} \tag{90}$$

where $B^c = \mathbb{R}^n\backslash\overline{B_1}$. We then define the weighted Sobolev space $W^{2,p}_{\beta,\gamma}(\mathbb{R}^n_\circ)$ to be those functions in $W^{2,p}_{\ell oc}(\mathbb{R}^n_\circ)$ for which the norm

$$\|u\|_{W^{2,p}_{\beta,\gamma}} = \sum_{|\alpha|\leq 2} \||x|^{|\alpha|}\partial^\alpha u\|_{L^p_{\beta,\gamma}(\mathbb{R}^n_\circ)} \tag{91}$$

is finite. Many authors have used similar weighted Sobolev spaces to study operators such as the Laplacian on $\mathbb{R}^n$, $\mathbb{R}^n_\circ$, and other noncompact manifolds with conical or cylindrical ends. Using the analysis in [**18**], [**19**] or [**16**], for example, it is easily verified that the bounded operator

$$\Delta : W^{2,p}_{\beta,\gamma}(\mathbb{R}^n_\circ) \to L^p_{\beta+2,\gamma+2}(\mathbb{R}^n_\circ) \tag{92}$$

is Fredholm (finite nullity and finite deficiency) for all values of β and γ *except* for the values $-2 + \frac{n}{q} + k$ and $-\frac{n}{p} - k$, where $q = p/(p-1)$ and k is any nonnegative integer. In fact, (92) is an isomorphism for $-n/p < \beta, \gamma < -2 + n/q$ (recall that we are assuming $n \geq 3$, so such β, γ exist). Since we are principally interested in the behavior of functions at the origin, we will fix $\gamma_0 \in (-n/p, -2+n/q)$. Then

$$\beta_- < \beta_+ \quad \Rightarrow \quad W^{2,p}_{\beta_-,\gamma_0}(\mathbb{R}^n_\circ) \subset W^{2,p}_{\beta_+,\gamma_0}(\mathbb{R}^n_\circ).$$

Moreover, for $\beta_+ \in (-2+n/q, -1+n/q)$, we find that (92) is surjective with a one-dimensional nullspace spanned by $|x|^{2-n} \in W^{2,p}_{\beta_+,\gamma_0}(\mathbb{R}^n_\circ)$; for $\beta_- \in (-n/p-2, -n/p-1)$, we find that (92) is injective with codimension equal to $n+1$.

Next we use perturbation theory (cf. [**10**], Ch. IV, Sec. 5) to conclude not only that the operator

$$\mathcal{L} : W^{2,p}_{\beta,\gamma}(\mathbb{R}^n_\circ) \to L^p_{\beta+2,\gamma+2}(\mathbb{R}^n_\circ) \tag{93}$$

is Fredholm for exactly the same values of β and γ as for (92), but the nullity and deficiency of (92) and (93) agree, provided δ is sufficiently small. So, in addition to the fixed $\gamma_0 \in (-n/p, -2+n/q)$, let us now choose $\beta_0 \in (-n/p, -2+n/q)$, as well as $\beta_- \in (-1-n/p, -n/p)$ and $\beta_+ \in (-2+n/q, -1+n/q)$. Then we assume that δ is so small that

$$\mathcal{L} : W^{2,p}_{\beta_0,\gamma_0}(\mathbb{R}^n_\circ) \to L^p_{\beta_0+2,\gamma_0+2}(\mathbb{R}^n_\circ) \quad \text{is an isomorphism,} \tag{94}$$

$$\mathcal{L} : W^{2,p}_{\beta_-,\gamma_0}(\mathbb{R}^n_\circ) \to L^p_{\beta_-+2,\gamma_0+2}(\mathbb{R}^n_\circ) \quad \text{is injective with codim} = n+1, \tag{95}$$

and

$$\mathcal{L} : W^{2,p}_{\beta_+,\gamma_0}(\mathbb{R}^n_\circ) \to L^p_{\beta_++2,\gamma_0+2}(\mathbb{R}^n_\circ) \quad \text{is surjective with nullity} = 1. \tag{96}$$

We claim that $Z \in W^{2,p}_{\beta_+,\gamma_0}(\mathbb{R}^n_\circ)$. In fact, this is quite simple to check given the explicit formulas (82) and (83), and the fact that $v \in X$, where X has the norm (54). Thus Z is a basis vector for the one-dimensional nullspace of (96).

Now suppose that $u \in W^{2,p}_{\ell oc}(\overline{B_1}\backslash\{0\})$ satisfies $\mathcal{L}u = 0$ and the growth estimate (41) with $\varepsilon_0 \in (0,1)$. Introduce a cutoff function $\chi \in C_0^\infty(B_1)$ equal to 1 on $B_{1/2}$. If we now specify that $\beta_+ \in (-1-\varepsilon_0+n/q, -1+n/q)$, then $\chi u \in W^{2,p}_{\beta_+,\gamma_0}(\mathbb{R}^n_\circ)$. Let $f = \mathcal{L}(\chi u)$. Since $f = 0$ for $|x| < 1/2$ and for $|x| > 1$, $f \in L^p_{\beta_0+2,\gamma_0+2}(\mathbb{R}^n_\circ)$. But (94) is an isomorphism, so we can find $v = \mathcal{L}^{-1} f \in W^{2,p}_{\beta_0,\gamma_0}(\mathbb{R}^n_\circ)$. Now $\chi u - v \in W^{2,p}_{\beta_+,\gamma_0}(\mathbb{R}^n_\circ)$ satisfies $\mathcal{L}(\chi u - v) = 0$. Since the nullspace of (96) is spanned by Z, there exists a constant C such that $\chi u - v = CZ$. But this means in particular that $u = CZ + v$ for $0 < |x| < 1/2$.

Now let us describe v asymptotically. Let $\zeta_0, \zeta_1, \ldots, \zeta_n$ denote a basis for the cokernel of (95); i.e., the ζ_i are linear functionals on $L^p_{\beta_-+2,\gamma_0+2}(\mathbb{R}^n_\circ)$ that are linearly independent and vanish on the image of (95). Now we want to choose $C_0, C_1, \ldots, C_n$ so that $\mathcal{L}(\chi(v - C_0 - \sum_{j=1}^n C_j x_j))$ is in the range of (95), i.e.

$$\zeta_i[\mathcal{L}(\chi(C_0 + \sum_{j=1}^n C_j x_j))] = \zeta_i[\mathcal{L}(\chi v)] \quad \text{for } i = 0, \ldots, n. \tag{97}$$

To be able to solve the linear system (97) for $C_0, \ldots, C_n$, we need to verify that the finite-dimensional linear map

$$(C_0, \ldots, C_n) \to \left(\zeta_0[\mathcal{L}(\chi(C_0 + \sum_{j=1}^n C_j x_j))], \ldots, \zeta_n[\mathcal{L}(\chi(C_0 + \sum_{j=1}^n C_j x_j))] \right) \tag{98}$$

is nonsingular. Suppose $(C_0,\ldots,C_n)$ is in the nullspace of (98). Then $\mathcal{L}(\chi(C_0+\sum_{j=1}^n C_j x_j))]$ is in the range of (95), and there exists $\psi\in W^{2,p}_{\beta_-,\gamma_0}(\mathbb{R}^n_\circ)$ such that

$$\mathcal{L}(\psi)=\mathcal{L}(\chi(C_0+\sum_{j=1}^n C_j x_j)),$$

i.e. $\mathcal{L}(\psi-\chi(C_0+\sum_{j=1}^n C_j x_j))=0$. But $\psi-\chi(C_0+\sum_{j=1}^n C_j x_j)\in W^{2,p}_{\beta_0,\gamma_0}(\mathbb{R}^n_\circ)$ and (94) is an isomorphism, so $\psi=\chi(C_0+\sum_{j=1}^n C_j x_j)$. However, χ and χx_j are *not* in $W^{2,p}_{\beta_-,\gamma_0}(\mathbb{R}^n_\circ)$, so the only way that we can have $\psi=\chi(C_0+\sum_{j=1}^n C_j x_j)$ is to have $C_0=C_1=\cdots=C_n=0$. Consequently, the linear map (98) is nonsingular.

Thus we can find $C_0,\ldots,C_n$ and $w\in W^{2,p}_{\beta_-,\gamma_0}(\mathbb{R}^n_\circ)$ so that $\mathcal{L}(w)=\mathcal{L}(\chi(v-C_0-C_1x_1\cdots C_nx_n))$. But (94) is an isomorphism, so we obtain $v=C_0+C_1x_1+\cdots+C_nx_n+w$ for $0<|x|<1/2$. This yields (42), and we only need to verify $M_{2,p}(w,r)\le c\,r^{2-\varepsilon_1}$. But recall that $w\in W^{2,p}_{\beta_-,\gamma_0}(\mathbb{R}^n_\circ)$ implies $M_{2,p}(w,r)\le c\,r^{-\beta_--n/p}$ for $0<r<1/2$, and we can then let $\beta_-=-2-n/p+\varepsilon_1$ for any $\varepsilon_1\in(0,1)$ to obtain the desired estimate. □

Now let us formulate the result for a general point $y\in\mathbb{R}^n$, where we do not assume $a_{ij}(y)=\delta_{ij}$. With y fixed and the same conditions (9) and (35) on ω, we now assume

$$\sup_{|x-y|=r}\|\mathbf{A}_x-\mathbf{A}_y\|\le\omega(r)\quad\text{for } 0<r<1, \tag{99}$$

and we want to construct a solution of

$$\mathcal{L}(x,\partial_x)Z_y(x)=0\quad\text{for } x\in B_\varepsilon(y)\backslash\{y\}, \tag{100}$$

for ε sufficiently small. (Unlike Theorem 1, in the Corollary below we need to assume that the coefficients are real-valued so that we can choose coordinates in which $a_{ij}(0)=\delta_{ij}$.)

COROLLARY 2. *For $n\ge 3$ and $p\in(1,\infty)$, fix $y\in\mathbb{R}^n$ and suppose that the constant coefficient operator $\mathcal{L}(y,\partial_x)$ is elliptic and the coefficients $a_{ij}(x)$ are bounded, measurable functions satisfying (99). For $\varepsilon>0$ sufficiently small, there exists a solution of (100) in the form*

$$Z_y(x)=h_y(|\mathbf{A}_y^{-1/2}(x-y)|)+v(x), \tag{101}$$

where h_y is defined by

$$h_y(r)=\int_r^\varepsilon s^{1-n}\,e^{I_y(s)}\,ds\,(1+\zeta_y(r)), \tag{102}$$

with $I_y(r)$ given by (13) and

$$M_{2,p}(\zeta_y,r;y)\le c\max\left(\omega(r),\sigma(r)\right), \tag{103}$$

and v in (101) satisfies

$$M_{2,p}(v,r;y)\le c\,r^{2-n}\,e^{I_y(r)}\,\omega(r). \tag{104}$$

Moreover, for any $u\in W^{2,p}_{\ell oc}(\overline{B_\varepsilon(y)}\backslash\{y\})$ that is a strong solution of $\mathcal{L}(x,\partial_x)u=0$ in $\overline{B_\varepsilon(y)}\backslash\{y\}$ subject to the growth condition

$$M_{2,p}(u,r;y)\le c\,r^{1-n+\varepsilon_0},$$

where $\varepsilon_0 > 0$, *there exist constants* $C, C_0, C_1, \ldots, C_n$ *(depending on* u*) such that*

$$u(x) = C\, Z_y(x) + C_0 + \sum_{j=1}^{n} C_j(x_j - y_j) + w(x) \quad \text{for } 0 < |x-y| < \varepsilon, \tag{105}$$

where w *satisfies*

$$M_{2,p}(w,r;y) \leq c\, r^{2-\varepsilon_1} \quad \text{for any } \varepsilon_1 > 0. \tag{106}$$

If we use integration by parts, we can write the solution in Corollary 1 as

$$Z_y(x) = \frac{\langle \mathbf{A}_y^{-1}(x-y),(x-y)\rangle^{\frac{2-n}{2}}}{(n-2)}\, e^{I_y\left(\sqrt{\langle \mathbf{A}_y^{-1}(x-y),(x-y)\rangle}\right)}\,(1+\xi_y(x)), \tag{107}$$

where $M_{1,\infty}(\xi_y, r; y) \leq c\, \max(\omega(r), \sigma(r))$ for $0 < r < \varepsilon$. Notice that, if $I_y(r)$ has a finite limit as $r \to 0$, then the leading term in (107) is just a constant times $\tilde{F}_y(x-y)$, the fundamental solution for $\mathcal{L}(y, \partial_x)$ at y (cf. (5)).

Proof of Corollary 2. For now we continue to assume $y = 0$, i.e.

$$\sup_{|x|=r} \|\mathbf{A}_x - \mathbf{A}_0\| \leq \omega(r), \tag{108}$$

and let us assume $\mathbf{A}_0$ is positive definite. Let $\mathbf{B} = (b_{ij}) = \mathbf{A}_0^{-1/2}$ so that $\mathbf{B}\mathbf{A}_0\mathbf{B} = \mathbf{I}$. Introduce new independent variables $\tilde{x} = \mathbf{B}x$ and the matrix $\tilde{\mathbf{A}}_{\tilde{x}} = (\tilde{a}_{ij}(\tilde{x})) = \mathbf{B}\mathbf{A}_x\mathbf{B}$, which satisfies $\tilde{a}_{ij}(0) = \delta_{ij}$. Using $\partial/\partial x_i = b_{ik}\partial/\partial\tilde{x}_k = b_{ki}\partial/\partial\tilde{x}_k$, we can then write

$$\mathcal{L}(x,\partial_x) = a_{ij}(x)\frac{\partial^2}{\partial x_i \partial x_j} = \tilde{a}_{k\ell}(\tilde{x})\frac{\partial^2}{\partial\tilde{x}_k\partial\tilde{x}_\ell} = \tilde{\mathcal{L}}(\tilde{x},\partial_{\tilde{x}}) \tag{109}$$

and apply Theorem 1 to $\tilde{\mathcal{L}}$ in the coordinates $\tilde{x}$. We conclude that for $\varepsilon > 0$ sufficiently small, there is a solution $\tilde{Z}$ of $\tilde{\mathcal{L}}(\tilde{x},\partial_{\tilde{x}})\tilde{Z}(\tilde{x}) = 0$ for $0 < |\tilde{x}| < \varepsilon$, of the form $\tilde{Z}(\tilde{x}) = \tilde{h}(|\tilde{x}|) + \tilde{v}(\tilde{x})$, where $M_{2,p}(\tilde{v}, r) \leq c\, r^{2-n} e^{I(r)}\omega(r)$ and $\tilde{h}(r)$ is of the form (38) with

$$I(r) = \frac{1}{|S^{n-1}|}\int_{r<|\tilde{z}|<\varepsilon}\left(\operatorname{tr}\tilde{\mathbf{A}}_{\tilde{z}} - n\langle\tilde{\mathbf{A}}_{\tilde{z}}\tilde{z},\tilde{z}\rangle|\tilde{z}|^{-2}\right)\frac{d\tilde{z}}{|\tilde{z}|^n}. \tag{110}$$

Expressed in terms of the original variables x, we obtain $Z(x) = \tilde{Z}(\tilde{x}) = \tilde{h}(|\mathbf{B}x|) + \tilde{v}(\mathbf{B}x)$, which satisfies $\mathcal{L}(x,\partial_x)Z(x) = 0$ for $0 < |\mathbf{B}x| < \varepsilon$; but choosing ε_0 sufficiently small, we conclude that $\mathcal{L}(x,\partial_x)Z(x) = 0$ for $0 < |x| < \varepsilon_0$.

Finally, if y is a general point in $\mathbb{R}^n$, then let $\tilde{x} = \mathbf{B}(x-y)$ with $\mathbf{B} = \mathbf{A}_y^{-1/2}$ and let $\tilde{\mathbf{A}}_{\tilde{x}} = (\tilde{a}_{ij}(\tilde{x})) = \mathbf{B}\mathbf{A}_x\mathbf{B}$; since $\tilde{x} = 0$ corresponds to $x = y$, we have $\tilde{a}_{ij}(0) = \delta_{ij}$, and we can apply Theorem 1 to $\tilde{\mathcal{L}}(\tilde{x},\partial_{\tilde{x}}) = \mathcal{L}(x,\partial_x)$. We obtain the solution $Z(x) = \tilde{h}(|\mathbf{B}(x-y)|) + \tilde{v}(\mathbf{B}(x-y))$, where $\tilde{h}(r)$ involves $I(r)$ as in (110). To transform (110) to the original variables, replace $\tilde{\mathbf{A}}_{\tilde{z}}$ by $\mathbf{A}_z$ and every other occurrence of $\tilde{z}$ by $z - y$; we find that (110) is of the desired form (13), so we may let $h_y(r) = \tilde{h}(r)$. Moreover, since $\tilde{v}$ satisfies $M_{2,p}(\tilde{v}, r) \leq c\, r^{2-n}e^{I(r)}\omega(r)$, it is clear that $v(x) = \tilde{v}(\mathbf{B}(x-y))$ satisfies $M_{2,p}(v,r;y) \leq c\, r^{2-n}e^{I(r)}\omega(r)$, completing the proof. □

3. Finding the Constant C_y so that $\mathcal{L}Z_y(x) = C_y\delta(x-y)$ in $B_\varepsilon(y)$

Let us now begin to discuss the role that $Z_y(x)$ plays in finding the fundamental solution. As before, we first consider $y = 0$ with $a_{ij}(0) = \delta_{ij}$ and then use a change of variables to consider a general fixed $y \in \mathbb{R}^n$; as in the proof of Theorem 1, we shall assume $\varepsilon = 1$. We first want to see whether the function $Z(x)$ found in Theorem 1 satisfies

$$-\mathcal{L}(x, \partial_x)Z(x) = C_0\delta(x) \quad \text{for } x \in B_1(0) \tag{111}$$

for some constant C_0.

It is not immediately clear how the left-hand side of (111) is defined. Recall from the proof of Theorem 1 the decomposition $Z(x) = h(|x|) + v(x)$, where h is given by (81) and $M_{2,p}(v, r) \le c\, r^{2-n}\, e^{I(r)}\, \omega(r)$. We can easily calculate $\partial_i\partial_j Z$ and show that for any $\mu > 0$ there is a constant C_μ so that

$$M_p(\partial_i\partial_j Z, r) \le C_\mu\, r^{-n-\mu}. \tag{112}$$

In fact, since the a_{ij} are bounded functions, we conclude that for each i, j the function

$$F_{ij}(x) = a_{ij}(x)\partial_i\partial_j Z(x) \quad \text{for } x \neq 0 \tag{113}$$

satisfies

$$M_p(F_{ij}, r) \le C_\mu\, r^{-n-\mu} \quad \text{for } 0 < r < 1. \tag{114}$$

But this estimate implies that F_{ij} can be regularized at $x = 0$ to give a distribution $\mathcal{F}_{ij}$, in particular as a continuous linear functional on the space of λ-Hölder continuous functions of compact support in $U = B_1(0)$:

$$\langle \mathcal{F}_{ij}, \phi\rangle = \int_{|x|<1} F_{ij}(x)\,[\phi(x) - \phi(0)\chi(|x|)]\, dx, \tag{115}$$

where $\chi(r)$ is a smooth cutoff function which is identically 1 near $r = 0$ but vanishes for $r > 1/2$. (Since $|\phi(x) - \phi(0)\chi(|x|)| \le C|x|^\lambda$, by choosing $0 < \mu < \lambda$ we see that the integral in (115) converges.) Now let us take the sum over all i, j:

$$\mathcal{F}_0 = \sum_{i,j} \mathcal{F}_{ij}. \tag{116}$$

Then $\mathcal{F}_0$ is a regularization of $\mathcal{L}(x, \partial_x)Z(x) = 0$, and so it vanishes:

$$\langle \mathcal{F}_0, \phi\rangle = \sum_{i,j}\langle \mathcal{F}_{ij}, \phi\rangle = \int \sum_{i,j} F_{ij}(x)\,[\phi(x) - \phi(0)\chi(|x|)]\, dx = 0.$$

Of course, regularization affects the distribution only at $x = 0$, so *if* we can interpret the expression

$$\mathcal{F} = \mathcal{L}(x, \partial_x)Z(x), \tag{117}$$

as a distribution, then $\mathcal{F}$ is supported only at $x = 0$. As such, it is a linear combination of the delta distribution and its derivatives. But since $\mathcal{F}$ is a continuous linear functional on λ-Hölder functions with $\lambda \in (0, 1)$, it must only involve the delta distribution itself; i.e., (111) must hold for some constant C_0. But we still have two problems: (i) how is $\mathcal{F}$ itself defined as a distribution? and (ii) how do we calculate the constant C_0?

The difficulty in defining $\mathcal{F}$ as a distribution in U is a consequence of the lack of regularity of the a_{ij}. In particular, there is no difficulty in defining the distributional derivatives of Z:

$$\langle \partial_i \partial_j Z, \phi \rangle = -\int_U \partial_j Z(x) \partial_i \phi(x)\, dx \tag{118}$$

for $\phi \in C_0^1(U)$, since the integral on the right converges. To handle the a_{ij}, we can write

$$\langle \mathcal{L}Z, \phi \rangle = \int_U \left[(a_{ij}(x) - \delta_{ij})\, \partial_i \partial_j Z(x)\, \phi(x) - \partial_i Z(x) \partial_i \phi(x) \right] dx. \tag{119}$$

Of course, the integral in (119) is actually an improper integral due to the singularity of $\partial_i \partial_j Z$ at $x = 0$; but provided this integral converges, we conclude that $-\mathcal{L}Z = C_0 \delta$ in U, and we can calculate C_0 by

$$C_0 = \lim_{\varepsilon \to 0} \int_{|x|<1} \left[(-a_{ij}(x) + \delta_{ij})\, \partial_i \partial_j Z(x)\, \phi_\varepsilon(|x|) + \partial_i Z(x) \partial_i \phi_\varepsilon(|x|) \right] dx, \tag{120}$$

where $\phi_\varepsilon(|x|) = \chi(|x|/\varepsilon)$ with the cutoff function χ introduced above. (Note that we are able to assume $\phi(x) = \phi(|x|)$ since we can write $\phi(x) = \phi_0(|x|) + \phi_1(x)$ with $|\phi_1(x)| + |x|\,|\nabla \phi_1(x)| \le c\,|x|$ for $|x| < 1$, which shows that $\langle \mathcal{L}Z, \phi_1 \rangle$ is well-defined and ϕ_1 contributes nothing to C_0.) We shall now prove the following.

THEOREM 2. *Suppose the conditions of Theorem 1 hold and Z is the function found there. If $I(0) = \lim_{r \to 0} I(r)$ exists and is finite, then the integral in (119) converges and we can calculate $C_0 = |S^{n-1}|\, e^{I(0)}$. If $I(r) \to -\infty$ as $r \to 0$, then the integral in (119) converges and we can calculate $C_0 = 0$.*

Proof of Theorem 2. Recall from the proof of Theorem 1 the decomposition $Z(x) = h(|x|) + v(x)$, where h is given by (81) and v satisfies (40). Since we always assume that $I(r)$ is bounded above, we obtain from (40) that

$$M_{2,p}(v, r) \le c\, r^{2-n}\, \omega(r) \quad \text{for } 0 < r < 1. \tag{121}$$

We shall separately consider the roles of v and h. In the estimates below, $\int_{|x|<\varepsilon}$ should actually be considered as an improper integral $\lim_{\eta \to 0} \int_{\eta<|x|<\varepsilon}$, but we avoid such cumbersome notation.

For v, we use (32) and (121) to conclude that

$$\left| \int_{|x|<\varepsilon} \partial_i v\, \partial_i \phi_\varepsilon\, dx \right| \le c\, \varepsilon^{-1} \int_0^\varepsilon \omega(r)\, dr \le c\, \omega(\varepsilon) \to 0$$

and

$$\left| \int_{|x|<\varepsilon} (a_{ij} - \delta_{ij})\, \partial_i \partial_j v\, \phi_\varepsilon\, dx \right| \le c \int_0^\varepsilon \omega^2(r)\, \frac{dr}{r} = c\, \sigma(\varepsilon) \to 0.$$

This shows that v makes no contribution to the value of C_0.

To determine the effect of h, let us write $h(r) = h_2(r) + h_3(r) + c$, where

$$h_2(r) = c_1 \int_r^1 s^{1-n} E_+(s)\, ds \tag{122}$$

with c_1 chosen as in the proof of Theorem 1 so that $c_1 E_+(r) = e^{I(r)} + o(1)$ as $r \to 0$, and

$$h_3(r) = \int_r^1 s^{1-n} E_+(s) \int_0^s \rho^{n-1} E_-(\rho) B[D^2 v](\rho)\, d\rho. \tag{123}$$

(Notice that h_2 and h_3 differ slightly from (82) and (83).) Let us consider h_3 first:

$$\partial_i h_3 = -x_i r^{-n} E_+(r) \int_0^r \rho^{n-1} E_-(\rho) B[D^2 v](\rho)\, d\rho$$

and

$$\partial_i \partial_j h_3 = -r^{-n} E_+(r) \left(\delta_{ij} - n\frac{x_i x_j}{r^2} + \frac{x_i x_j}{r^2} R(r)\right) \int_0^r \rho^{n-1} E_-(\rho) B[D^2 v](\rho)\, d\rho$$
$$- \frac{x_i x_j}{r^2} B[D^2 v](r).$$

The calculation of $\partial_i h_3$ combined with (74) and the assumption that $I(r)$ is bounded shows that

$$\left| \int_{|x|<\varepsilon} \partial_i h_3\, \partial_i \phi_\varepsilon\, dx \right| \le c\,\varepsilon^{-1} \int_0^\varepsilon \sigma(r)\, dr \le c\,\sigma(\varepsilon) \to 0. \tag{124}$$

On the other hand, the "worst" term in $(a_{ij} - \delta_{ij})\partial_i \partial_j h_3$ is

$$W(x) = -r^{-n} E_+(r) \Big(a_{ii} - n\frac{a_{ij} x_i x_j}{r^2}$$
$$- \left(\frac{a_{ij} x_i x_j}{r^2} - 1\right) R \Big) \int_0^r \rho^{n-1} E_-(\rho) B[D^2 v](\rho)\, d\rho.$$

Using the fact that ϕ_ε only depends on r and observing that

$$\int_{S^{n-1}} \left(a_{ii} - n\frac{a_{ij} x_i x_j}{r^2} - \left(\frac{a_{ij} x_i x_j}{r^2} - 1\right) R \right) d\theta = |S^{n-1}| R(r),$$

we can calculate

$$\langle W, \phi_\varepsilon \rangle = -|S^{n-1}| \int_0^\varepsilon E_+(r) \frac{R(r)}{r} \phi_\varepsilon(r) \int_0^r \rho^{n-1} E_-(\rho) B[D^2 v](\rho)\, d\rho\, dr,$$
$$= |S^{n-1}| \int_0^\varepsilon (E_+)'(r)\, \phi_\varepsilon(r) \int_0^r \rho^{n-1} E_-(\rho) B[D^2 v](\rho)\, d\rho\, dr.$$

Now we can integrate by parts to obtain

$$\langle W, \phi_\varepsilon \rangle = |S^{n-1}| \Big(\int_0^\varepsilon E_+(r)\, \phi_\varepsilon'(r) \int_0^r \rho^{n-1} E_-(\rho) B[D^2 v](\rho)\, d\rho\, dr$$
$$+ \int_0^\varepsilon \phi_\varepsilon(r)\, r^{n-1} B[D^2 v](r)\, dr \Big).$$

But, using $|\phi_\varepsilon'| \le c\,\varepsilon^{-1}$ and $|E_+| \le c$, we find

$$\left| \int_0^\varepsilon E_+(r)\, \phi_\varepsilon'(r) \int_0^r \rho^{n-1} E_-(\rho) B[D^2 v](\rho)\, d\rho\, dr \right| \le c\,\varepsilon^{-1} \int_0^\varepsilon \sigma(r)\, dr \le c\,\sigma(\varepsilon) \to 0$$

and

$$\left| \int_0^\varepsilon \phi_\varepsilon(r)\, r^{n-1} B[D^2 v](r)\, dr \right| \le c\,\sigma(\varepsilon) \to 0.$$

We conclude that h_3 makes no contribution to C_0.

Now consider h_2. We compute $\partial_i h_2 = -c_1 x_i r^{-n} E_+(r)$ and

$$\partial_i\partial_j h_2 = -c_1\, r^{-n}\, E_+(r)\left(\delta_{ij} - n\frac{x_i x_j}{r^2} - \frac{x_i x_j}{r^2}R(r)\right),$$

so

$$(a_{ij}-\delta_{ij})\partial_i\partial_j h_2 = -c_1\, r^{-n}\, E_+(r)\left(a_{ii} - n\frac{a_{ij}x_i x_j}{r^2} - \left(\frac{a_{ij}x_i x_j}{r^2}-1\right)R\right).$$

Notice that

$$\int_{|x|<\varepsilon} r^{-n}\, E_+(r)\left(a_{ii} - n\frac{a_{ij}x_i x_j}{r^2} - \left(\frac{a_{ij}x_i x_j}{r^2}-1\right)R\right)\chi(\frac{r}{\varepsilon})\,dx$$
$$= |S^{n-1}|\int_0^\varepsilon E_+(r)\frac{R(r)}{r}\chi(\frac{r}{\varepsilon})\,dr,$$

so we now may compute

$$-\langle \mathcal{L}h_2,\phi_\varepsilon\rangle = c_1|S^{n-1}|\int_0^\varepsilon \left(E_+(r)\frac{R(r)}{r}\chi(\frac{r}{\varepsilon}) - E_+(r)\left[\chi(\frac{r}{\varepsilon})\right]'\right)\,dr$$
$$= -c_1|S^{n-1}|\int_0^\varepsilon \left[E_+(r)\chi(\frac{r}{\varepsilon})\right]'\,dr.$$

Now, if $I(0)$ exists and is finite, then $c_1 E_+(0) = e^{I(0)}$, and we conclude that $-\langle \mathcal{L}h_2,\phi_\varepsilon\rangle = |S^{n-1}|e^{I(0)}$. On the other hand, if $I(r)\to-\infty$ as $r\to 0$, then $E_+(r)\to 0$, so $C_0 = 0$. This completes the proof of Theorem 2. □

Now let us consider a general fixed $y\in\mathbb{R}^n$ and try to find the constant C_y so that

(125) $$-\mathcal{L}(x,\partial_x)Z_y(x) = C_y\delta(x) \quad \text{for } x\in B_1(y).$$

We replace (119) by

(126) $$\langle \mathcal{L}Z_y,\phi\rangle = \int_U \left[(a_{ij}(x)-a_{ij}(y))\,\partial_i\partial_j Z_y(x)\,\phi(x) - a_{ij}(y)\partial_i Z_y(x)\partial_j\phi(x)\right]\,dx.$$

Corollary 3. *Suppose the conditions of Corollary 1 hold and Z_y is the function found there. If $I_y(0) = \lim_{r\to 0} I_y(r)$ exists and is finite, then the integral in (126) converges and Z_y satisfies (125) with C_y given by*

(127) $$C_y = |S^{n-1}|\sqrt{\det\mathbf{A}_y}\, e^{I_y(0)}.$$

If $I_y(r)\to-\infty$ as $r\to 0$, then the integral in (126) converges and Z_y satisfies (125) with $C_y = 0$.

Proof of Corollary 3. We need only show that

$$\langle -\mathcal{L}Z_y,\phi\rangle = |S^{n-1}|(\det\mathbf{A}_y)^{1/2}e^{I_y(0)}\phi(y)$$

for some $\phi\in C_0^\infty(B_{\varepsilon_y}(y))$. Let us recall the change of coordinates used in the proof of Theorem 1, namely $\tilde{x} = \mathbf{B}(x-y)$, where $\mathbf{B} = \mathbf{A}_y^{-1/2}$, and let $\tilde{\phi}(\tilde{x}) = \phi(x)$; then

$$-\int \mathcal{L}(x,\partial_x)Z_y(x)\phi(x)\,dx = -(\det\mathbf{A}_y)^{1/2}\int \tilde{\mathcal{L}}(\tilde{x},\partial_{\tilde{x}})\tilde{Z}_0(\tilde{x})\tilde{\phi}(\tilde{x})\,d\tilde{x}.$$

But Theorem 2 implies that $-\langle \tilde{\mathcal{L}}\tilde{Z}_0,\tilde{\phi}\rangle = |S^{n-1}|e^{I_y(0)}\tilde{\phi}(0)$. Since $\tilde{\phi}(0) = \phi(y)$, we obtain the desired result. □

4. Constructing the Fundamental Solution

Now we are in a position to construct the fundamental solution in a bounded open set $U \subset \mathbb{R}^n$ when $a_{ij} \in C^\omega(U)$, assuming that ω satisfies the square-Dini condition (9) and for every $y \in U$ we know that $I_y(0)$ exists and is finite. Given the results in the preceding section, it is natural to seek the fundamental solution in the form $F(x,y) = Z_y(x)/C_y + v(x,y)$, where $\mathcal{L}(x,\partial_x)v(x,y) = 0$. In fact, it will be convenient to construct $F(x,y)$ as the Green's function $G(x,y)$, in the sense of (8), for a smooth, bounded domain V that contains U.

But first let us observe that the additional assumption that $I_y(0)$ exists and is finite allows us to improve the asymptotic description of $Z_y(x)$. In fact, let us fix $y = 0$ with $a_{ij}(0) = \delta_{ij}$, and assume

$$|I(r) - I(0)| \le \theta(r), \tag{128}$$

where $\theta(r)$ is a positive, nondecreasing function for $0 < r < 1$ such that $\theta(0) = 0$; as with ω we additionally assume that for some $\nu \in (0,1)$ we have

$$\theta(r)\, r^{-1+\nu} \quad \text{is nonincreasing for } 0 < r < 1. \tag{129}$$

(If $\omega(r)$ satisfies the Dini condition, then we can take $\theta(r) = \omega(r)$.)

LEMMA 1. *Under the conditions of Theorem 1, let us additionally assume that (128) and (129) hold. Then the solution Z found in Theorem 1 satisfies*

$$Z(x) = \frac{|x|^{2-n} e^{I(0)}}{n-2}\,(1 + \xi(x)), \tag{130}$$

where for any $p \in (1,\infty)$ we can estimate ξ by

$$M_{2,p}(\xi, r) \le c \max(\omega(r), \sigma(r), \theta(r)) \quad \text{as } r \to 0, \tag{131}$$

where c depends only on ω, θ, n, and $I(0)$.

Proof of Lemma 1. In the decomposition $Z(x) = h(|x|) + v(x)$ as in (37), it is elementary to use (40) with (128) to show that (131) applies to $\xi = c\,|x|^{n-2}\, v$, so we focus on h. Recall the decomposition $h = h_0 + h_1$ using (82) and (83). Let us recall h_0'' from (87) and use (128) to estimate $e^{I(r)} \le (1 + 2\theta(r))e^{I(0)}$ for r sufficiently small; also recalling $|I'(r)| \le 2(n-1)\omega(r)/r$, we obtain

$$|h_0''(r) - (n-1)r^{-n}e^{I(0)}| \le c\, r^{-n} \max(\omega(r), \theta(r));$$

here (and henceforth) c depends only on ω, θ, n, and $I(0)$. Using (35) and (129), we can integrate this twice to obtain

$$\left| h_0(r) - \frac{r^{2-n} e^{I(0)}}{n-2} \right| \le c\, r^{2-n} \max(\omega(r), \theta(r)).$$

In fact, this argument has shown that $\xi_0(r) = (n-2)\, r^{n-2}\, e^{-I(0)}\, h_0(r) - 1$ satisfies the pointwise estimate

$$M_{2,\infty}(\xi_0, r) \le c \max(\omega(r), \theta(r)).$$

Similarly, let us recall h_1'' from (88); we can estimate most of the terms pointwise, but the term $B[D^2 v](r)$ can only be estimated in M_p, so $\xi_1 = (n-2)r^{n-2}e^{-I(0)}h_1(r)$ satisfies

$$M_{2,p}(\xi_1, r) \le c \max(\omega(r), \sigma(r)).$$

These may be combined to yield (131), so the lemma is proved. □

Now we are ready to construct our fundamental solution in a bounded open set.

THEOREM 3. *Suppose (1) is uniformly elliptic in a bounded open set $U \subset \mathbb{R}^n$ for $n \geq 3$, where the coefficients a_{ij} are continuous functions with uniform modulus of continuity ω satisfying (9). If $I_y(0)$ exists at every $y \in U$ and*

$$|I_y(r) - I_y(0)| \leq \theta(r) \quad \text{for all } y \in U, \tag{132}$$

where θ is a positive, nondecreasing function for $0 < r < 1$ with $\theta(0) = 0$ and (129), then there is a function $F(x, y)$ satisfying (2) for $x, y \in U$; moreover, $F(x, y)$ admits the asymptotic description

$$F(x,y) = \frac{\langle \mathbf{A}_y^{-1}(x-y), (x-y)\rangle^{\frac{2-n}{2}}}{(n-2)|S^{n-1}|\sqrt{\det \mathbf{A}_y}}\,(1 + H(x,y)), \tag{133}$$

where for any $p \in (1, \infty)$ and any compact set $K \subset U$ we have

$$M_{2,p}(H(\cdot, y), r; y) \leq c \max(\omega(r), \sigma(r), \theta(r)) \quad \text{as } r \to 0, \tag{134}$$

with constant c independent of $y \in K$.

Proof of Theorem 3. For each $y \in U$, denote the ε in Corollary 1 by ε_y. Note that the size of ε_y depends on the behavior of the coefficients a_{ij} near y through their ellipticity (i.e. the norm of $\mathbf{A}_y^{-1/2}$) and their continuity (i.e. ω). Since the ellipticity and modulus of continuity are uniform on U, we can find $\varepsilon > 0$ that is independent of $y \in U$. In fact, if we choose a smooth, bounded domain V with $V \supset \overline{U}$ and $\operatorname{dist}(U, \partial V) > \varepsilon$, then we can extend the coefficients a_{ij} to V in such a way as to maintain the uniform ellipticity as well as the modulus of continuity $\omega(r)$ (cf. [**20**]). Thus for all $y \in \overline{U}$ we can construct $Z_y(x)$ in $B_\varepsilon(y)\backslash\{y\}$. In fact, repeating this argument with a neighborhood of V, we may assume that $Z_y(x)$ is defined for all $y \in \overline{V}$.

For each $y \in \overline{V}$ let us use Corollary 2 to calculate C_y (which is positive since $I_y(0)$ is finite) and conclude that $-\mathcal{L}(x, \partial_x)Z_y(x)/C_y = \delta(x-y)$ for all $x, y \in \overline{V}$ with $|x - y| < \varepsilon$. We shall construct $G(x, y)$ as the Green's function for $\mathcal{L}$ in V. For fixed $y \in V$, let us introduce a smooth cutoff function $\eta_y(r)$ satisfying $\eta_y(r) = 1$ for sufficiently small $r > 0$ but such that $\eta_y(|\cdot - y|)$ has compact support in V. Then let us define

$$G(x,y) = \eta_y(|x-y|)Z_y(x)/C_y + v(x,y), \tag{135}$$

where $v(x, y)$ is to be determined. But if we apply $-\mathcal{L}(x, \partial_x)$ to $G(x, y)$ we obtain

$$-\mathcal{L}(x,\partial_x)G(x,y) = \delta(x-y) + \psi(x,y) - \mathcal{L}(x,\partial_x)v(x,y),$$

where $\psi(\cdot, y) \in L^p(V)$. So, for fixed $y \in V$, consider the Dirichlet problem for v:

$$\begin{aligned} \mathcal{L}(x,\partial_x)v(x,y) &= \psi(x,y) \quad \text{for } x \in V,\\ v(x,y) &= 0 \quad \text{for } x \in \partial V. \end{aligned} \tag{136}$$

It is well known (cf. Theorem 9.15 in [**9**]) that (136) has a unique solution $v(\cdot, y) \in W^{2,p}(V) \cap W_0^{1,p}(V)$, so using this $v(x, y)$ in (135) not only ensures that (8) holds for $x, y \in V$ (and hence for $x, y \in U$), but the following: for any $f \in C(\overline{V})$, $u(x) = -\int_V G(x, y)f(y)\,dy$ satisfies $\mathcal{L}u = f$ in V and $u = 0$ on ∂V. Now if we pick $\phi \in C^2(\overline{V})$ with $\phi = 0$ on ∂V and let $f = \mathcal{L}\phi$, then uniqueness of the solution shows that $u = \phi$ in V; i.e., (8) holds. In other words, the $G(x,y)$ that we have constructed is just the Green's function for $\mathcal{L}$ in V; in particular, (2) holds.

Now from (135) and Corollaries 1 and 2 we see that

$$G(x,y) = \frac{\langle \mathbf{A}_y^{-1}(x-y),(x-y)\rangle^{\frac{2-n}{2}}}{(n-2)|S^{n-1}|\sqrt{\det \mathbf{A}_y}}\, e^{J_y(\sqrt{\langle \mathbf{A}_y^{-1}(x-y),(x-y)\rangle})}\,(1+\xi_y(x)), \tag{137}$$

where $M_{1,\infty}(\xi_y, r; y) \le c \max(\omega(r), \sigma(r))$ for $0 < r < \varepsilon$ and $J_y(r) = I_y(r) - I_y(0)$ can also be written as

$$\frac{-1}{|S^{n-1}|}\int_{0<|z-y|<r}\left(\operatorname{tr}(\mathbf{A}_z\mathbf{A}_y^{-1}) - n\,\frac{\langle \mathbf{A}_z\mathbf{A}_y^{-1/2}(z-y), \mathbf{A}_y^{-1/2}(z-y)\rangle}{|z-y|^2}\right)\frac{dz}{|z-y|^n}. \tag{138}$$

However, if we choose coordinates in which $y = 0$ and $a_{ij}(0) = \delta_{ij}$, then we may apply Lemma 1 to absorb the exponential term in (137) into $1 + \xi_y$ and obtain

$$G(x,y) = \frac{\langle \mathbf{A}_y^{-1}(x-y),(x-y)\rangle^{\frac{2-n}{2}}}{(n-2)|S^{n-1}|\sqrt{\det \mathbf{A}_y}}\,(1+\xi_y(x)) \quad \text{as } x \to y, \tag{139}$$

where $M_{2,p}(\xi_y, r; y) \le c_y \max(\omega(r), \sigma(r), \theta(r))$ for $0 < r < \varepsilon$ with c_y depending on ω, θ, n, and $I_y(0)$. But if we select a compact subset $K \subset U$, then c_y may be taken to be independent of $y \in K$, so we can replace $\xi_y(x)$ by $H(x,y)$ and obtain (133), (134) as $|x-y| \to 0$.

Letting $F(x,y) = G(x,y)$ for $x, y \in U \subset V$, we have our fundamental solution in U. □

References

[1] P. Bauman, *Positive solutions of elliptic equations in nondivergence form and their adjoints,* Ark. Mat. **22** (1984), 153–173. MR765409 (86m:35008)

[2] ———, *Equivalence of the Green's function for diffusion operators in* $\mathbb{R}^n$*: A counterexample,* Proc. Amer. Math. Soc. **91** (1984), 64–68. MR735565 (85d:35026)

[3] ———, *A Wiener test for nondivergence structure, second-order elliptic equations,* Indiana Univ. Math. J. **34** (1985), 825–844. MR808829 (87b:35047)

[4] L. Escauriaza, *Bounds for the fundamental solution of elliptic and parabolic equations in nondivergence form,* Comm. Part. Diff. Equat. **25** (2000), 821–845. MR1759794 (2001i:35009)

[5] E. B. Fabes, D. Jerison, and C. E. Kenig, *Necessary and sufficient conditions for absolute continuity of elliptic harmonic measure,* Ann. of Math. (2) **119** (1984), 121–141. MR736563 (85h:35069)

[6] E. B. Fabes and D. Stroock, *The* L^p*-integrability of Green's functions and fundamental solutions for elliptic and parabolic equations,* Duke Math. J. **51** (1984), 977–1016. MR771392 (86g:35057)

[7] R. Fefferman, *A criterion for the absolute continuity for the harmonic measure associated with an elliptic operator,* J. American Mathematical Society **2** (1989), 127–135. MR955604 (90b:35068)

[8] D. Gilbarg and J. Serrin, *On isolated singularities of solutions of second-order elliptic equations,* J. Analyse Math. **4** (1955/56), 309-340. MR0081416 (18:399a)

[9] D. Gilbarg and N. S. Trudinger, *Elliptic Partial Differential Equations of Second Order,* 2nd Edition, Springer-Verlag, Berlin, 1983. MR737190 (86c:35035)

[10] T. Kato, *Perturbation Theory for Linear Operators,* Springer, New York, 1966. MR0203473 (34:3324)

[11] C. E. Kenig and J. Pipher, *The absolute continuity of elliptic measure revisited,* J. Fourier analysis and applications **4** (1998), 463–468. MR1658628 (99j:31010)

[12] V. Kozlov and V. Maz'ya, *The estimates for L_p-means and asymptotic behavior of solutions of elliptic boundary value problems in a cone.* I, Seminar Analysis Operator Equat. and Numer. Anal. Karl-Weierstrass-Institut für Mathematik (1985/86), 55–92. (Russian) MR0890526 (88e:47092)

[13] ———, *The estimates for L_p-means and asymptotic behavior of solutions of elliptic boundary value problems in a cone.* II, Math. Nachr. **137** (1988), 113–139. (Russian) MR968991 (90a:47124)

[14] ———, *Differential Equations with Operator Coefficients*, Springer Monographs in Mathematics, Springer-Verlag, 1999. MR1729870 (2001d:34090)

[15] ———, *Asymptotic formula for solutions to the Dirichlet problem for elliptic equations with discontinuous coefficients near the boundary* (English summary), Ann. Sc. Norm. Super. Pisa Cl. Sci. (5) **2** (2003), 551–600. MR2020860 (2005b:35056)

[16] R. Lockhart and R. McOwen, *Elliptic differential operators on noncompact manifolds,* Ann. Sc. Norm. Super. Pisa Cl. Sci. (4) **12** (1985), 409–447. MR837256 (87k:58266)

[17] V. Maz'ya and R. McOwen, *Asymptotics for solutions of elliptic equations in double divergent form,* Comm. Part. Diff. Equat. **32** (2007), 1–17. MR2304147 (2007m:35031)

[18] V. Maz'ya and B. Plamenevski, *Estimates in L_p and Hölder classes and the Miranda-Agmon maximum principle solutions of elliptic boundary problems in domains with singular points on the boundary*, Math. Nachr. **81** (1978), 25–82. (Russian) MR0492821 (58:11886)

[19] R. McOwen, *The behavior of the Laplacian on weighted Sobolev spaces,* Comm. Pure Appl. Math. **32** (1979), 783–795. MR539158 (81m:47069)

[20] V. Mil'man, *Extension of functions that preserve the modulus of continuity*, Math. Notes **61** (1997), 193–200. MR1620006 (99d:26002)

[21] C. Miranda, *Partial Differential Equations of Elliptic Type*, Springer-Verlag, Berlin, 1970. MR0284700 (44:1924)

[22] E. Stein, *Note on singular integral operators*, Proc. Amer. Math. Soc. **8** (1957), 250–254. MR0088606 (19:547b)

[23] E. Stein and A. Zygmund, *On the differentiability of functions*, Studia Math. **23** (1964), 247–283. MR0158955 (28:2176)

[24] M. Taylor, *Tools for PDE*, Amer. Math. Soc., Providence, RI, 2000. MR1766415 (2001g:35004)

LINKÖPING UNIVERSITY AND OHIO STATE UNIVERSITY AND UNIVERSITY OF LIVERPOOL
E-mail address: vlmaz@mai.liu.se

NORTHEASTERN UNIVERSITY
E-mail address: mcowen@neu.edu

Amer. Math. Soc. Transl.
Volume **229**, 2010

Boundary Regularity for Vectorial Problems

Giuseppe Mingione

To Nina Nikolaevna Uraltseva, with admiration

ABSTRACT. This paper surveys recent results concerning boundary regularity of solutions to elliptic and parabolic systems, and minima of integral functionals of the Calculus of Variations.

1. Foreword: Nina's jubilee

For all those of my generation involved in the regularity theory of solutions to elliptic, parabolic and variational problems, Nina Nikolaevna Uraltseva belongs to a restricted and distinguished group of mathematical heros whose work deeply influenced and shaped the subject. Indeed, although Nina is nowadays more strong and active than ever, the quality, the importance and the depth of her contributions already allow for an evaluation of her work from a historical perspective: the one reserved to those whose ideas are peaks that are not going to be smoothed out by the natural "homogenization process" of synthesis that scientific theories typically undergo through the passing of time. It is not the case here to recall all the change al posto di her major contributions she achieved while being one of the leaders of the celebrated St. Petersburg's school of Analysis, and that are partially reported in the classical books [**39, 42**], but, at least, I would like to recall a fundamental result, which certainly has a special meaning for my personal mathematical life: the proof of the $C^{1,\alpha}$ character of solutions to the p-Laplacian equation

$$\operatorname{div}\left(|Du|^{p-2}Du\right) = 0\,,$$

together with a counterexample disproving their further smoothness; see [**40**].

When Arina Arkhipova and Alexander Nazarov asked me to contribute to this volume, I felt extremely happy and honored. This modest essay is therefore dedicated to Nina with friendship, admiration and gratitude for her lifetime achievement.

2. General outlook and the case of systems

In this paper I will describe some of the results obtained in a series of papers [**5, 6, 7, 17, 36, 37, 38**], concerning the boundary regularity of solutions to Dirichlet elliptic, parabolic and variational problems. For this reason, the one you are reading now cannot be thought as a comprehensive survey on boundary regularity problems, and I do apologize in advance with those authors whose nevertheless significant work will not be properly described. I finally emphasize that the main fact here is that the problems treated are vectorial; that is, all solutions and minima considered here will be vector-valued maps.

2.1. General notation. In the following I shall denote by $\mathcal{H}^s$ the standard s-dimensional Hausdorff measure in $\mathbb{R}^n$, with $s \in [0, n]$, while the Hausdorff dimension of a set $A \subset \mathbb{R}^n$ will be denoted by $\dim_{\mathcal{H}}(A)$. In the following, if $h\colon A \to \mathbb{R}^k$ is an integrable map and $0 < |A| < \infty$, I will denote the integral average of $h(\cdot)$ over A by

$$(h)_A := \fint_A h(x)\, dx := \frac{1}{|A|} \int_A h(x)\, dx\,.$$

2.2. Partial regularity. First of all, in order to frame the problems in the right perspective I will give a very brief survey of the standard interior partial regularity theory. An important feature of elliptic and variational problems in the vectorial case is the one of exhibiting singularities: while under reasonable assumptions, solutions are everywhere regular in the scalar case, in the vectorial case they turn out to be continuous only outside a negligible, closed subset, called *the singular set, which can be indeed a very wild one.* To begin with, let me consider the classical case of nonlinear elliptic systems in divergence form:

$$\operatorname{div}\, a(x, u, Du) = 0\,, \tag{2.1}$$

where $a : \Omega \times \mathbb{R}^N \times \mathbb{R}^{nN} \to \mathbb{R}^{nN}$ is a continuous vector field of class C^1 in the third variable, satisfying

$$\begin{cases} |a(x,v,z)| \leq L(1+|z|^{p-1}), \\ \nu(1+|z|^2)^{\frac{p-2}{2}}|\lambda|^2 \leq \langle a_z(x,v,z)\lambda, \lambda\rangle \leq L(1+|z|^2)^{\frac{p-2}{2}}|\lambda|^2, \\ |a(x,u,z)-a(y,v,z)| \leq L\omega_\alpha(|x-y|+|u-v|)(1+|z|^{p-1})\,, \end{cases} \tag{2.2}$$

for all $x,\, y \in \Omega$, $u,\, v \in \mathbb{R}^N$ and $z,\, \lambda \in \mathbb{R}^{nN}$. Here $n \geq 2$, $N \geq 1$, Ω is a bounded open set in $\mathbb{R}^n$, while in the rest of the paper I assume $p \geq 2$. Finally

$$\omega_\alpha(s) := \min\{s^\alpha, 1\}, \qquad \alpha \in (0,1)\,. \tag{2.3}$$

I will also assume that $a_z(\cdot)$ is Carathéodory regular, which is obviously unavoidable to give meaning to the problem. The left-hand side in $(2.2)_2$ prescribes the strong ellipticity of the system considered, taking into account the convenient growth with respect to the gradient variable z, while the remaining ones in $(2.2)_{1,2}$ indeed fix the growth conditions of the systems and the well-posedness of the problems in the Sobolev space $W^{1,p}(\Omega, \mathbb{R}^N)$. Assumption $(2.2)_3$, together with (2.3), means that the map defined by

$$(x, v) \mapsto \frac{a(x,v,z)}{(1+|z|)^{p-1}}$$

is Hölder continuous with respect to (x, v) with exponent $\alpha \in (0,1)$, uniformly with respect to the gradient variable z.

DEFINITION 2.1. A weak solution to (2.1) is at this point of course a map $u \in W^{1,p}(\Omega, \mathbb{R}^N)$ such that

$$\int_\Omega \langle a(x,u,Du), D\varphi\rangle\, dx = 0 \text{ holds for every } \varphi \in C_c^\infty(\Omega, \mathbb{R}^N)\,. \tag{2.4}$$

In the following, when using the word regular for a map I will usually mean "Hölder continuous" or, more often, "with Hölder continuous gradient". The latter is the focal point of regularity theory, after which higher regularity of solutions can be obtained via standard bootstrap arguments.

Beside the case of systems, I will consider the case of variational integrals of the type

$$\mathcal{F}(v, A) := \int_A F(x, v, Dv)\, dx\,, \tag{2.5}$$

defined for Sobolev maps $v \in W^{1,p}_{\mathrm{loc}}(\Omega, \mathbb{R}^N)$ and open sets $A \subseteq \Omega$. I shall often denote $\mathcal{F} \equiv \mathcal{F}(v) \equiv \mathcal{F}(v, \Omega)$. Again, here I assume $n \geq 2$, $N \geq 1$, Ω is a bounded open set in $\mathbb{R}^n$ and $p \geq 2$. The integrand $F\colon \Omega \times \mathbb{R}^N \times \mathbb{R}^{nN} \to \mathbb{R}$ is of class C^2 in the third variable, satisfying

$$\begin{cases} \nu|z|^p \leq F(x, v, z) \leq L(1 + |z|^p), \\ \nu(1 + |z|^2)^{\frac{p-2}{2}}|\lambda|^2 \leq \langle F_{zz}(x, v, z)\lambda, \lambda\rangle \leq L(1 + |z|^2)^{\frac{p-2}{2}}|\lambda|^2, \\ |F(x, u, z) - F(y, v, z)| \leq L\omega_\alpha(|x - y| + |u - v|)(1 + |z|^p)\,, \end{cases} \tag{2.6}$$

for all $x, y \in \Omega$, $u, v \in \mathbb{R}^N$ and $z, \lambda \in \mathbb{R}^{nN}$, where $0 < \nu \leq L$ and $\omega_\alpha\colon \mathbb{R}^+ \to (0, 1)$ is as in (2.3). I will also assume that both $F_z(\cdot)$ and $F_{zz}(\cdot)$ are Carathéodory regular. Assumption $(2.6)_2$ describes a controlled, uniform convexity of the integrand F, via growth conditions imposed on the second derivatives F_{zz}, which are in turn prescribed according to the ones in $(2.6)_1$. Convexity in the gradient variable of F is the main assumption considered above. Assumption $(2.6)_3$, together with (2.3), means that the function

$$(x, v) \mapsto \frac{F(x, v, z)}{(1 + |z|)^p}$$

is Hölder continuous with respect to (x, v) with exponent $\alpha \in (0, 1)$, uniformly with respect to z.

DEFINITION 2.2. Local minimizers of $\mathcal{F}$ are maps $u \in W^{1,p}(\Omega, \mathbb{R}^N)$ such that

$$\mathcal{F}(u, A) \leq \mathcal{F}(v, A) \qquad \text{whenever } A \Subset \Omega \quad \text{and} \quad u - v \in W^{1,p}_0(A, \mathbb{R}^N)\,.$$

In the *scalar case* $N = 1$, when minima and solutions are real-valued functions, singularities do not show up and minimizers are everywhere regular in the interior:

THEOREM 2.1. *Let $u \in W^{1,p}(\Omega)$ be either a local minimizer of the functional $\mathcal{F}$ under the assumptions* (2.6)-(2.3), *or a solution to* (2.1) *under the assumptions* (2.2) *and* (2.3) *with $N = 1$. Then $Du \in C^{0,\alpha/2}_{\mathrm{loc}}(\Omega, \mathbb{R}^n)$ in the case of functionals, and $Du \in C^{0,\alpha}_{\mathrm{loc}}(\Omega, \mathbb{R}^n)$ in the case of equations.*

For the previous result, see for instance [**27**],[**43**] and related references. The main ingredient here is a suitable application of ideas going back to De Giorgi [**12**], and these are essentially scalar ones. In terms of pointwise regularity of solutions, the last two results are the best possible. Notice the decrease in the Hölder exponent as we move from solutions to minima: counterexamples show that minimizers are not $C^{0,\alpha}$-regular in general; they are only $C^{0,\alpha/2}$-regular.

In the *vectorial case* $N > 1$ the situation drastically changes and minima and solutions generally lose a lot of regularity properties. This was first shown by De Giorgi [**13**] and Maz'ja [**44**]; then came a series of counterexamples, including the one of Nečas [**48**] and that of Šverák and Yan [**52**], who showed that even when considering the favorable case of smooth and quadratic-growth functionals as Nečas, there exist unbounded minimizers. In general the singular set is nonempty, but not when considering very special structures [**53**], [**31**].

After realizing that solutions to vectorial problems can be irregular, the next step is the so-called partial regularity; i.e., *proving that the regularity properties of the scalar case are preserved at least on large subsets.* Indeed the so-called partial regularity holds.

THEOREM 2.2. *Let $u \in W^{1,p}(\Omega, \mathbb{R}^N)$ be a weak solution to the system* (2.1), *under the assumptions* (2.2) *and* (2.3). *Then there exists an open subset $\Omega_u \subset \Omega$ such that $|\Omega \setminus \Omega_u| = 0$ and $Du \in C^{0,\alpha}_{\mathrm{loc}}(\Omega_u, \mathbb{R}^{nN})$.*

THEOREM 2.3. *Let $u \in W^{1,p}(\Omega, \mathbb{R}^N)$ be a local minimizer of the functional $\mathcal{F}$, under the assumptions* (2.6) *and* (2.3). *Then there exists an open subset $\Omega_u \subset \Omega$ such that $|\Omega \setminus \Omega_u| = 0$ and $Du \in C^{0,\alpha/2}_{\mathrm{loc}}(\Omega_u, \mathbb{R}^{nN})$.*

Theorems 2.2 and 2.3 are the vectorial counterpart of the scalar Theorem 2.1; Hölder continuity is preserved only on an open subset $\Omega_u \subset \Omega$; the set

$$\Sigma_u := \Omega \setminus \Omega_u \tag{2.7}$$

is called the singular set of Du (or of u, since we have in general $C^{1,\sigma}$-regularity in mind). The proofs of Theorems 2.3 and 2.2 automatically provide a characterization of Σ_u as the set of "non-Lebesgue points" of Du in the sense that

$$\Sigma_u = \Sigma_u^0 \cup \Sigma_u^1\,, \tag{2.8}$$

where

$$\begin{aligned}\Sigma_u^0 := \Bigg\{ x \in \Omega : \liminf_{\rho \downarrow 0} \fint_{B(x,\rho)} &|Du(y) - (Du)_{B(x,\rho)}|^p \, dy > 0 \\ &\text{or } \limsup_{\rho \downarrow 0} |(Du)_{B(x,\rho)}| = \infty \Bigg\}\end{aligned} \tag{2.9}$$

and

$$\begin{aligned}\Sigma_u^1 := \Bigg\{ x \in \Omega : \liminf_{\rho \downarrow 0} \fint_{B(x,\rho)} &|u(y) - (u)_{B(x,\rho)}|^p \, dy > 0 \\ &\text{or } \limsup_{\rho \downarrow 0} |(u)_{B(x,\rho)}| = \infty \Bigg\}.\end{aligned} \tag{2.10}$$

Indeed the proofs are based on the following fact: a point $x_0 \in \Omega$ is regular, that is, Du is Hölder continuous in a neighborhood of x_0, if

$$\fint_{B(x_0,\varrho)} |Du(x) - (Du)_{B(x,\rho)}|^p \, dx \leq \varepsilon \quad \text{and} \quad \limsup_{\varrho \downarrow 0} |(Du)_{B(x,\rho)}| \leq \infty\,, \tag{2.11}$$

where $\varepsilon > 0$ is a suitably small number. The proofs also require a similar condition for u. Since such conditions are satisfied almost everywhere by Lebesgue's differentiation theory, partial regularity, with the characterization in (2.8), follows. As a matter of fact all the proofs indeed yield local results in the sense that conditions (2.11) ensure that x_0 is a regular point. This is clarified by the following:

DEFINITION 2.3. A point $x_0 \in \Omega$ is a regular point for Du iff there exists a neighborhood $U \subset \Omega$ of x_0 such that $Du \in C^{0,\alpha}(\Omega \cap U)$. Accordingly, the regular set of Du is

$$\Omega_u := \{x \in \partial\Omega : u \in C^{1,\alpha}(\Omega \cap U, \mathbb{R}^N) \text{ for some neighborhood } U \text{ of } x\}.$$

In the form given here, Theorems 2.3 and 2.2 are the result of the efforts of several authors, and their first versions were originally obtained by Giaquinta, Giusti and Modica [**24**], [**26**]; see also [**32**], [**33**]; the methods of such authors partially rely on suitable freezing techniques, originally pioneered by Campanato [**8**] in the case of linear elliptic equations. For the present optimal form, I refer to Duzaar and Grotowski [**15**], who introduced an interesting new technique coming from Geometric Measure Theory problems: the A-harmonic approximation method [**20**]. For more on partial regularity the reader may look at the classical treatises [**22, 27**], and at the more recent survey [**47**] for a more updated presentation.

2.3. Boundary regularity. Let me now switch to the boundary regularity, and therefore consider the following Dirichlet problem:

$$\begin{cases} \operatorname{div} a(x,u,Du) = 0 & \text{in } \Omega, \\ u = u_0 & \text{on } \partial\Omega\,, \end{cases} \tag{2.12}$$

where

$$u_0 \in C^{1,\alpha}(\overline{\Omega}, \mathbb{R}^N) \qquad \text{and} \qquad \partial\Omega \text{ is } C^{1,\alpha}\text{-regular}\,, \tag{2.13}$$

and the assumptions (2.2)-(2.3) are in force. In order to describe the boundary partial regularity we start by

DEFINITION 2.4. A point $x \in \partial\Omega$ is a regular boundary point for Du iff there exists a neighborhood $U \subset \mathbb{R}^n$ of x such that $Du \in C^{0,\alpha}(\overline{\Omega \cap U})$. Accordingly, the boundary regular set of Du is

$$\Omega_u^b := \{x \in \partial\Omega : u \in C^{1,\alpha}(\overline{\Omega \cap U}, \mathbb{R}^N) \text{ for some neighborhood } U \text{ of } x\}. \tag{2.14}$$

As a matter of fact, the local interior regularity characterization given in (2.11) extends to the boundary. Indeed, as proved in [**28, 30**], a boundary point $x_0 \in \partial\Omega$ is regular iff the following conditions are satisfied:

$$\begin{cases} \liminf_{\rho\downarrow 0} \fint_{B(x,\rho)\cap\Omega} |Du(y) - (Du)_{B(x,\rho)}|^p \, dy = 0, \\ \\ \limsup_{\rho\downarrow 0} \left| \fint_{B(x_0,\rho)\cap\Omega} Du(y)\, dy \right| < \infty \end{cases} \tag{2.15}$$

and

$$\begin{cases} \liminf_{\rho\downarrow 0} \fint_{B(x,\rho)\cap\Omega} |u(y) - (u)_{B(x_0,\rho)}|^p \, dy = 0, \\ \\ \limsup_{\rho\downarrow 0} \left| \fint_{B(x_0,\rho)\cap\Omega} u(y)\, dy \right| < \infty\,. \end{cases} \tag{2.16}$$

A completely similar result holds in the case of minima of variational Dirichlet problems of the type

$$\begin{cases} u \mapsto \min_{w} \mathcal{F}[w], \\ w \in u_0 + W_0^{1,p}(\Omega, \mathbb{R}^N) , \end{cases} \tag{2.17}$$

where $\mathcal{F}[\cdot]$ is defined in (2.5). The crucial point now is that, *exactly as in the case of Σ_u^0 and Σ_u^1 defined in* (2.9)-(2.10), *conditions* (2.15)-(2.16) *are satisfied only almost everywhere with respect to the Lebesgue measure* and therefore, the whole boundary $\partial\Omega$ being a null set, it follows that *conditions* (2.15)-(2.16) *alone do not yield the existence of even one regular boundary point.* Moreover, and even for a case of systems with a simpler structure, Giaquinta has proved in [**21**] the existence of boundary irregular points, although both u_0 and $\partial\Omega$ are assumed to be smooth; this also confirms the fact that the occurrence of singularities is a genuine peculiarity of vectorial problems. Summarizing, while boundary singularities are known to exist, even the existence of one regular boundary point remains at this stage an open problem. This situation is in striking contrast to what happens in the scalar case $N = 1$, when every boundary point is regular [**25**]. In the rest of the paper I am going to describe approaches leading to the existence of regular boundary points for large classes of vectorial problems.

2.4. A dual approach: L^p-Hölder continuity and Sobolev traces. Here I would like to describe a circle of related ideas leading to establish, simultaneously, singular set estimates in the interior and existence of boundary regular points on the one hand, and a weak form of global Hölder continuity replacing the partial one together with the existence of gradient boundary traces on the other hand. For simplicity I will refer to systems with only x-dependence as

$$\text{div } a(x, Du) = 0 \, , \tag{2.18}$$

with related boundary value problems

$$\begin{cases} \text{div } a(x, Du) = 0 & \text{in } \Omega, \\ u = u_0 & \text{on } \partial\Omega \, , \end{cases} \tag{2.19}$$

where u_0 and Ω are as in (2.13). In [**45**] it is possible to find

THEOREM 2.4 ([**45**]). *Let $u \in W^{1,p}(\Omega, \mathbb{R}^N)$ be a weak solution to the system* (2.18), *under the assumptions* (2.2) *and* (2.3). *Then, denoting by Σ_u the singular set of u in the sense of* (2.7) *it follows that*

$$\dim_{\mathcal{H}}(\Sigma_u) \le n - 2\alpha \, . \tag{2.20}$$

The previous result allows us to bound the Hausdorff dimension of the singular set according to the regularity of the coefficients. There is a dual version of this result, which actually implies it. We have seen that solutions are $C^{0,\alpha}$ locally regular, only outside the singular set Σ_u. On the other hand, if we are allowed to measure Hölder continuity in a weaker way *by performing an integral averaging procedure over* Ω, it turns out that Du is still Hölder continuous with exponent α locally in Ω in such a weaker sense, while no singular sets appear in this respect. For this, more definitions are needed.

DEFINITION 2.5. A measurable map $w : \Omega \to \mathbb{R}^k$ belongs to the *fractional Sobolev space* $W^{\sigma,p}(\Omega,\mathbb{R}^k)$ for parameters $\sigma \in (0,1)$ and $p \in [1,\infty)$, provided the following Gagliardo-type norm is finite:

$$\|w\|_{W^{\sigma,p}(\Omega)} = \left(\int_\Omega |w(x)|^p \, dx\right)^{\frac{1}{p}} + \left(\int_\Omega\int_\Omega \frac{|w(x)-w(y)|^p}{|x-y|^{n+\sigma p}}\, dx\, dy\right)^{\frac{1}{p}} . \tag{2.21}$$

The map w belongs to the Nikolski space $\mathcal{N}^{\sigma,p}(\Omega,\mathbb{R}^k)$ if and only if

$$\|w\|_{\mathcal{N}^{\sigma,q}(\Omega)} := \|w\|_{L^p(\Omega)} + \sup_{h,e}\left(\int_{A_h} \frac{|w(x+he)-w(x)|^p}{|h|^{\sigma p}}\right)^{\frac{1}{p}} < \infty , \tag{2.22}$$

where the supremum is taken considering $h \in R\setminus\{0\}$, $e \in R^n$ such that $|e| = 1$, and $A_h := \{x \in \Omega : \operatorname{dist}(x,\partial\Omega) > |h|\}$. The local variants $W^{\sigma,p}_{\mathrm{loc}}(\Omega)$ and $\mathcal{N}^{\sigma,p}_{\mathrm{loc}}(\Omega)$ are defined in the obvious way.

The relations between the spaces introduced in the previous definition are related by the following strict inclusions:

$$W^{\sigma,p}(\Omega,\mathbb{R}^k) \subset \mathcal{N}^{\sigma,p}(\Omega,\mathbb{R}^k) \subset W^{\sigma-\varepsilon,p}(\Omega,\mathbb{R}^k) , \tag{2.23}$$

which hold for every $\varepsilon \in (0,\sigma)$. Such spaces are particular instances of another family of spaces, the so-called Besov spaces $B^\sigma_{p,q}$ [**50**]: indeed $B^\sigma_{p,p} \equiv W^{\sigma,p}$, $B^\sigma_{p,\infty} \equiv \mathcal{N}^{\sigma,p}$, and $B^\sigma_{\infty,\infty} \equiv C^{0,\sigma}$. At this point it follows that

THEOREM 2.5 ([**45**]). *Let $u \in W^{1,p}(\Omega,\mathbb{R}^N)$ be a weak solution to the system* (2.18), *under the assumptions* (2.2) *and* (2.3). *It follows that*

$$Du \in W^{\sigma,p}_{\mathrm{loc}}(\Omega) \qquad \text{for every } \sigma < \frac{2\alpha}{p} . \tag{2.24}$$

Now, let me read $C^{0,\sigma}$-regularity as σ-Hölder continuity in the L^∞-norm:

$$\|Du(\cdot+h) - Du(\cdot)\|_{L^\infty(\Omega')} \le [Du]_{0,\sigma;\Omega'}|h|^\sigma, \qquad \Omega' \Subset \Omega , \tag{2.25}$$

for any $h \in \mathbb{R}^n$ such that $|h| \le \operatorname{dist}(\Omega',\partial\Omega)$. Whilst (2.25) is in general lost in the vectorial case due to the presence of the singular set, it happens that the gradient belongs to some Nikolski space, and so L^p-Hölder continuity holds:

$$\|Du(\cdot+h) - Du(\cdot)\|_{L^p(\Omega')} \le \|u\|_{\mathcal{N}^{\sigma,p}(\Omega')}|h|^\sigma, \quad \Omega' \Subset \Omega, \quad 1 < p < \infty . \tag{2.26}$$

In other words, *when passing from the scalar to the vectorial case, Hölder continuity is preserved modulo being read in the right form.*

The linkage between Theorem 2.4 and Theorem 2.5 is very simple; as a consequence of general properties of fractional Sobolev spaces (see Proposition 2.1 below) we have that

$$(2.24) \Longrightarrow (2.20) .$$

Now, let me switch to the case of boundary regularity, and let me denote by Σ^b_u the set of boundary irregular points, that is, according to Definition 2.4,

$$\Sigma^b_u := \partial\Omega \setminus \Omega^b_u .$$

Therefore the question is now whether or not $\Sigma^b_u \ne \partial\Omega$, while the boundary analog of interior boundary regularity would clearly be

$$\mathcal{H}^{n-1}(\Sigma^b_u) = 0 . \tag{2.27}$$

Now, on the one hand, since the final aim is to prove the boundary regularity of the gradient of solutions, we expect that Du is in at least in some weak sense defined on $\partial\Omega$. In other words, it should be possible to define the trace of Du. Therefore let me recall the actions of the classical trace operator,

$$\text{Tr}\colon W^{\sigma,p}(\Omega) \to W^{\sigma-\frac{1}{p},p}(\partial\Omega), \qquad \sigma > \frac{1}{p}. \tag{2.28}$$

Keeping the previous result in mind, together with the result of Theorem 2.5, we realize that extending (2.24) up to the boundary, and eventually requiring that

$$\alpha > \frac{1}{2} \tag{2.29}$$

would yield

$$Du|_{\partial\Omega} \in W^{\frac{2\alpha-1-\varepsilon}{p},p}(\partial\Omega, \mathbb{R}^{Nn}) \qquad \text{for every } \varepsilon \in (0, 2\alpha-1)\,,$$

and therefore the existence of the boundary trace of Du. On the other hand, the dual aspect between Theorems 2.4 and 2.5 is revealed here as follows: using general properties of fractional Sobolev functions (see Proposition 2.2 below) and again assuming that (2.24) holds globally, we would conclude with

$$\mathcal{H}^{n-2\alpha+\varepsilon}(\Sigma_u^b) = 0 \qquad \text{for every } \varepsilon > 0,$$

assuming (2.29) at this point yields the desired partial boundary regularity, that is, (2.27). As a matter of fact, partial boundary regularity holds exactly in the way outlined above:

THEOREM 2.6 ([**17**]). *Let $u \in W^{1,p}(\Omega, \mathbb{R}^N)$ be the unique weak solution to the problem* (2.18), *under the assumptions* (2.2), (2.3) *and* (2.13). *It follows that*

$$Du \in W^{\sigma,p}(\Omega) \qquad \textit{for every } \sigma < \frac{2\alpha}{p}\,.$$

Moreover, assuming (2.29), *it follows also that*

$$\mathcal{H}^{n-1}(\Sigma_u^b) = 0; \tag{2.30}$$

that is, almost every boundary point, in the sense of the usual surface measure, is a regular point for Du.

REMARK 2.1 (The two-dimensional case). In the following, when referring to the boundary regularity, I shall mostly consider the case

$$n \geq 3$$

since in the two-dimensional case, the singular set is in general empty, and this type of global regularity extends up to the boundary. For this I refer to the interesting work of Arkhipova [**1**]; see also [**36**, Theorem 1.7] for the interior variational case.

2.5. Dimension reduction and fractional differentiability. The dimension reduction effect of fractional differentiability is displayed in the following propositions; the results below are actually standard, and for the proof we refer to [**17, 45**] and the references therein.

PROPOSITION 2.1. *Suppose that $w \in W^{\theta,q}_{\text{loc}}(\Omega, \mathbb{R}^k)$ with $\theta \in (0,1]$, $q \geq 1$, $k \in \mathbb{N}$. Moreover, let*

$$A := \left\{ x \in \Omega \ : \ \limsup_{\varrho \downarrow 0} \fint_{B(x,\varrho)} |w(y) - (w)_{B(x,\varrho)}|^q \, dy > 0 \right\},$$

$$B := \left\{ x \in \Omega \ : \ \limsup_{\varrho \downarrow 0} |(w)_{B(x,\varrho)}| = \infty \right\} .$$

Then

$$\dim_{\mathcal{H}}(A) \leq n - \theta q \qquad \textit{and} \qquad \dim_{\mathcal{H}}(B) \leq n - \theta q .$$

PROPOSITION 2.2. *Suppose that $w \in W^{\theta,q}(\Omega, \mathbb{R}^k)$ with $\theta \in (0,1]$, $q \geq 1$, $k \in \mathbb{N}$, and where Ω is a Lipschitz domain. Moreover, let*

$$A := \left\{ x \in \bar{\Omega} \ : \ \limsup_{\varrho \downarrow 0} \fint_{B(x,\varrho)\cap\Omega} |v(y) - (v)_{B(x,\varrho)\cap\Omega}|^q \, dy > 0 \right\} ,$$

$$B := \left\{ x \in \bar{\Omega} \ : \ \limsup_{\varrho \downarrow 0} |(w)_{B(x,\varrho)\cap\Omega}| = \infty \right\} .$$

Then

$$\dim_{\mathcal{H}}(A) \leq n - \theta q \qquad \textit{and} \qquad \dim_{\mathcal{H}}(B) \leq n - \theta q .$$

3. More on elliptic systems

3.1. Systems with u-dependence. Here I will discuss the boundary regularity results available for solutions to Dirichlet problems involving complete systems as in (2.12). In the previous section we have seen that conditions such as (2.29) play a decisive role in the boundary regularity in the context of Theorem 2.6. When dealing with systems of the type (2.1), the presence of $u(x)$ acts as a measurable coefficient when considered in $a(\cdot)$, and therefore $x \to a(x, u(x), \cdot)$ globally looks like a measurable map, therefore making a condition such as (2.29) immaterial. In fact, the analog of Theorem 2.4 features a weaker dimension reduction, actually based on the use of Gehring's lemma. Let me recall the background: both for solutions to systems as in (2.1) and for minima of functionals of the type in (2.5), it can be proved that there exists a number q, depending only on $n, L/\nu, p, N$ such that

$$Du \in L^q_{\text{loc}}(\Omega, \mathbb{R}^{Nn}), \qquad q > p . \tag{3.31}$$

THEOREM 3.1 ([**46**]). *Let $u \in W^{1,p}(\Omega, \mathbb{R}^N)$ be a weak solution to the system (2.1), under the assumptions (2.2) and (2.3). Then, denoting by Σ_u the singular set of u in the sense of (2.7), it follows that*

$$\dim_{\mathcal{H}}(\Sigma_u) \leq n - \min\{2\alpha, q - p\} . \tag{3.32}$$

Note that in [**46**] the previous result is proved under the additional assumption that solutions are bounded, but this only depends on the fact that the systems considered are not homogeneous; moreover, the precise exponent can be computed following [**36**]. For minima of integral functionals it follows that

THEOREM 3.2 ([**36**]). *Let $u \in W^{1,p}(\Omega, \mathbb{R}^N)$ be a local minimizer of the functional $\mathcal{F}$, under the assumptions (2.6) and (2.3). Then, denoting by Σ_u the singular set of u in the sense of (2.7), it follows that*

$$\dim_{\mathcal{H}}(\Sigma_u) \leq n - \min\{\alpha, q - p\} . \tag{3.33}$$

Note that, comparing (3.33) with (3.32), we find α instead of 2α, with a loss of regularity similar to the one we already found by passing from Theorem 2.2 to Theorem 2.3.

It is clear that, also when pushing a condition (3.32) up to the boundary, we cannot get the existence of regular boundary points since the estimates on the exponent q in (3.31) do not ensure in general that $q - p > 2\alpha$. On the one hand, in low dimensions, solutions are in general a bit more regular, and the idea is to use the higher regularity of $u(x)$ to get a better control of the oscillations of $x \mapsto a(x, u(x), \cdot)$. A first result in this direction is

THEOREM 3.3 ([**46**]). *Let $u \in W^{1,p}(\Omega, \mathbb{R}^N)$ be a weak solution to the system* (2.1), *under the assumptions* (2.2) *and* (2.3), *and assume also that*

$$n \leq p + 2 . \tag{3.34}$$

Then, denoting by Σ_u the singular set of u in the sense of (2.7), *it follows that*

$$\dim_{\mathcal{H}}(\Sigma_u) \leq n - 2\alpha . \tag{3.35}$$

Estimate (3.34) is strong enough to allow a dimension reduction when pushed up to the boundary. Indeed we have

THEOREM 3.4 ([**17**]). *Let $u \in W^{1,p}(\Omega, \mathbb{R}^N)$ be a weak solution to the problem* (2.12), *under the assumptions* (2.2), (2.3) *and* (2.13); *assume also that*

$$n \leq p + 2 \qquad \text{and} \qquad \alpha > \frac{1}{2}$$

hold. Then almost every boundary point, in the sense of the usual surface measure, is a regular point for Du; that is, (2.30) *holds.*

The last two theorems are again a consequence of a fractional differentiability result, this time obtained on open subsets of properly large Hausdorff codimension, where the solution is proved to be Hölder continuous; this is essentially the effect of the low-dimensional assumption (3.34). Further boundary results are still available; for this I would recommend to look at [**17**], and also at the last section of this paper.

3.2. Additional results. The boundary regularity results presented up to now are the first and, up to now, the only ones applying to general nonlinear systems of the type in (2.12) and (2.19). The literature anyway contains several papers dedicated to the issue in certain special situations. A remarkable one is provided by those called, with some terminological ambiguity, "quasilinear" systems, i.e. problems of the type

$$\begin{cases} \operatorname{div}\,(A(x,u)Du) = 0 & \text{in } \Omega, \\ u = u_0 & \text{on } \partial\Omega\,, \end{cases} \tag{3.36}$$

where $A \colon \Omega \times \mathbb{R}^n \to \mathbb{R}^{N^2 n^2}$ is a an elliptic matrix with Hölder continuous entries:

$$\nu\langle \lambda, \lambda\rangle \leq \langle A(x,y)\lambda, \lambda\rangle, \qquad |A(x,u)| \leq L\,,$$

are satisfied for every choice of $x \in \Omega$, $y \in \mathbb{R}^N$ and $\lambda \in \mathbb{R}^{Nn}$. In this case, partial boundary regularity results can be found in the papers [**1, 2, 3, 10, 29**]. Further boundary estimates for solutions to systems without coefficients can be found in [**9**].

4. Variational problems

The case of boundary regularity of solutions to variational problems of the type in (2.17) is, for a number of important reasons, rather different from the one involving elliptic systems; let me just mention two. The first, which is intrinsic to the problem, is that when considering functionals such as the one in (2.5), these typically do not possess the associated Euler-Lagrange system; indeed, this formally looks like

$$\operatorname{div} F_z(x, u, Du) = F_u(x, u, Du) \tag{4.37}$$

and the right-hand side of the previous equations clearly loses meaning when $F(\cdot)$ is not differentiable with respect to the second variable. In (2.6) only Hölder continuity is assumed. This immediately tells us that the techniques developed for systems are not applicable, at least without genuinely new ingredients. Moreover, let me observe that also in the favorable case in which (4.37) makes sense (assume $F(\cdot)$ to be as smooth as you like) the one in (4.37) turns out to be a system with right-hand side of critical growth

$$|F_u(x, u, Du)| \leq c(|Du|^p + 1)$$

so that even partial regularity holds only when assuming that solutions are bounded; on the other hand, the counterexample in [**52**] shows that even in the smoothest case of totally autonomous analytic integrands, $F(Du)$ minima can be unbounded. This suggests that in order to gain regularity, a stronger use of the minimality properties must be made, with a situation in a certain sense analogous to the one on harmonic mappings into spheres: while stationary mappings, that is, solutions to the system can be everywhere discontinuous (as shown in [**49**]) for minimum points,m a satisfying partial regularity theory can be derived; see [**51**]. A second difficulty in tackling the case of minima, which this time is more linked to the type of technique employed here, is that while trying to proceed as for the case of systems as in the previous two sections, that is, trying to push estimate (3.33) up to the boundary, even in the most favorable case that q is large, the estimate gives an upper bound of $n - \alpha$, which is still too large in order to get boundary partial regularity; this would force us to have $\alpha > 1$, while here it is $\alpha \in (0, 1)$. For these and other reasons, the problems of establishing boundary regularity have remained essentially untouched for several years, while a first result has been obtained by Jost and Meier [**35**], who proved everywhere boundary regularity of minima (not considering the gradient) of quadratic functionals with special structure such as

$$w \mapsto \int_\Omega c(x, w)|Dw|^2 \, dx\,. \tag{4.38}$$

The main feature of the functional in (4.38) is that its energy density depends on the gradient variable only via its euclidean norm $|Dw|$. This renders the situation very special and allows the authors to use powerful techniques that are not available for more general functionals. In fact, full boundary regularity follows in such cases, a result that is known to fail in the general vectorial case; see [**22**]. These special techniques are in turn related to the result of Uhlenbeck [**53**]. As a matter of fact, a nontrivial extension of the result of Jost and Meier covers degenerate cases such as

$$w \mapsto \int_\Omega c(x, w)|Dw|^p \, dx\,, \qquad p > 1\,,$$

and can be found in [**16**].

Here I would like to present some of the results obtained in [**38**], which seems to be, on the other hand, the first paper to address in large generality the problem of gradient boundary regularity for minima of integral functionals; see also [**37**] for a related announcement. In [**38**], there is actually a number of cases presented, and here I will confine myself to present only some of them, not the most general perhaps, but those which are probably the most suitable to give a flavor of the viewpoints adopted. The results considered will anyway be general enough to cover important model cases such as

$$w \mapsto \int_\Omega c(x,w)f(Dw)\,dx \qquad \text{and} \qquad w \mapsto \int_\Omega c(x)f(Dw) + h(x,w)\,dx\,. \tag{4.39}$$

As mentioned above, the first problem to tackle is that in the best possible case the estimate in (3.33) provides an upper bound as $n-\alpha$, which is insufficient. Therefore the first point in [**38**] is to identify classes of functionals for which we can get an estimate of the type $n-2\alpha$. Then, imposing (2.29) gives the desired boundary partial regularity. A first important outcome is that such classes include functionals as in (4.39). I will present here three significant model results; further cases can be found in [**38**]. The first concerns the significant model case

$$\mathcal{F}_1[w] := \int_\Omega c(x)f(Dw) + d(w)(1+|Dw|^2)^{\frac{\gamma}{2}}\,dx\,. \tag{4.40}$$

I shall assume the following standard Hölder continuity of the coefficients:

$$0 < \nu \le c(x) \le L\,, \qquad\qquad c(\cdot) \in C^{0,\alpha}(\Omega)\,, \tag{4.41}$$

whenever $x \in \Omega$, and

$$0 \le d(v) \le L\,, \qquad\qquad d(\cdot) \in C^{0,\beta}(\Omega)\,, \tag{4.42}$$

whenever $v \in \mathbb{R}^N$, while, as usual, the following strong convexity of the C^2 density $f(\cdot)$ is assumed to hold whenever $z, \lambda \in \mathbb{R}^{Nn}$:

$$\nu|z|^p \le f(z)\,, \qquad \nu(1+|z|^2)^{\frac{p-2}{2}}|\lambda|^2 \le \langle f_{zz}(z)\lambda,\lambda\rangle \le L(1+|z|^2)^{\frac{p-2}{2}}|\lambda|^2\,. \tag{4.43}$$

THEOREM 4.1 ([**38**]). *Under the assumptions* (4.41)-(4.43) *with*

$$\alpha > \frac{1}{2}, \qquad\qquad \beta > \frac{2}{3}, \qquad\qquad \gamma \le \frac{2p}{3}\,, \tag{4.44}$$

let $u \in u_0 + W^{1,p}_0(\Omega,\mathbb{R}^N)$ *be a solution to the Dirichlet problem* (2.17) *with* (2.13) *in force and* $\mathcal{F}[\cdot] \equiv \mathcal{F}_1[\cdot]$. *Then* $\mathcal{H}^{n-1}$*-almost every boundary point is regular for* Du*; that is,* (2.30) *holds. Moreover, the global higher differentiability*

$$Du \in W^{\frac{1}{p}+\varepsilon,p}(\Omega,\mathbb{R}^{Nn}) \qquad \textit{and} \qquad Du|_{\partial\Omega} \in W^{\varepsilon,p}(\partial\Omega,\mathbb{R}^{Nn}) \tag{4.45}$$

hold for some $\varepsilon \equiv \varepsilon(\alpha,\beta) > 0$.

The main point in the previous theorem is the interaction between the regularity of $v \to F(\cdot,v,\cdot)$, described in (4.42), and the growth, with respect to the gradient variable Dw, of the part of the integrand depending directly on w. This is explained by the fact that in (4.44) we in particular have $\gamma < p$; moreover, the third inequality in (4.44) tells us that γ cannot be too close to p. Theorem 4.1 is actually a particular case of a more general phenomenon described in [**38**]. Indeed, the theory presented in [**38**] extends to more general functionals, including those of the type

$$\mathcal{F}_2[w] := \int_\Omega f(Dw) + h(x, w, Dw)\, dx\,, \tag{4.46}$$

where an essential feature is that the function $h(\cdot)$ grows at a suitably lower rate

$$0 \leq h(x, y, z) \leq L(1 + |z|^2)^{\gamma/2}, \qquad \gamma < p\,. \tag{4.47}$$

Specifically, let me consider the following set of assumptions:

$$\begin{cases} \nu|z|^p \leq F(x, v, z) \leq L(1 + |z|^2)^{\frac{p}{2}}, \\ \nu(1 + |z|^2)^{\frac{p-2}{2}}|\lambda|^2 \leq \langle F_{zz}(x, v, z)\lambda, \lambda\rangle \leq L(1 + |z|^2)^{\frac{p-2}{2}}|\lambda|^2, \\ |F(x, u, z) - F(x, v, z)| \leq L\omega_\beta(|u - v|)(1 + |z|^2)^{\frac{\gamma}{2}}, \\ |F_z(x, v, z) - F_z(y, v, z)| \leq L|x - y|^\alpha(1 + |z|^2)^{\frac{p-1}{2}}, \\ |F_z(x, u, z) - F_z(x, v, z)| \leq L\omega_\beta(|u - v|)(1 + |z|^2)^{\frac{\gamma-1}{2}} \end{cases} \tag{4.48}$$

where $\omega_\alpha(\cdot), \omega_\beta(\cdot)$ are as in (2.3), and the rest are as in (2.6). Except for $(4.48)_4$, these assumptions are standard in the context of criteria for partial interior regularity and for boundary regularity. Condition $(4.48)_4$ is anyway not particularly restrictive and automatically satisfied in several model cases as those in (4.39) and (4.46). Now it follows that

THEOREM 4.2 ([**38**]). *Under the assumptions* (4.48) *with* $\gamma < p$, *take*

$$\frac{2}{3} \leq s \leq \frac{p}{p+1} \tag{4.49}$$

and assume that

$$\alpha > \frac{1}{2}, \qquad \beta > s, \qquad \gamma \leq ps + \frac{2ps}{n-2}\,. \tag{4.50}$$

Let $u \in u_0 + W^{1,p}_0(\Omega, \mathbb{R}^N)$ *be a solution to the Dirichlet problem* (2.17) *with* (2.13) *in force. Then* $\mathcal{H}^{n-1}$*-almost every boundary point is regular for* Du, *and* (4.45) *holds.*

The previous statement should be understood as follows: (α, β, γ) characterize the structure of the integrand $F(\cdot)$ via (4.48), while the parameter s, when varying in the range (4.49), can be considered to parametrize the various results. Condition (4.50) states that *the less regularity we assume on* $y \mapsto F(\cdot, y, \cdot)$, *the less it can be allowed to grow with respect to* z. The extreme cases of assumptions (4.49)-(4.50) are given by $s = 2/3$ corresponding to

$$\alpha > \frac{1}{2}, \qquad \beta > \frac{2}{3}, \qquad \gamma \leq \frac{2p}{3} + \frac{4p}{3(n-2)}\,,$$

and by $s = p/(p+1)$ corresponding to

$$\alpha > \frac{1}{2}, \qquad \beta > \frac{p}{p+1}, \qquad \gamma \leq \frac{p^2}{p+1} + \frac{2p^2}{(n-2)(p+1)}\,.$$

Thus, as expected, strong growth of $z \mapsto F(\cdot, \cdot, z)$ and $z \mapsto F_z(\cdot, \cdot, z)$ must be compensated by a higher degree of Hölder continuity of $y \mapsto F(\cdot, y, \cdot)$ and $y \mapsto F_z(\cdot, y, \cdot)$.

At this point, dealing with the borderline case $\gamma = p$ is possible, but, as suggested by Theorem 3.4, provided by considering a low-dimensional assumption.

THEOREM 4.3 ([**38**]). *Under the assumptions* (2.6) *with* $\gamma = p$ *and*

$$\alpha > \frac{1}{2}, \qquad \beta > \max\left\{1 - \frac{2}{n}, \frac{2}{3}\right\}, \qquad n \leq p+2, \tag{4.51}$$

let $u \in u_0 + W_0^{1,p}(\Omega, \mathbb{R}^N)$ *be a solution to the Dirichlet problem* (2.17) *with* (2.13) *in force. Then* $\mathcal{H}^{n-1}$*-almost every boundary point is regular for* Du*; that is,* (2.30) *holds.*

Further results can again be found in [**38**]; in particular, an approach to relax the condition (2.29) is proposed through the consideration of certain types of rough coefficients. For more on this aspect the reader is referred to the last section of this paper, where such an approach is presented in the case of systems.

5. Nonlinear parabolic systems

In this section, I will describe some of the boundary regularity results recently obtained in [**5, 6**]; further results will appear in [**7**]. For the sake of simplicity, I will confine myself to systems with linear growth, and without explicit dependence on u, again referring to [**5, 6, 7**] for results concerning more general cases. Therefore this means that I am going to deal with parabolic systems of the type

$$u_t - \operatorname{div} a(x, t, Du) = 0, \tag{5.52}$$

where the vector field $a(\cdot)$ will in general satisfy assumptions of the type (2.2) with $p = 2$.

5.1. Setting of the problem. I will consider Cauchy-Dirichlet problems of the type

$$\begin{cases} u_t - \operatorname{div} a(x, t, Du) = 0 & \text{in } \Omega_T, \\ u = g & \text{on } \partial_{\mathcal{P}}\Omega_T, \end{cases} \tag{5.53}$$

with a Carathéodory vector field $a \colon \Omega_T \times \mathbb{R}^N \times \mathbb{R}^{Nn} \to \mathbb{R}^{Nn}$ and $g(\cdot) \in L^2(\Omega, \mathbb{R}^N)$. The problem (5.53) will be considered in a cylindrical domain

$$\Omega_T := \Omega \times (0, T),$$

where $\Omega \subset \mathbb{R}^n$, $n \geq 2$ is a bounded domain in $\mathbb{R}^n$ and $T > 0$. The parabolic boundary of Ω_T consists of the lateral and initial boundary and will be denoted by

$$\partial_{\mathcal{P}}\Omega_T = \big(\partial\Omega \times (0, T)\big) \cup \big(\Omega \times \{0\}\big) \cup \big(\partial\Omega \times \{0\}\big).$$

The set $\big(\partial\Omega \times \{0\}\big)$ is called the set of edge points; moreover, in the following it is

$$\Omega_0 := \Omega \times \{0\}.$$

When dealing with parabolic problems it is convenient to equip $\mathbb{R}^{n+1}$ with the parabolic metric given by

$$d_{\mathcal{P}}((x,t),(y,s)) := \max\left\{|x-y|, \sqrt{|t-s|}\right\}, \qquad x, y \in \mathbb{R}^n,\ s, t \in \mathbb{R}. \tag{5.54}$$

Accordingly, the parabolic cylinder centered at (x_0, t_0) with radius $\varrho > 0$ is defined as the ball of radius ϱ, centered at (x_0, t_0), with respect to the previous parabolic metric, that is,

$$Q_\varrho(x_0, t_0) \equiv B_\varrho(x_0) \times (t_0 - \varrho^2, t_0 + \varrho^2). \tag{5.55}$$

First let me specify here the notion of solutions suited to the problem (5.53).

DEFINITION 5.1. A map $u \in L^2(0,T;W^{1,2}(\Omega,\mathbb{R}^N))$ is called a (weak) solution to $(5.53)_1$ if and only if

$$\int_{\Omega_T} u \cdot \varphi_t - \langle a(x,t,u,Du), D\varphi\rangle \, dx\, dt = 0$$

holds for every test-function $\varphi \in C_0^\infty(\Omega_T,\mathbb{R}^N)$, and the following boundary conditions hold:

$$u(\cdot,t) - g(\cdot,t) \in W_0^{1,2}(\Omega,\mathbb{R}^N) \qquad \text{for a.e. } t \in (0,T)$$

in the sense that

$$\lim_{h\downarrow 0} \frac{1}{h}\int_0^h \int_\Omega |u(x,t)-g(x,0)|^2\, dx\, dt = 0\,. \tag{5.56}$$

Here I shall assume that the vector field $a\colon \Omega_T \times \mathbb{R}^{Nn} \to \mathbb{R}^{Nn}$ is of class C^1 with respect to the gradient variable and fulfills the following growth and ellipticity assumptions:

$$\begin{cases} |a(x,t,z)| + (1+|z|)|a_z(x,t,z)| \le L\,(1+|z|), \\ \langle a_z(x,t,z)\lambda,\lambda\rangle \ge \nu\,|\lambda|^2, \\ |a(x,t,z)-a(x_0,t,v,z)| \le L\;\omega_\alpha(|x-x_0|)(1+|z|), \\ |a(x,t,z)-a(x,t_0,v,z)| \le L\;\omega_\alpha(\sqrt{|t-t_0|})(1+|z|) \end{cases} \tag{5.57}$$

for every choice of $x, x_0 \in \Omega$, $t, t_0 \in (0,T)$ and $z, \lambda \in \mathbb{R}^{Nn}$. The structure constants will as usual satisfy $0 < \nu \le 1 \le L < \infty$. Concerning the regularity of the boundary datum the assumptions are, initially,

$$\partial\Omega \text{ is } C^{1,\alpha}, \quad Dg \in C^{\alpha,\alpha/2}(\overline{\Omega}\times[0,T),\mathbb{R}^{Nn}), \quad g_t \in L^{2,2-2\alpha}(\Omega_T,\mathbb{R}^N). \tag{5.58}$$

The second inclusion above means that the spatial derivative Dg is Hölder continuous in Ω_T with exponent α with respect to the parabolic metric defined in (5.54). The last inclusion in (5.58) refers to Morrey spaces; the definition is

DEFINITION 5.2. With $q \ge 1$, $\theta \in [0,n+2]$, a measurable map $v\colon \Omega_T \to \mathbb{R}^k$, $k \ge 1$ belongs to the (parabolic) Morrey space $L^{q,\theta}(Q;\mathbb{R}^k)$ if and only if

$$\|v\|^q_{L^{q,\theta}(\Omega_T;\mathbb{R}^k)} := \sup_{z_0\in\Omega_T,\, 0<\rho<\operatorname{diam}(\Omega_T)} \varrho^{\theta-(n+2)} \int_{\Omega_T\cap Q_\rho(x_0,t_0)} |v|^q dx\, dt < \infty$$

holds for every parabolic cylinder $Q_\rho(x_0,t_0) \subset \mathbb{R}^{n+1}$.

Let me also finally observe that assumptions $(5.57)_{3,4}$ actually amount to requiring that the map

$$(x,t) \mapsto \frac{a(x,t,z)}{(1+|z|)^{p-1}}$$

is α-Hölder continuous with respect to the parabolic metric defined in (5.54), uniformly with respect to the spatial gradient variable z.

5.2. Partial regularity in the interior. I will first give here a short overview on the interior results, therefore considering weak solutions to $(5.53)_1$, not taking into account the boundary datum $g(\cdot)$. Exactly as in the elliptic case, when dealing with general nonlinear parabolic systems in general the results achievable do not go beyond the partial regularity, i.e. again regularity outside a closed negligible set. For results in this direction I refer to [**18, 19**] and the references therein. A typical interior partial regularity result in this direction is

THEOREM 5.1 ([**18**]). *Let* $u \in L^2(0,T;W^{1,2}(\Omega;\mathbb{R}^N))$ *be a weak solution of the nonlinear parabolic system* $(5.53)_1$ *in* Ω_T *under the assumptions* (5.57). *Then there exists an open subset* $\Omega_u \subset \Omega_T$ *such that*

$$Du \in C^{\alpha,\alpha/2}(\Omega_u,\mathbb{R}^{Nn}) \qquad \text{and} \qquad |\Omega_T \setminus \Omega_u| = 0\,.$$

Moreover, defining $\Sigma_u := \Omega_T \setminus \Omega_u$ *as the singular set of* u, *it follows that*

$$\Sigma_u \subset \Sigma_u^1 \cup \Sigma_u^2\,,$$

where

$$\Sigma_u^1 := \left\{(x_0,t_0) \in \Omega_T : \liminf_{\varrho \downarrow 0} \fint_{Q_\varrho(x_0,t_0)} |Du - (Du)_{Q_\varrho(x_0,t_0)}|^2\, dx\, dt > 0\right\}$$

and

$$\Sigma_u^2 := \left\{(x_0,t_0) \in \Omega_T \ : \ \limsup_{\varrho \downarrow 0} |(Du)_{Q_\varrho(x_0,t_0)}| = \infty\right\}.$$

As in the elliptic case, one would like to know a bit more on the size of the singular set Σ_u. To this aim, since the setting is $\mathbb{R}^{n+1}$ equipped with the parabolic metric, it is appropriate to introduce the so-called parabolic Hausdorff measure, with a related concept of dimension. Such a measure is obtained in a standard way using, in the usual Hausdorff-Carathéodory measures construction, the parabolic balls generated by the parabolic metric defined in (5.54), that is, the standard parabolic cylinders defined in (5.55).

DEFINITION 5.3. For $s \in [0, n+2]$ and $F \subset \mathbb{R}^{n+1}$, let

$$\mathcal{H}^s_{\mathcal{P},\delta}(F) := \inf\left\{\sum_{i=1}^{\infty} R_i^s \ : \ F \subset \bigcup_{i=1}^{\infty} Q((x_i,t_i),R_i), \quad R_i \leq \delta\right\}.$$

The s-dimensional parabolic Hausdorff measure is defined by

$$\mathcal{H}^s_{\mathcal{P}}(F) := \sup_{\delta>0} \mathcal{H}^s_{\mathcal{P},\delta}(F)\,.$$

Accordingly, the parabolic Hausdorff dimension is defined by

$$\dim_{\mathcal{P}}(F) := \inf\{s > 0 \ : \ \mathcal{H}^s_{\mathcal{P}}(F) = 0\} = \sup\{s > 0 \ : \ \mathcal{H}^s_{\mathcal{P}}(F) = \infty\}\,. \tag{5.59}$$

Let me observe that, due to the faster shrinking in the time direction with respect to the spatial one of the parabolic cylinders defined in (5.55), the limit parabolic Hausdorff dimension in $\mathbb{R}^{n+1}$ is $n+2$: $\dim_{\mathcal{P}}(F) \leq n+2$ for every $F \subset \mathbb{R}^{n+1}$, *while* $\mathcal{H}^{n+2}_{\mathcal{P}}$ *is equivalent to the Lebesgue measure in* $\mathbb{R}^{n+1}$. At this point, nontrivial information about the smallness of a set F in terms of parabolic Hausdorff dimension amounts to proving that

$$\dim_{\mathcal{P}}(F) < n+2$$

rather than $\dim_{\mathcal{P}}(F) < n+1$. As a matter of fact, the following analog of Theorem 2.4 holds:

THEOREM 5.2 ([**18**]). *Let* $u \in L^2(0,T;W^{1,2}(\Omega;\mathbb{R}^N))$ *be a weak solution of the nonlinear parabolic system* $(5.53)_1$ *in* Ω_T *under the assumptions* (5.57). *Denote by* Σ_u *his singular set of* u *in the sense of Theorem 5.3; then it follows that*

$$\dim_{\mathcal{P}}(F) < n+2-2\alpha\,. \tag{5.60}$$

Towards the existence of regular boundary points, following the elliptic strategy, the previous result will suggest a criterium for establishing the reasonable singular sets estimates at the boundary, with a related criterium to establish the existence of regular boundary points.

5.3. Boundary regularity. I will now concentrate on the boundary situation. As in the elliptic case, in order to study the boundary regularity of solutions, we recall the definition of the set of regular boundary points:

$$\mathrm{Reg}_{\mathcal{P}} u \equiv \big\{(x_0,t_0) \in \partial_{\mathcal{P}}\Omega_T : \\ Du \in C^0(\overline{U \cap \Omega_T}, \mathbb{R}^{Nn}) \text{ for some neighborhood } U \text{ of } (x_0,t_0)\big\}.$$

The first result to mention is concerned with the characterization of regular boundary points:

THEOREM 5.3 ([**5**]). *Let $u \in L^2(0,T;W^{1,2}(\Omega;\mathbb{R}^N))$ be the unique weak solution of the Cauchy-Dirichlet problem* (5.53) *under the assumptions* (5.57) (5.58). *Then,*

$$\partial_{\mathcal{P}}\Omega_T \setminus \mathrm{Reg}_{\mathcal{P}} u \subset \Sigma_u^1 \cup \Sigma_u^2$$

where

$$\Sigma_u^1 = \left\{(x_0,t_0) \in \partial_{\mathcal{P}}\Omega_T : \liminf_{\rho\downarrow 0} \fint_{\Omega_T\cap Q_\rho(x_0,t_0)} \left|Du - (Du)_{\Omega_T\cap Q_\rho(x_0,t_0)}\right|^2 dx\,dt > 0\right\}$$

and

$$\Sigma_u^2 = \left\{(x_0,t_0) \in \partial_{\mathcal{P}}\Omega_T : \limsup_{\rho\downarrow 0} \left|(Du)_{\Omega_T\cap Q_\rho(x_0,t_0)}\right| = \infty\right\}.$$

Furthermore, if $(x_0,t_0) \in \mathrm{Reg}_{\mathcal{P}} u$, then

$$Du \in C^{\alpha,\alpha/2}(\overline{U \cap \Omega_T};\mathbb{R}^{Nn}) \tag{5.61}$$

for some neighborhood U of (x_0,t_0).

Theorem 5.3 is the starting point for the almost everywhere boundary regularity result described at the beginning of this Introduction. Indeed, to prove that a boundary point $(x_0,t_0) \in \partial_{\mathcal{P}}\Omega_T$ is regular, it suffices to prove that the following conditions hold:

$$\liminf_{\rho\downarrow 0} \fint_{\Omega_T\cap Q_\rho(x_0,t_0)} \left|Du - (Du)_{\Omega_T\cap Q_\rho(x_0,t_0)}\right|^2 dx\,dt = 0 \tag{5.62}$$

and

$$\limsup_{\rho\downarrow 0} \left|(Du)_{\Omega_T\cap Q_\rho(x_0,t_0)}\right| < \infty\,. \tag{5.63}$$

The strategy now consists of proving an up-to-the-boundary fractional differentiability result for Du which in turn implies that conditions (5.62) and (5.63) are satisfied at almost every point $(x_0,t_0) \in \partial_{\mathcal{P}}\Omega_T$, where "almost everywhere" refers to the standard boundary surface measure on $\partial_{\mathcal{P}}\Omega_T$.

The natural quantity to measure the size of the singular sets

$$\mathrm{Sing}_{\mathrm{lat}} u = \partial_{\mathrm{lat}}\Omega_T \setminus \mathrm{Reg}_{\mathcal{P}} u \text{ and } \quad \mathrm{Sing}_{\mathrm{ini}} u = \Omega_0 \setminus \mathrm{Reg}_{\mathcal{P}} u$$

is obviously the parabolic Hausdorff dimension defined in Definition 5.3. Taking into account that $\dim_{\mathcal{P}}(\partial_{\mathrm{lat}}\Omega_T) = n+1$, respectively $\dim_{\mathcal{P}}(\Omega_0) = n$, we are looking for conditions ensuring a bound of the form

$$\dim_{\mathcal{P}}(\mathrm{Sing}_{\mathrm{lat}} u) < n+1, \quad \text{resp.} \quad \dim_{\mathcal{P}}(\mathrm{Sing}_{\mathrm{ini}} u) < n.$$

In turn these will ensure that almost every point, in $\partial_{\mathrm{lat}}\Omega_T$ and Ω_0 respectively, and with respect to the related surface measure, will be regular points. Note that, in order to conclude that almost every boundary point is regular with respect to the surface measure, it is not necessary to take into account edge points, i.e. those lying on $\partial\Omega \times \{0\}$, since we already have $\dim_{\mathcal{P}}(\partial\Omega \times \{0\}) = n-1$. As a final consequence, almost every (parabolic) boundary point will be regular when considering the associated surface measure. The main attention is anyway obviously to give the lateral boundary points.

The first result, which actually holds under assumptions weaker than those presented here, is

THEOREM 5.4 ([**6**], Lateral boundary existence). *Let $u \in L^2(0,T;W^{1,2}(\Omega;\mathbb{R}^N))$ be the unique weak solution of the Cauchy-Dirichlet problem* (5.53) *under the assumptions* (5.57)-(5.58). *Moreover, assume that g_t is Hölder continuous in the time variable for some positive exponent. If*

$$\alpha > \frac{1}{2}, \tag{5.64}$$

then $\mathcal{H}^{n+1}_{\mathcal{P}}$-almost every lateral boundary point is a regular point of Du.

Further results can be found in [**6, 7**]; I would just like to remark that the assumptions on the boundary datum $g(\cdot)$ considered in the previous theorem can be weakened. Anyway, the main point here is not the regularity of $g(\cdot)$, which could also be taken to be a smooth function without in my opinion compromising too much the nature of the result, but rather the existence of boundary regular points, as irregular ones already appear when $g(\cdot)$ is indeed smooth. Again, let me observe that condition (5.64) naturally follows from a suitable "boundaryzation" of (5.60), requiring that

$$n+2-2\alpha < n-1 = \text{ dimension of the lateral boundary}.$$

The second result concerns the *existence of regular initial boundary points.* As mentioned before, the initial boundary Ω_0 only has parabolic Hausdorff dimension equal to n. Therefore, contrary to the lateral-boundary situation, we now have to reduce the dimension of the singular set below n. This yields a positive result only when α is close to 1. To be precise, at the initial boundary, the following holds:

THEOREM 5.5 ([**6**], Initial time existence). *Let $u \in L^2(0,T;W^{1,2}(\Omega;\mathbb{R}^N))$ be the unique weak solution of the nonlinear parabolic Cauchy-Dirichlet problem* (5.53) *under the assumptions* (5.57)-(5.58). *Then there exists a positive constant $\delta = \delta(n,\nu,L,\|g\|_{C^{1;\beta,0}})$ such that if*

$$\alpha > 1-\delta \tag{5.65}$$

holds, then $\mathcal{H}^{n}_{\mathcal{P}}$-almost every initial boundary point is a regular point of Du.

This time condition (5.65) serves to make effective the "boundaryzation" of (5.60) in the sense that

$$n+2-2\alpha < n = \text{ dimension of the initial boundary}.$$

6. Rough coefficients

The purpose of this section is to present an approach allowing for considering a condition less strong than (2.29) when dealing with boundary regularity problems. This approach has been proposed in [**38**] both for variational integrals and for elliptic systems, but for the sake of brevity, I will confine myself to presenting this last case here. The viewpoint can now be described as follows: since the existence of regular boundary points is linked to the possibility of proving up-to-the-boundary higher fractional differentiability of Du, that is, L^p-Hölder continuity, and since this is in turn linked to the Hölder continuity of the coefficients as described in Section 2.4, it is therefore natural to consider more "rough" coefficients which are themselves only fractional differentiable rather than Hölder continuous. A model problem to keep in mind is

$$\begin{cases} \operatorname{div}(c(x)a(Du)) = 0 & \text{in } \Omega \\ u = u_0 & \text{on } \partial\Omega \end{cases} \qquad 0 < \nu \leq c(x) \leq L\,, \tag{6.66}$$

where it is desirable to replace the standard condition $c(\cdot) \in C^{0,\alpha}$ and $\alpha > 1/2$ with $c(\cdot) \in W^{\alpha,q} \cap C^{0,\sigma}$ for some suitable exponent q and for arbitrary small σ, therefore allowing only for a modest degree of Hölder continuity. This is actually possible and indeed the following holds.

THEOREM 6.1 ([**38**]). *Let $u \in W^{1,p}(\Omega, \mathbb{R}^N)$ be the unique weak solution to the problem* (6.66), *under the assumptions* (2.2) *and* (2.13); *assume also that*

$$c(\cdot) \in W^{\alpha,n}(\Omega) \cap C^{0,\sigma}(\Omega) \qquad \textit{with } \alpha > \frac{1}{2} \textit{ and } \sigma > 0 \tag{6.67}$$

holds. Then almost every boundary point, in the sense of the usual surface measure, is a regular point for Du and moreover (4.45) *holds.*

In other words it is possible to require $(1/2)$-Hölder continuity only in the L^n scale, rather than in the traditional L^∞ one.

The previous result, again presented for expository purposes, can be extended in several ways. The first direction to take now is to find a way to formulate conditions of the type (6.67) for general nonlinear systems as (2.18), where "coefficients do not split"; i.e., there is no function $c(\cdot)$ to prescribe fractional differentiability. A possible way to do this is provided by the work of DeVore and Sharpley [**14**], who noticed that for $0 < s < 1$ and $1 < q < \infty$, if $c(\cdot) \in W^{s,q}(\mathbb{R}^n)$, then there exists $g \in L^q(\mathbb{R}^n)$ (actually they canonically take the s-fractional sharp maximal function of $c(\cdot)$) such that, for almost all $x, y \in \mathbb{R}^n$ it follows that

$$|c(x) - c(y)| \leq (g(x) + g(y))|x - y|^s\,. \tag{6.68}$$

Observe that in the case $s = 1$, so that $c(\cdot) \in W^{1,q}$, one can take $g \approx M(|Dc|)$. Here $M(\cdot)$ of course denotes the standard Hardy-Littlewood maximal operator on $\mathbb{R}^n$. The authors of [**14**] thereby define a new function space C^s_q, saying that

$$c(\cdot) \in C^s_q(\mathbb{R}^n) \Longleftrightarrow c(\cdot) \in L^q(\mathbb{R}^n) \text{ and (6.68) holds for some } g \in L^q(\mathbb{R}^n)\,.$$

These spaces are generally not Besov spaces, but are nevertheless comparable to them and to the usual fractional Sobolev spaces in the sense that

$$W^{\alpha,q}(\mathbb{R}^n) \subset C^\alpha_q(\mathbb{R}^n) \subset W^{\alpha-\varepsilon,q}(\mathbb{R}^n)$$

holds for every $\varepsilon \in (0, \alpha)$. At this point it is rather obvious how to formulate the fact that systems of the type (2.18) have $C^{s,q}$-coefficients; it is sufficient to prescribe that

$$\frac{|a(x,z) - a(y,z)|}{(1+|z|)^{p-1}} \leq (g(x) + g(y))|x-y|^s$$

holds whenever $x, y \in \Omega$, $z \in \mathbb{R}^{Nn}$, for a function $g(\cdot) \in L^q(\Omega)$. Accordingly, I will switch to describing the assumptions to treat problems of the type (2.12); these are

$$\begin{cases} |a(x,v,z)| + |a_z(x,v,z)|(1+|z|^2)^{\frac{1}{2}} \leq L(1+|z|^2)^{\frac{p-1}{2}}, \\ \nu(1+|z|^2)^{\frac{p-2}{2}}|\lambda|^2 \leq \langle a_z(x,v,z)\lambda, \lambda\rangle, \\ |a(x,v,z) - a(y,v,z)| \leq (g(x)+g(y))|x-y|^\alpha(1+|z|^2)^{\frac{p-1}{2}}, \\ |a(x,u,z) - a(x,v,z)| \leq L\omega_\alpha(|u-v|)(1+|z|^2)^{\frac{\gamma-1}{2}}, \end{cases} \tag{6.69}$$

with the same meaning as in (2.2), but now it is

$$\gamma \leq p \qquad \text{and} \qquad g(\cdot) \in L^{n_s}(\Omega), \qquad n_s \leq n;$$

obviously $\omega_\alpha(\cdot)$ is as in (2.3). In order to have partial boundary regularity and therefore a singular set to estimate, I again assume that

$$\begin{cases} |a(x,u,z) - a(y,v,z)| \leq L\omega_\sigma(|x-y| + |u-v|)(1+|z|^2)^{\frac{p-1}{2}}, \\ \omega_\sigma(s) = \min\{s^\sigma, 1\}, \qquad \text{for some } \sigma > 0. \end{cases} \tag{6.70}$$

Let me remark again that the difference between the "traditional" assumptions (2.2) with $\alpha > 1/2$ and the ones in (6.69)-(6.70) is that here the usual "L^∞ Hölder continuity" is required only with an arbitrarily small exponent $\sigma > 0$.

Now we give two results which parallel, and actually extend already in the case of Hölder continuous coefficients, Theorems 2.6 and 3.4.

THEOREM 6.2 ([**38**]). *Under the assumptions* (6.69)-(6.70) *with* $\gamma < p$, *assume that*

$$\alpha > \frac{1}{2}, \qquad \gamma \leq p - \frac{1}{2} + \frac{p}{n-2}, \tag{6.71}$$

and let $u \in u_0 + W_0^{1,p}(\Omega, \mathbb{R}^N)$ *be a solution to the Dirichlet problem* (2.12) *with* (2.13) *in force. Then* $\mathcal{H}^{n-1}$*-almost every boundary point is regular for* Du *and* (4.45) *holds.*

The case $\gamma = p$ again needs a dimensional restriction.

THEOREM 6.3 ([**38**]). *Under the assumptions* (6.69)-(6.70) *with*

$$\alpha > \frac{1}{2}, \qquad n \leq p+2,$$

let $u \in u_0 + W_0^{1,p}(\Omega, \mathbb{R}^N)$ *be a solution to the Dirichlet problem* (2.12) *with* (2.13) *in force. Then* $\mathcal{H}^{n-1}$*-almost every boundary point is regular for* Du.

A parabolic version of the above result will soon be available in [**7**].

6.1. Acknowledgment. The author is supported by the ERC grant 207573 "Vectorial Problems".

References

[1] Arkhipova A., *Partial regularity up to the boundary of weak solutions of elliptic systems with nonlinearity q greater than two.* J. Math. Sci. (N. Y.) **115** (2003), 2735–2746. MR1810609 (2002a:35061)

[2] ———, *New a priori estimates for q-nonlinear elliptic systems with strong nonlinearities in the gradient,* $1 < q < 2$. J. Math. Sci. (N. Y.) **132**(2006), 255–273. MR2120183 (2005j:35042)

[3] Arkhipova A. and Ladyzhenskaya O., *On a modification of Gehring's lemma.* J. Math. Sci. (N. Y.) **109** (2002), 1805–1813. MR1754355 (2001e:35101)

[4] Beck L., *Partial regularity for weak solutions of nonlinear elliptic systems: The subquadratic case.* Manuscripta Math. **123** (2007), 453–491. MR2320739 (2008k:351216)

[5] Bögelein V., Duzaar F., and Mingione G., *The boundary regularity of non-linear parabolic systems.* I. *Ann. Inst. Henri Poincaré, Anal. Non Linéaire*, 27 (2010), 201-255.

[6] ———, *The boundary regularity of non-linear parabolic systems.* II. *Ann. Inst. Henri Poincaré, Anal. Non Linéaire*, 27 (2010), 145-200.

[7] ———, *The boundary regularity of non-linear parabolic systems.* III. In preparation.

[8] Campanato S., *Equazioni ellittiche del* II *ordine e spazi* $L^{(2,\lambda)}$. Ann. Mat. Pura Appl. (IV) **69** (1965), 321–381. MR0192168 (33:395)

[9] ———, *Elliptic systems with non-linearity q greater than or equal to two. Regularity of the solution of the Dirichlet problem.* Ann. Mat. Pura Appl. (IV) **147** (1987), 117–150. MR916705 (89c:35056)

[10] Colombini F., *Un teorema di regolarità alla frontiera per soluzioni di sistemi ellittici quasi lineari.* Ann. Scuola Norm. Sup. Pisa (3) **25** (1971), 115–161. MR0289939 (44:7124)

[11] Dacorogna B., *Direct methods in the calculus of variations.* Applied Mathematical Sciences, 78. Springer-Verlag, Berlin, 1989. MR990890 (90e:49001)

[12] De Giorgi E., *Sulla differenziabilità e l'analiticità delle estremali degli integrali multipli regolari.* Mem. Accad. Sci. Torino Cl. Sci. Fis. Mat. Nat. (III) **125** (1957), no. 3, 25–43. MR0093649 (20:172)

[13] ———, *Un esempio di estremali discontinue per un problema variazionale di tipo ellittico.* Boll. Un. Mat. Ital. (4) **1** (1968), 135–137. MR0227827 (37:3411)

[14] DeVore R. A. and Sharpley R. C., *Maximal functions measuring smoothness.* Mem. Amer. Math. Soc. **47** (1984), no. 293. MR727820 (85g:46039)

[15] Duzaar F. and Grotowski J. F., *Optimal interior partial regularity for nonlinear elliptic systems: The method of A-harmonic approximation.* Manuscripta Math. **103** (2000), 267–298. MR1802484 (2001j:35079)

[16] Duzaar F., Grotowski J. F., and Kronz M., *Partial and full boundary regularity for minimizers of functionals with nonquadratic growth.* J. Convex Anal. **11** (2004), 437–476. MR2158914 (2006e:49081)

[17] Duzaar F., Kristensen J., and Mingione G., *The existence of regular boundary points for nonlinear elliptic systems.* J. Reine Angew. Math. (Crelles J.) **602** (2007), 17–58. MR2300451 (2008b:35057)

[18] Duzaar F. and Mingione G., *Second order parabolic systems, optimal regularity, and singular sets of solutions.* Ann. Inst. Henri Poincaré Anal. Non Linéaire, **22** (2005), 705–751. MR2172857 (2008h:35139)

[19] Duzaar F., Mingione G., and Steffen K., *Parabolic systems with polynomial growth and regularity. Memoirs of the AMS*, to appear.

[20] Duzaar F. and Steffen K., *Optimal interior and boundary regularity for almost minimizers to elliptic variational integrals.* J. Reine Angew. Math. (Crelles J.) **546** (2002), 73–138. MR1900994 (2003b:49038)

[21] Giaquinta M., *A counter-example to the boundary regularity of solutions to quasilinear systems.* Manuscripta Math. **24** (1978), 217–220. MR492658 (80a:35041)

[22] Giaquinta M., *Multiple integrals in the calculus of variations and nonlinear elliptic systems.* Annals of Math. Studies, 105. Princeton Univ. Press, Princeton, NJ, 1983. MR717034 (86b:49003)

[23] Giaquinta M. and Giusti E., *On the regularity of the minima of variational integrals.* Acta Math. **148** (1982), 31–46. MR666107 (84b:58034)

[24] ———, *Differentiability of minima of nondifferentiable functionals.* Invent. Math. **72** (1983), 285–298. MR700772 (84f:58037)

[25] ———, *Global $C^{1,\alpha}$-regularity for second order quasilinear elliptic equations in divergence form.* J. Reine Angew. Math. (Crelles J.) **351** (1984), 55–65. MR749677 (85k:35077)

[26] Giaquinta M. and Modica G., *Almost-everywhere regularity for solutions of nonlinear elliptic systems.* Manuscripta Math. **28** (1979), 109–158. MR535699 (80m:35033)

[27] Giusti E. *Direct methods in the calculus of variations.* World Scientific, River Edge, NJ, 2003. MR1962933 (2004g:49003)

[28] Grotowski J. F., *Boundary regularity for nonlinear elliptic systems.* Calc. Var. Partial Differential Equations **27** (2002), 2491–2512. MR1944037 (2003j:35061)

[29] Grotowski J.F., *Boundary regularity for quasilinear elliptic systems.* Comm. Partial Differential Equations **27** (2002), 2491–2512, 353–388. MR1944037 (2003j:35061)

[30] Hamburger C., *Partial boundary regularity of solutions of nonlinear superelliptic systems.* Boll. UMI.(B) (8) **10** (2007), 63–81. MR2310958 (2008g:35038)

[31] Hildebrandt S. and Widman K.O., *Some regularity results for quasilinear elliptic systems of second order.* Math. Z. **142** (1975), 67–86. MR0377273 (51:13446)

[32] Ivert P.A., *Regularitätsuntersuchungen von Lösungen elliptischer Systeme von quasilinearen Differentialgleichungen zweiter Ordnung.* Manuscripta Math. **30** (1979/80), 53–88. MR552363 (82d:35045)

[33] ———, *Partial regularity of vector valued functions minimizing variational integrals.* Preprint Univ. Bonn, 1982.

[34] Iwaniec T., *The Gehring lemma.* Quasiconformal mappings and analysis (Ann Arbor, MI, 1995), 181–204, Springer, New York, 1998. MR1488451 (99e:30012)

[35] Jost J. and Meier M., *Boundary regularity for minima of certain quadratic functionals.* Math. Ann. **262** (1983), 549–561. MR696525 (84i:35051)

[36] Kristensen J. and Mingione G., *The singular set of minima of integral functionals.* Arch. Ration. Mech. Anal. **180** (2006), 331–398. MR2214961 (2007d:49059)

[37] ———, *Boundary regularity of minima.* Rend. Lincei Mat. Appl. **19** (2008), 265–277. MR2465679 (2009m:49069)

[38] ———, *Boundary regularity in variational problems. Arch. Ration. Mech. Anal.*. doi: 10.1007/s00205-010-0294-x.

[39] Ladyzhenskaya O. A. and Ural'tseva N. N., *Linear and quasilinear elliptic equations.* Academic Press, New York–London, 1968. MR0244627 (39:5941)

[40] ———, *Linear and quasilinear elliptic equations.* Second edition "Nauka", Moscow, 1973. MR0509265 (58:23009)

[41] ———, *Local estimates for gradients of solutions of non-uniformly elliptic and parabolic equations.* Comm. Pure Appl. Math. **23** (1970), 677–703. MR0265745 (42:654)

[42] Ladyzhenskaya O. A., Solonnikov V. A., and Ural'tseva N. N., *Linear and Quasilinear Equations of Parabolic Type.* Amer. Math. Soc., Providence, RI, 1967. MR0241822 (39:3159b)

[43] Manfredi J. J., *Regularity of the gradient for a class of nonlinear possibly degenerate elliptic equations.* Ph.D. Thesis, University of Washington, St. Louis.

[44] Maz'ja V. G.: *Examples of nonregular solutions of quasilinear elliptic equations with analytic coefficients.* (Russian) *Funckcional. Anal. i Priložen.* 2 (1968), 53–57.

[45] Mingione G., *The singular set of solutions to non-differentiable elliptic systems.* Arch. Ration. Mech. Anal. **166** (2003), 287–301. MR1961442 (2004b:35070)

[46] ———, *Bounds for the singular set of solutions to non linear elliptic systems.* Calc. Var. Partial Differential Equations **18** (2003), 373–400. MR2020367 (2004i:35121)

[47] ———, *Regularity of minima: An invitation to the Dark side of the Calculus of Variations. Applications of Mathematics* **51** (2006), 355–425. MR2291779 (2007k:35086)

[48] Nečas J., *Example of an irregular solution to a nonlinear elliptic system with analytic coefficients and conditions for regularity.* Theory of nonlinear operators (Proc. Fourth Internat. Summer School), Acad. Sci., Berlin, 1975, pp. 197–206. MR0509483 (58:23038)

[49] Rivière T., *Everywhere discontinuous harmonic maps into spheres.* Acta Math. **175** (1995), 197–226. MR1368247 (96k:58059)

[50] Runst T. and Sickel W., *Sobolev spaces of fractional order, Nemytskij operators, and nonlinear partial differential equations.* Walter de Gruyter, Berlin, 1996. MR1419319 (98a:47071)

[51] Schoen R. and Uhlenbeck K., *A regularity theory for harmonic maps.* J. Differential Geom. **17** (1983), 307–336. MR664498 (84b:58037a)

[52] Šverák V. and Yan X., *Non-Lipschitz minimizers of smooth uniformly convex variational integrals.* Proc. Nat. Acad. Sci. USA **99/24** (2002), 15269–15276. MR1946762 (2003h:49066)

[53] Uhlenbeck K., *Regularity for a class of non-linear elliptic systems.* Acta Math. **138** (1977), 219–240. MR0474389 (57:14031)
[54] Ural'tseva N. N., *Degenerate quasilinear elliptic systems.* Zap. Nauch. Sem. Leningrad. Otdel. Mat. Inst. Steklov. (LOMI) **7** (1968), 184–222. MR0244628 (39:5942)

DIPARTIMENTO DI MATEMATICA, UNIVERSITÀ DI PARMA, VIALE G. P. USBERTI 53/A, CAMPUS, 43100 PARMA, ITALY
E-mail address: giuseppe.mingione@unipr.it

Amer. Math. Soc. Transl.
(2) Vol. **229**, 2010

Attainability of Infima in the Critical Sobolev Trace Embedding Theorem on Manifolds

Alexander Nazarov and Alexander Reznikov

Dedicated to Nina Nikolaevna Ural'tseva with admiration

ABSTRACT. Sufficient conditions for the existence of extremal functions in the trace Sobolev inequality and the trace Sobolev–Poincaré inequality on Riemannian manifolds are established. It is shown that some of these conditions are sharp.

1. Introduction

Let $n \geqslant 2$, and let Ω be an n-dimensional ($\mathcal{C}^3$-smooth compact Riemannian) manifold with $\mathcal{C}^3$-smooth boundary. For $1 < p < n$, we denote by $p^\star = \frac{(n-1)p}{n-p}$ the trace Sobolev exponent for p, that is, the critical exponent for the trace embedding $W_p^1(\Omega) \hookrightarrow L_q(\partial\Omega)$.

Since the embedding operator $W_p^1(\Omega) \hookrightarrow L_{p^\star}(\partial\Omega)$ is noncompact, the problem of attainability of the norm of this operator (i.e., the problem of existence of an extremal function in the trace embedding theorem) is nontrivial. The corresponding problem for the conventional embedding $W_p^1(\Omega) \hookrightarrow L_{p^*}(\Omega)$ (here $p^* = \frac{np}{n-p}$ is the Sobolev conjugate of p) was treated in many papers; see, e.g., the recent survey [**N**] and further references therein.

The problem for the trace embedding is considerably less investigated. Let us consider the inequality

$$K(n,p) \equiv \inf_{v \in \dot{\mathcal{C}}^\infty(\mathbb{R}^n_+)\setminus\{0\}} \frac{\|\nabla v\|_{p,\mathbb{R}^n_+}}{\|v(\cdot,0)\|_{p^\star,\mathbb{R}^{n-1}}} > 0, \tag{1}$$

where $\dot{\mathcal{C}}^\infty(\mathbb{R}^n_+)$ is the set of functions on $\mathbb{R}^n_+$ with bounded support.

Obviously, the functional in (1) is invariant with respect to translations and dilations of v. It was proved in the remarkable paper [**Nt**] that the infimum in (1) is attained on the function with *unbounded* support

$$\widetilde{w}_\varepsilon(x) = |x - x^\varepsilon|^{-\frac{n-p}{p-1}}, \tag{2}$$

This paper was partially supported by grant NSh.227.2008.1. The first author was also supported by RFBR grant 08-01-00748.

with $x^\varepsilon = (0,\dots,0,-\varepsilon)$ (for the case $p=2$ this was established earlier in [**E**]). The result of [**Nt**] implies that

$$K(n,p) = \Big(\frac{n-p}{p-1}\Big)^{\frac{1}{p'}} \Big(\frac{\omega_{n-2}}{2}\cdot \mathcal{B}\Big(\frac{n-1}{2},\frac{n-1}{2(p-1)}\Big)\Big)^{\frac{1}{(n-1)p'}}.$$

In the paper [**NR**], the authors considered the critical trace embedding in the bounded domain $\Omega \Subset \mathbb{R}^n$, i.e. the inequality

$$(I) \qquad \lambda_1(n,p,\Omega) = \inf_{v\in W^1_p(\Omega)\setminus\{0\}} \frac{\|v\|_{W^1_p(\Omega)}}{\|v\|_{p^\star,\partial\Omega}} > 0$$

(the norm of the numerator is defined as $\|v\|^p_{W^1_p(\Omega)} = \|\nabla v\|^p_{p,\Omega} + \|v\|^p_{p,\Omega}$). In [**NR**, Remark 4] it is noted that the main result holds true for an arbitrary manifold Ω with a smooth boundary, provided $\partial\Omega$ contains a point with positive mean curvature (with respect to the inner normal). Namely, in this case, the infimum in (I) is attained for $1 < p < \frac{n+1}{2} + \beta$, where $\beta > 0$ depends on Ω.

This paper deals with the more complicated case where the mean curvature is non-positive. Our first result reads as follows.

THEOREM 1. *Let $n \geqslant 5$, and let Ω be an n-dimensional manifold with $\mathcal{C}^3$-smooth boundary. Suppose that the mean curvature of $\partial\Omega$ with respect to the inner normal is nonpositive everywhere. Finally, suppose that there exists a point $y^0 \in \partial\Omega$ such that $\partial\Omega$ is a totally geodesic submanifold in Ω at y^0 and the scalar curvature of Ω is positive at y^0. Then for some $\beta > 0$, for $2 < p < \frac{n+2}{3} + \beta$, the infimum in* (I) *is attained.*

REMARK 1. By a standard argument it follows that under suitable normalization the extremal function in (I), if it exists, is a positive solution to the nonlinear Neumann problem

$$-\Delta_p u + u^{p-1} = 0 \quad \text{in } \Omega, \qquad |\nabla u|^{p-2}\frac{\partial u}{\partial \mathbf{n}} = u^{p^\star-1} \quad \text{on } \partial\Omega$$

(here $\Delta_p u = \operatorname{div}(|\nabla u|^{p-2}\nabla u)$).

Observe that, in contrast to [**NR**], the assumptions of Theorem 1 contain a nontrivial left border of the interval for p. We show that this left border is sharp.

THEOREM 2. *Let $n \geqslant 2$, and let Ω be an n-dimensional hemisphere*

$$\Omega = \{(\theta,\phi_1,\dots,\phi_{n-1}) \in \mathbb{S}^n : \; 0 < \theta < \frac{\pi}{2}\}.$$

Then for any $1 < p < 2$, there exists $\varkappa^* > 0$ such that for $\varkappa > \varkappa^*$ the infimum in (I) *is not attained on $\varkappa\Omega$. For $n \geqslant 3$ this is true also for $p = 2$.*

REMARK 2. It is easy to see that the nonattainability of the infimum in (I) on $\varkappa\Omega$ for large values of $\varkappa$ is equivalent to the validity of the "optimal trace Sobolev inequality"

$$\|v\|^p_{p^\star,\partial\Omega} \leqslant K^{-p}(n,p)\cdot \|\nabla v\|^p_{p,\Omega} + C(p,\Omega)\cdot \|v\|^p_{p,\Omega}, \qquad v \in W^1_p(\Omega).$$

For the conventional optimal Sobolev inequality, see [**Dr1**] and the references therein.

REMARK 3. Note that the fact that Ω is a hemisphere is used only for symmetrization arguments. Using the techniques of [**Dr1**], one can obtain the estimate (42) and prove Theorem 2 in the general case under the assumption that $\partial\Omega$ is a totally geodesic submanifold in Ω.

Finally, we consider the trace Sobolev–Poincaré inequality

$$(II) \qquad \lambda_2(n,p,\Omega) = \inf_{v\in W_p^1(\Omega)\setminus\{c\}} \frac{\|\nabla v\|_{p,\Omega}}{\|v-\overline{v}\|_{p^\star,\partial\Omega}} > 0$$

(here we use the notation $\overline{v} = |\partial\Omega|^{-1}\int_{\partial\Omega} v\, d\Sigma$).

THEOREM 3. *Let $n \geqslant 4$, and let Ω be as in Theorem 1. Then for some $\beta(\Omega) > 0$ and for $1 < p < \frac{n+2}{3} + \beta$, the infimum in* (II) *is attained.*

The structure of our paper is as follows. In §2 we establish required integral estimates and prove Theorem 1. Theorem 2 is proved in §3. Inequality (II) is considered in §4.

We are grateful to Prof. S.V. Ivanov for important advice.

Let us recall some notation. $x = (x_1,\dots,x_{n-1},x_n) = (x',x_n)$ is a point in $\mathbb{R}^n$. Put $|x'| = \sqrt{x_1^2+\dots+x_{n-1}^2}$ and $Q_\rho = \{x : |x'| < \rho, |x_n| < \rho\}$. By y we denote points in Ω.

Consider the exponential map at $y^0 \in \partial\Omega$ and recall (see, e.g., [**BZ**, §11]) that there exists $\rho_0 > 0$ such that $\exp_{y^0}^{-1}$ is a diffeomorphism from the Riemannian ball $B_{2\rho_0}(y^0)\cap\Omega$ onto the set $B_{2\rho_0}(0)\cap\{x_n > F(x')\} \subset \mathbb{R}^n$. Moreover, one can suppose that $F(0) = \nabla' F(0) = 0$. In what follows we always assume $\rho < \rho_0$. For a function f on $B_{2\rho_0}(y^0)\cap\Omega$ we define the "transplanted" function $\widetilde{f}(x) \equiv f(\exp_{y^0}(x))$.

We denote by $G = (g_{ij}(y))$ the Riemannian metric tensor on Ω and by $(g^{ij}(y))$ the inverse tensor. It is well known that for $f : \Omega \to \mathbb{R}$ the Riemannian length of ∇f is given by $|\nabla f|^2 = \sum\limits_{ij} \widetilde{g}^{ij}\widetilde{f}_{x_i}\widetilde{f}_{x_j}$. Also $dy = \sqrt{\det(\widetilde{G})}dx$.

$\omega_{n-1} = \frac{2\pi^{n/2}}{\Gamma(\frac{n}{2})}$ stands for the area of the unit sphere in $\mathbb{R}^n$. We denote by $p' = \frac{p}{p-1}$ the Hölder conjugate exponent to p, and put $q = \frac{n-1}{2}p'$. Γ is the Euler gamma function; $\mathcal{B}$ is the Euler beta function.

We denote by $o_\rho(1)$ a quantity which tends to zero as $\rho \to 0$. We use the letter C to denote various positive constants. To indicate that C depends on some parameters, we write $C(\dots)$.

2. Existence of the minimizer

PROOF OF THEOREM 1. Our main tool is the concentration-compactness principle of Lions ([**Ls**]). It is used in various forms; for the problem (I) it can be reformulated as follows (the proof is a verbatim repetition of [**NR**, Proposition 1]).

PROPOSITION 1. *Let the infimum in* (I) *satisfy the inequality*

$$\lambda_1(n,p,\Omega) < K(n,p).$$

Then the infimum is attained.

Thus, to prove the attainability of the infimum in (I), it is sufficient to present a function such that the quotient in (I) is less than $K(n,p)$. Following [**DN**], see also [**LPT**], we succeed, constructing a function with a small support, in simulating the behavior of $\widetilde{w}_\varepsilon$ in Riemannian normal coordinates. Namely, for sufficiently small $\varepsilon > 0$ and $\rho > 0$ we introduce the function $u(y)$ such that

$$\widetilde{u}(x) = \widetilde{\varphi}(|x'|, x_n)\widetilde{w}_\varepsilon(x). \tag{3}$$

Here $\widetilde{w}_\varepsilon$ is defined in (2), while $\widetilde{\varphi}$ is a smooth cutoff function such that

$$\widetilde{\varphi} = 1 \quad \text{in} \quad Q_{\frac{\rho}{2}}, \qquad \widetilde{\varphi} = 0 \quad \text{in} \quad \mathbb{R}^n \setminus Q_\rho,$$

and $|\nabla \widetilde{\varphi}| \leqslant \frac{C}{\rho}$.

2.1. Auxiliary relations. It is well known (see, e.g., [**BZ**, §12.6]) that

$$\widetilde{g}_{ij}(0) = \delta_{ij}, \ \frac{\partial \widetilde{g}_{ij}}{\partial x_k}(0) = 0. \tag{4}$$

Further, from [**BZ**, §§14.5 and 19.5] we conclude that for $i \neq j$,

$$\frac{\partial^2 \widetilde{g}_{ii}}{\partial x_j^2}(0) = -\frac{\mathcal{R}_{ij}}{3}, \qquad \frac{\partial^2 \widetilde{g}_{ij}}{\partial x_i \partial x_j}(0) = \frac{2\mathcal{R}_{ij}}{3}. \tag{5}$$

Here $\mathcal{R}_{ij} = K_\sigma(e_i, e_j)$ is a sectional curvature of Ω at y^0. We write $\mathcal{R}_{ii} = 0$, so the formulas (5) hold true even for $i = j$.

Also we denote by

$$\mathcal{R}_i = \sum_j \mathcal{R}_{ij} = Ric(e_i)$$

a Ricci curvature at y^0 and introduce the notation

$$\mathcal{R}'_i = \sum_{j<n} \mathcal{R}_{ij}, \qquad i < n.$$

Next, consider the function F giving the local representation of $\partial\Omega$ at y^0. Since the boundary is $\mathcal{C}^3$-smooth, we have

$$F(x') = \sum_{i,j<n} a_{ij} x_i x_i + \sum_{i,j,k<n} b_{ijk} x_i x_j x_k + o(|x'|^3) \quad \text{as} \quad x' \to 0. \tag{6}$$

Moreover, the assumption on $\partial\Omega$ to be totally geodesic at y^0 means that $a_{ij} = 0$.

2.2. Estimate of $\|\nabla u\|_{p,\Omega}$ for $1 < p \leqslant \frac{n+2}{3}$. We apply the estimate

$$|\nabla(fg)|^p \leqslant |f\, \nabla g|^p + C\, \left(|\nabla f\, g|^p + |\nabla f\, g| \cdot |f\, \nabla g|^{p-1}\right) \tag{7}$$

to ∇u. Since $\nabla \widetilde{\varphi}$ does not vanish only in $Q_\rho \setminus Q_{\frac{\rho}{2}}$, one obtains

$$\int_\Omega \left(|\nabla\varphi \cdot w_\varepsilon|^p + |\nabla\varphi \cdot w_\varepsilon| \cdot |\varphi\, \nabla w_\varepsilon|^{p-1}\right) dy \tag{8}$$

$$\leqslant C\left(\int_0^{\rho/2}\int_{\rho/2}^{\rho} + \int_{\rho/2}^{\rho}\int_0^{\rho}\right) \frac{\rho^{-p} r^{n-2}\, dx_n dr}{(r^2 + x_n^2)^{\frac{p(n-p)}{2(p-1)}}} \cdot \left[1 + \left(\frac{\rho}{\sqrt{r^2 + x_n^2}}\right)^{p-1}\right] \leqslant C\, \rho^{n-2q}.$$

The inequalities (7) and (8) imply that

$$\int_\Omega |\nabla u|^p dy \leqslant \int_{\Omega \cap \exp_{y^0}(Q_\rho)} |\nabla w_\varepsilon|^p dy + C\rho^{n-2q}. \tag{9}$$

We have

$$\int_{\Omega\cap\exp_{y^0}(Q_\rho)} |\nabla w_\varepsilon|^p dy = \int_{|x'|<\rho}\int_{F(x')}^{\rho} \Big(\sum_{ij} \widetilde{g}^{ij}\frac{\partial\widetilde{w_\varepsilon}}{\partial x_i}\frac{\partial\widetilde{w_\varepsilon}}{\partial x_j}\Big)^{\frac{p}{2}}\cdot\sqrt{\det(\widetilde{G})}\,dx_n dx'.$$

Expanding the metric tensor by the Taylor formula and using (4), we obtain

$$\begin{aligned}(10)\quad &\int_{\Omega\cap\exp_{y^0}(Q_\rho)} |\nabla w_\varepsilon|^p dy \\ &= \mathcal{I}_1 + \frac{1}{4}\sum_{ijk}\frac{\partial^2\widetilde{g}_{kk}}{\partial x_i\partial x_j}(0)\cdot\mathcal{I}_2^{(ij)} - \frac{p}{4}\sum_{ijkl}\frac{\partial^2\widetilde{g}_{ij}}{\partial x_k\partial x_l}(0)\cdot\mathcal{I}_3^{(ijkl)} + \mathcal{I}_4 - \mathcal{I}_5,\end{aligned}$$

where

$$\mathcal{I}_1 = \int_{|x'|<\rho}\int_0^\rho |\nabla\widetilde{w}_\varepsilon|^p dx_n dx',$$

$$\mathcal{I}_2^{(ij)} = \int_{|x'|<\rho}\int_0^\rho x_i x_j|\nabla\widetilde{w}_\varepsilon|^p dx_n dx',$$

$$\mathcal{I}_3^{(ijkl)} = \int_{|x'|<\rho}\int_0^\rho x_k x_l\frac{\partial\widetilde{w}_\varepsilon}{\partial x_i}\frac{\partial\widetilde{w}_\varepsilon}{\partial x_j}|\nabla\widetilde{w}_\varepsilon|^{p-2} dx_n dx',$$

$$\mathcal{I}_4 = \int_{|x'|<\rho}\int_0^\rho |x|^2|\nabla\widetilde{w}_\varepsilon|^p dx_n dx'\cdot o_\rho(1),$$

$$\mathcal{I}_5 = \int_{|x'|<\rho}\int_0^{F(x')} |\nabla\widetilde{w}_\varepsilon|^p\cdot\big(1+o(\rho^2)\big)\,dx_n dx'.$$

It is easy to see that

$$(11)\qquad \mathcal{I}_1 = \Big(\frac{n-p}{p-1}\Big)^p\cdot\big(\varepsilon^{n-2q}\mathcal{E}_1 - O(\rho^{n-2q})\big),$$

where

$$(12)\qquad \mathcal{E}_1 = \omega_{n-2}\int_0^\infty\int_0^\infty \frac{r^{n-2}dr dx_n}{(r^2+(x_n+1)^2)^q} = \frac{\omega_{n-2}}{2(2q-n)}\cdot\mathcal{B}\big(\tfrac{n-1}{2},\tfrac{n-1}{2(p-1)}\big).$$

In a similar way we get

$$(13)\qquad \mathcal{I}_2^{(ij)} = \Big(\frac{n-p}{p-1}\Big)^p\delta_{ij}\cdot\begin{cases}\frac{\varepsilon^{n-2q+2}}{2q-n-2}\cdot\mathcal{E}_2^{(j)} - O(\rho^{n-2q+2}), & p<\frac{n+2}{3},\\ \ln(\frac{1}{\varepsilon})\cdot\mathcal{E}_2^{(j)} - O(\ln(\frac{1}{\rho})), & p=\frac{n+2}{3},\end{cases}$$

where

$$(14)\qquad \mathcal{E}_2^{(j)} = \frac{\omega_{n-2}}{2q-n-1}\cdot\mathcal{B}\big(\tfrac{n-1}{2},\tfrac{n-1}{2(p-1)}\big)\cdot\begin{cases}\frac{1}{2}, & j<n,\\ \frac{1}{2q-n}, & j=n.\end{cases}$$

To calculate $\mathcal{I}_3$, we consider three cases.

1. $i<n$, $j<n$. Then

$$\begin{aligned}(15)\quad \mathcal{I}_3^{(ijkl)} &= \Big(\frac{n-p}{p-1}\Big)^p\int_{|x'|<\rho}\int_0^\rho \frac{x_i x_j x_k x_l dx}{(|x'|^2+(x_n+\varepsilon)^2)^{q+1}}dx \\ &= (\delta_{ik}\delta_{jl}+\delta_{il}\delta_{jk})\mathcal{F}^{(ij)} + \delta_{ij}\delta_{kl}\mathcal{F}^{(ik)},\end{aligned}$$

where, for $i \neq k$,

$$(16)\quad \mathcal{F}^{(ik)} = \Big(\frac{n-p}{p-1}\Big)^p \int_{|x'|<\rho}\int_0^\rho \frac{x_i^2 x_k^2 dx}{(|x'|^2+(x_n+\varepsilon)^2)^{q+1}}dx$$

$$= \Big(\frac{n-p}{p-1}\Big)^p \cdot \frac{1}{2q} \cdot \begin{cases} \frac{\varepsilon^{n-2q+2}}{2q-n-2}\cdot \mathcal{E}_2^{(k)} + O(\rho^{n-2q+2}), & p<\frac{n+2}{3},\\ \ln(\frac{1}{\varepsilon})\cdot \mathcal{E}_2^{(k)} + O(\ln(\frac{1}{\rho})), & p=\frac{n+2}{3}\end{cases}$$

(note that the case $i = k$ is not interesting for us due to $\mathcal{R}_{ii} = 0$).

2. $i < n$, $j = n$. Then

$$(17)\quad \mathcal{I}_3^{(ijkl)} = \Big(\frac{n-p}{p-1}\Big)^p \int_{|x'|<\rho}\int_0^\rho \frac{x_i(x_n+\varepsilon)x_k x_l dx}{(|x'|^2+(x_n+\varepsilon)^2)^{q+1}}dx = (\delta_{ik}\delta_{ln}+\delta_{il}\delta_{kn})\mathcal{F}_1^{(i)},$$

where

$$(18)\quad \mathcal{F}_1^{(i)} = \Big(\frac{n-p}{p-1}\Big)^p \int_{|x'|<\rho}\int_0^\rho \frac{x_i^2(x_n+\varepsilon)x_n dx}{(|x'|^2+(x_n+\varepsilon)^2)^{q+1}}dx$$

$$= \Big(\frac{n-p}{p-1}\Big)^p \cdot \frac{1}{2q} \cdot \begin{cases} \frac{\varepsilon^{n-2q+2}}{2q-n-2}\cdot \mathcal{E}_2^{(i)} + O(\rho^{n-2q+2}), & p<\frac{n+2}{3},\\ \ln(\frac{1}{\varepsilon})\cdot \mathcal{E}_2^{(i)} + O(\ln(\frac{1}{\rho})), & p=\frac{n+2}{3}.\end{cases}$$

Obviously, a similar result is true for $i = n$, $j < n$.

3. $i = j = n$. Then

$$(19)\quad \mathcal{I}_3^{(ijkl)} = \Big(\frac{n-p}{p-1}\Big)^p \int_{|x'|<\rho}\int_0^\rho \frac{(x_n+\varepsilon)^2 x_k x_l dx}{(|x'|^2+(x_n+\varepsilon)^2)^{q+1}}dx = \delta_{kl}\mathcal{F}_2^{(k)},$$

where, for $k \neq n$,

$$(20)\quad \mathcal{F}_2^{(k)} = \Big(\frac{n-p}{p-1}\Big)^p \int_{|x'|<\rho}\int_0^\rho \frac{x_k^2(x_n+\varepsilon)^2 dx}{(|x'|^2+(x_n+\varepsilon)^2)^{q+1}}dx$$

$$= \Big(\frac{n-p}{p-1}\Big)^p \cdot \frac{2q-n-1}{2q} \cdot \begin{cases} \frac{\varepsilon^{n-2q+2}}{2q-n-2}\cdot \mathcal{E}_2^{(k)} + O(\rho^{n-2q+2}), & p<\frac{n+2}{3},\\ \ln(\frac{1}{\varepsilon})\cdot \mathcal{E}_2^{(k)} + O(\ln(\frac{1}{\rho})), & p=\frac{n+2}{3}.\end{cases}$$

Further, it is obvious that

$$(21)\quad \mathcal{I}_4 = \mathcal{I}_2^{(nn)}\cdot o_\rho(1).$$

Finally, the last term in (10) can be estimated using (6):

$$(22)\quad \mathcal{I}_5 = \Big(\frac{n-p}{p-1}\Big)^p \int_{|x'|<\rho} \sum_{i,j,k<n} \frac{b_{ijk}x_i x_j x_k}{(|x'|^2+\varepsilon^2)^q}\,dx' + \int_{|x'|<\rho}\frac{|x'|^3\,dx'}{(|x'|^2+\varepsilon^2)^q}\cdot o_\rho(1) = \mathcal{I}_2^{(nn)}\cdot o_\rho(1).$$

Substituting relations (10)-(22) into (9), we obtain with regards to (5),

$$(23)\quad \int_\Omega |\nabla u|^p dy \leqslant \Big(\frac{n-p}{p-1}\Big)^p \times \begin{cases} \varepsilon^{n-2q}\mathcal{E}_1 - \mathcal{E}_2\frac{\varepsilon^{n-2q+2}}{2q-n-2} + O(\rho^{n-2q}), & p<\frac{n+2}{3},\\ \varepsilon^{n-2q}\mathcal{E}_1 - \mathcal{E}_2\ln(\frac{1}{\varepsilon}) + O(\rho^{n-2q}), & p=\frac{n+2}{3},\end{cases}$$

where

$$\mathcal{E}_2 = \frac{\mathcal{E}_2^{(1)}}{12(2q-n+1)} \cdot \big(12\mathcal{R}_n + (2q-n+4)\sum_{i<n} \mathcal{R}'_i + o_\rho(1)\big). \tag{24}$$

2.3. Estimate of $\|u\|_{p^\star,\partial\Omega}$ for $1 < p < \frac{n+1}{2}$. It is obvious that the area on $\partial\Omega$ satisfies the relation

$$d\Sigma = \sqrt{1+|\nabla F(x')|^2} \cdot \sqrt{\det(\widetilde{G'})}\, dx' \geqslant \sqrt{\det(\widetilde{G'})}\, dx',$$

where $\widetilde{G'} = (\widetilde{g}_{ij})_{i,j<n}$. Thus, we have

$$\begin{aligned} \int_{\partial\Omega} u^{p^\star}\, d\Sigma &\geqslant \int_{|x'|<\frac{\rho}{2}} \frac{\sqrt{\det(\widetilde{G'})}\, dx'}{(|x'|^2 + (\varepsilon + F(x'))^2)^q} \\ &= \int_{|x'|<\frac{\rho}{2}} \frac{\sqrt{\det(\widetilde{G'})}\, dx'}{(|x'|^2+\varepsilon^2)^q} \times \left[1 - 2q\frac{\varepsilon F(x')}{|x'|^2+\varepsilon^2} - \frac{F^2(x')}{|x'|^2+\varepsilon^2}\cdot O(1)\right]. \end{aligned} \tag{25}$$

Expanding the metric tensor by the Taylor formula and using (4), we obtain

$$\int_{\partial\Omega} u^{p^\star}\, d\Sigma \geqslant \mathcal{J}_1 + \frac{1}{4}\sum_{i,j,k<n} \frac{\partial^2 \widetilde{g}_{kk}}{\partial x_i \partial x_j}(0) \cdot \mathcal{J}_2^{(ij)} - \mathcal{J}_3 - 2q\mathcal{J}_4 - \mathcal{J}_5, \tag{26}$$

where

$$\begin{aligned} \mathcal{J}_1 &= \int_{|x'|<\frac{\rho}{2}} \frac{dx'}{(|x'|^2+\varepsilon^2)^q}, \\ \mathcal{J}_2^{(ij)} &= \int_{|x'|<\frac{\rho}{2}} \frac{x_i x_j dx'}{(|x'|^2+\varepsilon^2)^q}, \\ \mathcal{J}_3 &= \int_{|x'|<\frac{\rho}{2}} \frac{|x'|^2\, dx'}{(|x'|^2+\varepsilon^2)^q} \cdot o_\rho(1), \\ \mathcal{J}_4 &= \int_{|x'|<\frac{\rho}{2}} \frac{\varepsilon F(x')\big(1+o(\rho^2)\big)\, dx'}{(|x'|^2+\varepsilon^2)^{q+1}}, \\ \mathcal{J}_5 &= \int_{|x'|<\frac{\rho}{2}} \frac{F^2(x')\, dx'}{(|x'|^2+\varepsilon^2)^{q+1}} \cdot O(1). \end{aligned}$$

It is easy to see that

$$\mathcal{J}_1 = \varepsilon^{n-2q-1}\mathcal{E}'_1 - O(\rho^{n-2q-1}), \tag{27}$$

where

$$\mathcal{E}'_1 = \omega_{n-2}\int_0^\infty \frac{r^{n-2}dr}{(r^2+1)^q} = \frac{\omega_{n-2}}{2}\cdot \mathcal{B}\big(\tfrac{n-1}{2}, \tfrac{n-1}{2(p-1)}\big). \tag{28}$$

Similarly,

$$\mathcal{J}_2^{(ij)} = \delta_{ij}\varepsilon^{n-2q+1}\mathcal{E}_2'^{(i)} - O(\rho^{n-2q+1}), \tag{29}$$

where

$$\mathcal{E}_2'^{(i)} = \int_{\mathbb{R}^{n-1}} \frac{x_i^2\, dx'}{(|x'|^2+1)^q} = \frac{\omega_{n-2}}{2(2q-n-1)}\cdot \mathcal{B}\big(\tfrac{n-1}{2}, \tfrac{n-1}{2(p-1)}\big). \tag{30}$$

Further, using (6), we obtain

$$(31)\quad \mathcal{J}_4 = \int_{|x'|<\rho} \sum_{i,j,k<n} \frac{\varepsilon b_{ijk} x_i x_j x_k}{(|x'|^2+\varepsilon^2)^{q+1}}\, dx' + \int_{|x'|<\rho} \frac{\varepsilon |x'|^3\, dx'}{(|x'|^2+\varepsilon^2)^{q+1}} \cdot o_\rho(1) = \mathcal{J}_2^{(11)} \cdot o_\rho(1).$$

Finally, it is obvious that

$$(32)\quad \mathcal{J}_3 = \mathcal{J}_2^{(11)} \cdot o_\rho(1), \qquad \mathcal{J}_5 = \mathcal{J}_2^{(11)} \cdot o_\rho(1).$$

Substituting relations (27)-(32) into (26), we obtain with regards to (5),

$$(33)\quad \int_{\partial\Omega} u^{p^\star}\, d\Sigma \geqslant \varepsilon^{n-2q-1}\mathcal{E}_1' - \varepsilon^{n-2q+1}\mathcal{E}_2' - O(\rho^{n-2q-1}),$$

where

$$(34)\quad \mathcal{E}_2' = \frac{\mathcal{E}_2'^{(1)}}{12} \cdot \Big(\sum_{i<n} \mathcal{R}_i' + o_\rho(1)\Big).$$

2.4. Estimate of $\|u\|_{p,\Omega}$. Since $dy \leqslant C dx$ and $F(x') = O(|x'|^3)$ for $|x'| < \rho_0$, we have

$$(35)\quad \int_\Omega u^p dy \leqslant C \int_{|x'|<\rho} dx' \int_{-C|x'|^3}^{\infty} \frac{dx_n}{(|x'|^2 + (x_n+\varepsilon)^2)^{\frac{p(n-p)}{2(p-1)}}} \leqslant \begin{cases} C\varepsilon^{\frac{p^2-n}{p-1}}, & p < \sqrt{n}, \\ C(1+\ln\frac{\rho}{\varepsilon}), & p = \sqrt{n}, \\ C, & p > \sqrt{n}. \end{cases}$$

2.5. Completion of the proof. Estimates (23) and (35) imply that

$$\|u\|^p_{W^1_p(\Omega)} \leqslant \begin{cases} \left(\frac{n-p}{p-1}\right)^p \varepsilon^{n-2q}\mathcal{E}_1 - \varepsilon^{n-2q+2}\frac{\mathcal{E}_2}{2q-n-2} + C\,\varepsilon^{\frac{p^2-n}{p-1}} + C(\rho), & 1<p<\sqrt{n}, \\ \left(\frac{n-p}{p-1}\right)^p \varepsilon^{n-2q}\mathcal{E}_1 - \varepsilon^{n-2q+2}\frac{\mathcal{E}_2}{2q-n-2} + C\ln(\frac{1}{\varepsilon}) + C(\rho), & p=\sqrt{n}, \\ \left(\frac{n-p}{p-1}\right)^p \varepsilon^{n-2q}\mathcal{E}_1 - \varepsilon^{n-2q+2}\frac{\mathcal{E}_2}{2q-n-2} + C(\rho), & \sqrt{n}<p<\frac{n+2}{3}, \\ \left(\frac{n-p}{p-1}\right)^p \varepsilon^{n-2q}\mathcal{E}_1 - \mathcal{E}_2\ln(\frac{1}{\varepsilon}) + C\,\rho^{\frac{p-n}{p-1}}, & p=\frac{n+2}{3}, \end{cases}$$

where $\mathcal{E}_1$, $\mathcal{E}_2$ are defined in (12) and (24).

Since $n-2q+2 < \frac{p^2-n}{p-1}$ for $p>2$, we get for $2<p<\frac{n+2}{3}$, with regard to (33),

$$\frac{\|u\|^p_{W^1_p(\Omega)}}{\|u\|^p_{p^\star,\partial\Omega}} \leqslant \left(\frac{n-p}{p-1}\right)^p \frac{\mathcal{E}_1}{(\mathcal{E}_1')^{\frac{p}{p^\star}}} \cdot \left(1 + \varepsilon^2\left(\frac{p}{p^\star}\frac{\mathcal{E}_2'}{\mathcal{E}_1'} - \frac{\mathcal{E}_2}{(2q-n-2)\mathcal{E}_1}\right) + o(\varepsilon^2)\right).$$

Direct computation shows that for $2<p<\frac{n+2}{3}$,

$$(36)\quad \frac{p}{p^\star}\frac{\mathcal{E}_2'}{\mathcal{E}_1'} - \frac{\mathcal{E}_2}{(2q-n-2)\mathcal{E}_1} = -\frac{(2q-n)\big(2\mathcal{R}_n + \sum_{i<n}\mathcal{R}_i' + o_\rho(1)\big)}{2(2q-n+1)(2q-n-1)(2q-n-2)}.$$

The last quantity is negative for sufficiently small ρ since

$$2\mathcal{R}_n + \sum_{i<n}\mathcal{R}_i' = \sum_i \mathcal{R}_i = \mathcal{R}(y^0)$$

is the scalar curvature of Ω at y^0, which is positive by assumption.

Hence, for ε and ρ sufficiently small, one has

$$\frac{\|u\|_{W_p^1(\Omega)}}{\|u\|_{p^\star,\partial\Omega}} < \frac{n-p}{p-1}\frac{\mathcal{E}_1^{\frac{1}{p}}}{(\mathcal{E}_1')^{\frac{1}{p^\star}}}. \tag{37}$$

Similarly, in the case $p = \frac{n+2}{3}$, the relation (37) holds true for ε and ρ sufficiently small. By continuity, for some $\beta > 0$ and $p < \frac{n+2}{3} + \beta$,

$$\lambda_1(n,p,\Omega) < \frac{n-p}{p-1}\frac{\mathcal{E}_1^{\frac{1}{p}}}{(\mathcal{E}_1')^{\frac{1}{p^\star}}} = K(n,p).$$

The application of Proposition 1 completes the proof. □

3. Nonexistence of the minimizer

Proof of Theorem 2. We proceed by contradiction. Suppose there exists an unbounded sequence of $\varkappa$'s such that the infimum in (I) is attained on $\varkappa\Omega$. Then, by a standard argument, the corresponding minimizers $u_\varkappa$ are positive in $\varkappa\Omega$.

Consider the functions $v_\varkappa(y) = C(\varkappa)u_\varkappa(\varkappa y)$ in Ω, with $C(\varkappa)$ given by the normalization condition $\|v_\varkappa\|_{p^\star,\partial\Omega} = 1$. Then the weak Euler equation for the $v_\varkappa$ reads as follows:

$$\int_\Omega |\nabla v_\varkappa|^{p-2}\langle \nabla v_\varkappa, \nabla h\rangle\, dy + \varkappa^p \int_\Omega v_\varkappa^{p-1} h\, dy = \lambda_\varkappa^p \int_{\partial\Omega} v_\varkappa^{p^\star-1} h\, d\Sigma, \tag{38}$$

where $\lambda_\varkappa = \lambda_1(n,p,\varkappa\Omega)$. Since, given $\varkappa$, for the sequence considered in the proof of Theorem 1 the quotient (I) tends to $K(n,p)$, one has

$$\lambda_\varkappa \leqslant K(n,p). \tag{39}$$

Put $h = v_\varkappa$ in (38). Then

$$\|\nabla v_\varkappa\|_{p,\Omega}^p + \varkappa^p \|v_\varkappa\|_{p,\Omega}^p = \lambda_\varkappa^p. \tag{40}$$

Without loss of generality, $v_\varkappa \rightharpoonup v$ in $W_p^1(\Omega)$. From (40) we conclude that $v = 0$. Arguing as in the proof of Theorem 2.2 in [**LPT**], one deduces that there exists a point $y^0 \in \partial\Omega$ such that $|v_\varkappa|^{p^\star} \rightharpoonup \delta(y - y^0)$ in the sense of measures on $\partial\Omega$. Therefore, for a given $\rho > 0$,

$$\|v_\varkappa\|_{p^\star,\partial\Omega\setminus B_{\frac{\rho}{2}}(y^0)} \to 0. \tag{41}$$

Remark 4. In fact, the stronger property $\sup_{\Omega\setminus B_\rho(y^0)} v_\varkappa \to 0$ can be proved, but we do not need it.

Now we use spherical symmetrization (see, for instance, [**PS**]) in "planes" $\theta = const$. Since this operation retains the denominator in (I) and does not increase the numerator, we can assume that $v_\varkappa\big|_{\partial\Omega}$ is symmetrically decreasing around the point y^0. Since the normalization of $v_\varkappa$ implies that $\|v_\varkappa\|_{p^\star,weak,\partial\Omega} \leqslant C$, this symmetry provides the estimate

$$\operatorname{dist}(y,y^0)^{\frac{n}{p}-1} \cdot v_\varkappa(y) \leqslant C, \qquad y \in \partial\Omega. \tag{42}$$

By η we denote a smooth cutoff function such that

$$\eta = 0 \quad \text{in} \quad B_{\frac{\rho}{2}}(y^0); \qquad \eta = 1 \quad \text{in} \quad \Omega \setminus B_\rho(y^0).$$

We substitute $h = \eta^p v_\varkappa$ into the equation (38):

$$\int_\Omega \eta^p |\nabla v_\varkappa|^p dy + p\int_\Omega \eta^{p-1} v_\varkappa |\nabla v_\varkappa|^{p-2}\langle \nabla v_\varkappa, \nabla\eta\rangle\, dy + \varkappa^p \int_\Omega \eta^p v_\varkappa^p\, dy = \lambda_\varkappa^p \int_{\partial\Omega} \eta^p v_\varkappa^{p^\star}\, d\Sigma.$$

Recalling (39), we arrive at

$$\|\eta\nabla v_\varkappa\|_{p,\Omega}^p + \varkappa^p \|\eta v_\varkappa\|_{p,\Omega}^p \leqslant K^p(n,p)\int_{\partial\Omega} \eta^p v_\varkappa^{p^\star}\, d\Sigma + C\|v_\varkappa\|_{p,\Omega}\cdot \|\eta\nabla v_\varkappa\|_{p,\Omega}^{p-1}. \tag{43}$$

The Hölder inequality and (41) give us

$$\int_{\partial\Omega} \eta^p v_\varkappa^{p^\star}\, d\Sigma \leqslant \|\eta v_\varkappa\|_{p^\star,\partial\Omega}^p \cdot \|v_\varkappa\|_{p^\star,\partial\Omega\setminus B_{\frac{\rho}{2}}(x^0)}^{p^\star - p} = \|\eta v_\varkappa\|_{p^\star,\partial\Omega}^p \cdot o_\varkappa(1).$$

We apply the trace embedding theorem to the function $\eta v_\varkappa$:

$$\|\eta v_\varkappa\|_{p^\star,\partial\Omega}^p \leqslant C(\|\nabla(\eta v_\varkappa)\|_{p,\Omega}^p + \|\eta v_\varkappa\|_{p,\Omega}^p) \leqslant C(\|\eta\nabla v_\varkappa\|_{p,\Omega}^p + \|v_\varkappa\|_{p,\Omega}^p). \tag{44}$$

Substituting these relations into (43), we obtain

$$\left(\frac{\|\eta\nabla v_\varkappa\|_{p,\Omega}}{\|v_\varkappa\|_{p,\Omega}}\right)^p \leqslant C\left(\frac{\|\eta\nabla v_\varkappa\|_{p,\Omega}}{\|v_\varkappa\|_{p,\Omega}}\right)^{p-1} + o_\varkappa(1)\cdot\left(1 + \left(\frac{\|\eta\nabla v_\varkappa\|_{p,\Omega}}{\|v_\varkappa\|_{p,\Omega}}\right)^p\right).$$

Therefore, for $\varkappa$ sufficiently large one has

$$\frac{\|\eta\nabla v_\varkappa\|_{p,\Omega}}{\|v_\varkappa\|_{p,\Omega}} \leqslant C. \tag{45}$$

To obtain accurate estimates of $v_\varkappa$ at the neighborhood of y^0, we use the idea of [**Dr1**]. Recall that $\exp_{y^0}$ is a diffeomorphism from the half-ball $B_{2\rho_0}^+ \subset \mathbb{R}^n$ onto the Riemannian half-ball $B_{2\rho_0}^+(y^0)$. Without loss of generality, one can assume that $\rho < \rho_0$.

Denote by η_1 a smooth cutoff function such that

$$\eta_1 = 1 \quad \text{in} \quad B_\rho(y^0); \qquad \eta_1 = 0 \quad \text{in} \quad \Omega\setminus B_{2\rho}(y^0).$$

By (1),

$$\|\widetilde{\eta}_1 \widetilde{v}_\varkappa\|_{p^\star,\mathbb{R}^{n-1}}^p \leqslant \frac{1}{K^p(n,p)}\cdot \|\nabla(\widetilde{\eta}_1 \widetilde{v}_\varkappa)\|_{p,\mathbb{R}_+^n}^p. \tag{46}$$

By [**BZ**, §19.5],

$$|\nabla(\widetilde{\eta}_1 \widetilde{v}_\varkappa)|(x) = |\nabla(\eta_1 v_\varkappa)|(y)\cdot\big(1 + O(\mathrm{dist}^2(y,y^0))\big)$$

and

$$dx = dy\cdot\big(1 + O(\mathrm{dist}^2(y,y^0))\big).$$

This gives

$$\begin{aligned} \|\nabla(\widetilde{\eta}_1 \widetilde{v}_\varkappa)\|_{p,\mathbb{R}_+^n}^p &= \int_{B_{2\rho}(0)} |\nabla(\widetilde{\eta}_1 \widetilde{v}_\varkappa)|^p(x)\, dx \\ &\leqslant \int_{B_{2\rho}(y^0)} |\nabla(\eta_1 v_\varkappa)|^p(y)\cdot\big(1 + C\mathrm{dist}^2(y,y^0)\big)\, dy \end{aligned} \tag{47}$$

and

$$\begin{aligned}\|\widetilde{\eta}_1\widetilde{v}_\varkappa\|^{p^\star}_{p^\star,\mathbb{R}^{n-1}} &= \int_{B_{2\rho}(0)\cap\mathbb{R}^{n-1}} (\widetilde{\eta}_1\widetilde{v}_\varkappa)^{p^\star}(x')\,dx' \\ &\geqslant \int_{B_{2\rho}(y^0)\cap\partial\Omega} (\eta_1 v_\varkappa)^{p^\star}(y)\cdot\big(1 - C\operatorname{dist}^2(y,y^0)\big)\,dy \\ &\geqslant \|v_\varkappa\|^{p^\star}_{p^\star,B_\rho(y^0)\cap\partial\Omega} - C\int_{\partial\Omega} v_\varkappa^{p^\star}(y)\cdot\operatorname{dist}^2(y,y^0)\,dy.\end{aligned}$$

Moreover, with regard to the normalization of $v_\varkappa$ and to $p < p^\star$,

$$\|\widetilde{\eta}_1\widetilde{v}_\varkappa\|^{p}_{p^\star,\mathbb{R}^{n-1}} \geqslant \|v_\varkappa\|^{p^\star}_{p^\star,B_\rho(y^0)\cap\partial\Omega} - C\int_{\partial\Omega} v_\varkappa^{p^\star}(y)\cdot\operatorname{dist}^2(y,y^0)\,dy. \tag{48}$$

Substituting (47) and (48) into (46), we obtain by (7)

$$\begin{aligned}&\|v_\varkappa\|^{p^\star}_{p^\star,\partial\Omega\cap B_\rho(y^0)} \\ &\quad\leqslant \frac{1}{K^p(n,p)}\cdot\|\nabla v_\varkappa\|^p_{p,\Omega} + C\Big[\|v_\varkappa\|^p_{p,\Omega} + \|v_\varkappa\|_{p,\Omega}\cdot\|\nabla v_\varkappa\|^{p-1}_{p,\Omega\setminus B_\rho(y^0)}\Big] \\ &\qquad + C\int_\Omega |\nabla v_\varkappa|^p\operatorname{dist}^2(y,y^0))\,dy + C\int_{\partial\Omega} v_\varkappa^{p^\star}\operatorname{dist}^2(y,y^0)\,dy.\end{aligned} \tag{49}$$

We estimate the last term in the right-hand side of (49) by (42):

$$\begin{aligned}&\int_{\partial\Omega} v_\varkappa^{p^\star}\operatorname{dist}^2(y,y^0)\,dy \\ &\qquad= \int_{\partial\Omega}\big(v_\varkappa^{\frac{p}{n-p}}\operatorname{dist}(y,y^0)\big)^{p-1}\cdot v_\varkappa^p\operatorname{dist}^{3-p}(y,y^0)\,dy \\ &\qquad\qquad\leqslant C\int_{\partial\Omega} v_\varkappa^p\operatorname{dist}^{3-p}(y,y^0)\,dy.\end{aligned}$$

Using the trace embedding theorem and the Hölder inequality, we obtain

$$\begin{aligned}&\int_{\partial\Omega} v_\varkappa^{p^\star}\operatorname{dist}^2(y,y^0)\,dy \\ &\quad\leqslant C\Big(\int_\Omega |\nabla(v_\varkappa^p\operatorname{dist}^{3-p}(y,y^0))|\,dy + \int_\Omega v_\varkappa^p\operatorname{dist}^{3-p}(y,y^0)\,dy\Big) \\ &\leqslant C\Big(\int_\Omega |\nabla v_\varkappa|^p\operatorname{dist}^{(3-p)p}(y,y^0)\,dy\Big)^{\frac{1}{p}}\cdot\|v_\varkappa\|^{p-1}_{p,\Omega} + C\int_\Omega v_\varkappa^p\operatorname{dist}^{2-p}(y,y^0)\,dy \\ &\qquad\leqslant C\Big(\int_\Omega |\nabla v_\varkappa|^p\operatorname{dist}^2(y,y^0)\,dy\Big)^{\frac{1}{p}}\cdot\|v_\varkappa\|^{p-1}_{p,\Omega} + C\|v_\varkappa\|^p_{p,\Omega}\end{aligned} \tag{50}$$

(the last inequality is due to $1 < p \leqslant 2$).

Further, we substitute $h = v_\varkappa\operatorname{dist}^2(y,y^0)$ into (2). Canceling a positive term results in

$$\int_\Omega |\nabla v_\varkappa|^p\operatorname{dist}^2(y,y^0)\,dy \leqslant K^p(n,p)\int_{\partial\Omega} v_\varkappa^{p^\star}\operatorname{dist}^2(y,y^0)\,dy.$$

Substituting this relations into (50), we arrive at

$$\int_\Omega |\nabla v_\varkappa|^p\operatorname{dist}^2(y,y^0)\,dy \leqslant C\Big(\int_\Omega |\nabla v_\varkappa|^p\operatorname{dist}^2(y,y^0)\,dy\Big)^{\frac{1}{p}}\|v_\varkappa\|^{p-1}_{p,\Omega} + C\|v_\varkappa\|^p_{p,\Omega},$$

which gives

$$\int_\Omega |\nabla v_\varkappa|^p \operatorname{dist}^2(y,y^0)\,dy \leqslant C\|v_\varkappa\|^p_{p,\Omega}, \qquad \int_{\partial\Omega} v_\varkappa^{p^\star} \operatorname{dist}^2(y,y^0)\,dy \leqslant C\|v_\varkappa\|^p_{p,\Omega}.$$

We substitute these relations into (49). Expressing $\|\nabla v_\varkappa\|^p_{p,\Omega}$ from (40) and recalling (39), we get

$$\frac{\varkappa^p}{\gamma^p}\cdot\|v_\varkappa\|^p_{p,\Omega} \leqslant 1 - \|v_\varkappa\|^{p^\star}_{p^\star,\partial\Omega\cap B_\rho(y^0)} + C\Big[\|v_\varkappa\|^p_{p,\Omega} + \|v_\varkappa\|_{p,\Omega}\cdot\|\nabla v_\varkappa\|^{p-1}_{p,\Omega\setminus B_\rho(y^0)}\Big].$$

Thus,

$$\frac{\varkappa^p}{\gamma^p} \leqslant \frac{\|v_\varkappa\|^{p^\star}_{p^\star,\partial\Omega\setminus B_\rho(y^0)}}{\|v_\varkappa\|^p_{p,\Omega}} + C\left(1+\left(\frac{\|\nabla v_\varkappa\|_{p,\Omega\setminus B_\rho(y^0)}}{\|v_\varkappa\|_{p,\Omega}}\right)^{p-1}\right). \tag{51}$$

Using normalization of $v_\varkappa$, (44) and (45), we obtain

$$\frac{\|v_\varkappa\|^{p^\star}_{p^\star,\partial\Omega\setminus B_\rho(y^0)}}{\|v_\varkappa\|^p_{p,\Omega}} \leqslant \frac{\|v_\varkappa\|^{p}_{p^\star,\partial\Omega\setminus B_\rho(y^0)}}{\|v_\varkappa\|^p_{p,\Omega}} \leqslant \frac{\|\eta v_\varkappa\|^{p}_{p^\star,\partial\Omega}}{\|v_\varkappa\|^p_{p,\Omega}} \leqslant C\left(1+\frac{\|\eta\nabla v_\varkappa\|^p_{p,\Omega}}{\|v_\varkappa\|^p_{p,\Omega}}\right) \leqslant C,$$

while the second term in (51) is bounded by (45):

$$\frac{\|\nabla v_\varkappa\|_{p,\Omega\setminus B_\rho(y^0)}}{\|v_\varkappa\|_{p,\Omega}} \leqslant \frac{\|\eta\nabla v_\varkappa\|_{p,\Omega}}{\|v_\varkappa\|_{p,\Omega}} \leqslant C.$$

Therefore, the right-hand side in (51) is bounded. This gives a contradiction for $\varkappa$ large enough. □

4. The trace Sobolev–Poincaré inequality

PROOF OF THEOREM 3. We use the following analog of Proposition 1 (see also [**NR**, Proposition 2]).

PROPOSITION 2. *Let the infimum in* (II) *satisfy the inequality* $\lambda_2(n,p,\Omega) < K(n,p)$. *Then the infimum is attained.*

To construct a function with zero trace mean-value having the quotient (II) less than $K(n,p)$ we modify the function (3). Similarly to [**DN**, Theorem 5.1], for large p we proceed by subtracting a suitable function with a small support while for small p we will subtract a constant.

We need the follownig estimate for the function u defined in (3):

$$\int_{\partial\Omega} u(y)d\Sigma \leqslant C\int_{|x'|<\rho}\frac{dx'}{(|x'|^2+\varepsilon^2)^{\frac{n-p}{2(p-1)}}} \leqslant \begin{cases} C\varepsilon^{n-1-\frac{n-p}{p-1}}, & p<\frac{2n-1}{n},\\ C\ln(\frac{1}{\varepsilon})+C(\rho), & p=\frac{2n-1}{n},\\ C, & p>\frac{2n-1}{n}.\end{cases} \tag{52}$$

1. First, consider the case $p>\frac{2n-1}{n}$. Let u_1 be a smooth nonnegative function such that $u_1\big|_{\partial\Omega}\not\equiv 0$ and $\operatorname{supp} u_1\cap\operatorname{supp} u=\emptyset$. We define the function with zero trace mean-value

$$\widehat{u}(y) = u(y) - u_1(y)\,\frac{\int_{\partial\Omega} u\,d\Sigma}{\int_{\partial\Omega} u_1 d\Sigma}.$$

The estimates (23), (33) and (36) yield for $p<\frac{n+2}{3}$,

$$\frac{\|\nabla u\|^p_{p,\Omega}}{\|u\|^p_{p^\star,\partial\Omega}} \leqslant K^p(n,p)\cdot\left(1-\varepsilon^2\frac{(2q-n)\mathcal{R}(y^0)}{2(2q-n+1)(2q-n-1)(2q-n-2)}+o(\varepsilon^2)\right)$$

and therefore, for small ρ and ε,

$$\frac{\|\nabla \widehat{u}\|_{p,\Omega}}{\|\widehat{u}\|_{p^\star,\partial\Omega}} < K(n,p). \tag{53}$$

For $p = \frac{n+2}{3}$ this relation also holds true for ε and ρ sufficiently small. By continuity, for some $\beta > 0$ and $p < \frac{n+2}{3} + \beta$,

$$\lambda_2(n,p,\Omega) < K(n,p). \tag{54}$$

2. Now we consider the case $1 < p \leqslant \frac{2n-1}{n}$. Define u by (3) and put $\rho = \varepsilon^\gamma$ with some $\gamma \in (0,\gamma^*)$, where $\gamma^* = \frac{n+2-3p}{n-p} \in (0,1)$.

The estimate (33) yields

$$\int_{\partial\Omega} u^{p^\star}\, d\Sigma \geqslant \varepsilon^{n-2q-1}\mathcal{E}_1' - \varepsilon^{n-2q+1}\mathcal{E}_2' - C\varepsilon^{\gamma(n-2q-1)}.$$

Immediate computation shows that

$$\gamma^*(n-2q-1) > n-2q+1.$$

Therefore, for $\gamma \in (0,\gamma^*)$,

$$\int_{\partial\Omega} u^{p^\star}\, d\Sigma \geqslant \varepsilon^{n-2q-1}\mathcal{E}_1' - \varepsilon^{n-2q+1}(\mathcal{E}_2' + o_\varepsilon(1)).$$

Furthermore, the estimate (23) implies that

$$\int_\Omega |\nabla u|^p\, dy \leqslant \Big(\frac{n-p}{p-1}\Big)^p \cdot \big(\varepsilon^{n-2q}\mathcal{E}_1 - \varepsilon^{n-2q+2}\frac{\mathcal{E}_2}{2q-n-2} + C\varepsilon^{\gamma(n-2q)}\big).$$

Direct calculations result in

$$\gamma^*(2q-n) = 2q-n-2.$$

Hence, for $\gamma \in (0,\gamma^*)$,

$$\int_\Omega |\nabla u|^p\, dy \leqslant \Big(\frac{n-p}{p-1}\Big)^p \cdot \Big(\varepsilon^{n-2q}\mathcal{E}_1 - \varepsilon^{n-2q+2}\frac{\mathcal{E}_2 - o_\varepsilon(1)}{2q-n-2}\Big). \tag{55}$$

Now we define the function with zero trace mean-value

$$\widehat{u}(y) = u(y) - \frac{1}{|\partial\Omega|}\int_{\partial\Omega} u\, d\Sigma.$$

Using (52) and the Minkowski inequality we conclude that

$$\begin{aligned} \|\widehat{u}\|_{p^\star,\partial\Omega} &\geqslant \Big(\int_{\operatorname{supp} u\cap\partial\Omega} |\widehat{u}|^{p^\star}\, d\Sigma\Big)^{\frac{1}{p^\star}} \\ &\geqslant \|u\|_{p^\star,\partial\Omega} - \frac{|\operatorname{supp} u\cap\partial\Omega|^{\frac{1}{p^\star}}}{|\partial\Omega|}\int_{\partial\Omega} u d\Sigma \geqslant \varepsilon^{\frac{n-2q-1}{p^\star}}(\mathcal{E}_1')^{\frac{1}{p^\star}} \\ &\qquad \times \Big(1 - \varepsilon^2\frac{\mathcal{E}_2' + o_\varepsilon(1)}{p^\star\mathcal{E}_1'} - C\varepsilon^{\gamma\frac{n-1}{p^\star} - \frac{n-2q-1}{p^\star}}\Phi(\varepsilon)\Big), \end{aligned} \tag{56}$$

where

$$\Phi(\varepsilon) = \begin{cases} \varepsilon^{n-1-\frac{n-p}{p-1}}, & \text{if } 1 < p < \frac{2n-1}{n}, \\ 1 + \ln(\frac{1}{\varepsilon}), & \text{if } p = \frac{2n-1}{n}. \end{cases}$$

Immediate calculations show that for $n \geqslant 4$ and $1 < p \leqslant \frac{2n-1}{n}$, the inequality

$$\gamma^*\frac{n-1}{p^\star} + \frac{n-2q-1}{p^\star} + n - 1 - \frac{n-p}{p-1} > 2$$

holds true for $\gamma = \gamma^*$ and, therefore, for some $\gamma \in (0, \gamma^*)$. Choose such a γ; then (56) implies that

$$\|\widehat{u}\|^p_{p^\star,\partial\Omega} \geqslant \varepsilon^{n-2q}(\mathcal{E}_1')^{\frac{p}{p^\star}} \left(1 - \varepsilon^2 \frac{p}{p^\star} \frac{\mathcal{E}_2' + o_\varepsilon(1)}{\mathcal{E}_1'}\right). \tag{57}$$

The relations (55), (57) and (36) imply (53) for ε small enough. Thus, the inequality (54) is valid in this case. The application of Proposition 2 completes the proof. □

References

[BZ] Yu. D. Burago and V. A. Zalgaller, *Introduction to Riemannian geometry*, St. Petersburg, Nauka, 1995 (Russian). MR1356465 (96h:53002)

[DN] A. V. Demyanov and A. I. Nazarov, *On the existence of an extremal function in Sobolev embedding theorems with critical exponents*, Algebra Anal. **17** (2005), N5, 105–140 (Russian). English transl. in: St. Petersburg Math. J. **17** (2006), N5, 108-142. MR2241425 (2007e:42016)

[Dr] O. Druet, *Optimal Sobolev inequalities of arbitrary order on compact Riemannian manifolds*, J. Funct. Anal. **159** (1998), 217–242. MR1654123 (99m:53076)

[Dr1] O. Druet, *The best constants problem in Sobolev inequalities*, Math. Annalen **314** (1999), 327–346. MR1697448 (2000d:58033)

[E] J. F. Escobar, *Sharp constant in a Sobolev trace inequality*, Indiana Univ. Math. J. **37** (1988), 687–698. MR962929 (90a:46071)

[Ls] P.-L. Lions, *The concentration-compactness principle in the Calculus of Variations. The limit case*, Parts 1 and 2, Rev. Mat. Iberoamericana **1** (1985), 45–121 (1), 145–201 (2). MR850686 (87j:49012); MR0834360 (87c:49007)

[LPT] P.-L. Lions, F. Pacella, and M. Tricarico, *Best constants in Sobolev inequalities for functions vanishing on some part of the boundary and related questions*, Indiana Univ. Math. J. **37** (1988), 301–324. MR963504 (89i:46036)

[Nt] B. Nazaret, *Best constant in Sobolev trace inequalities on the half-space*, Nonlin. Analysis, **65** (2006), N10, 1977–1985. MR2258478 (2007m:46047)

[N] A. I. Nazarov, *Dirichlet and Neumann problems to critical Emden–Fowler type equations*, J. Global Optim. **40** (2008), 289–303. MR2373558 (2009c:35150)

[NR] A. I. Nazarov and A. B. Reznikov, *On the existence of an extremal function in critical Sobolev trace embedding theorem*, J. Funct. Anal. **258** (2010), N11, 3906–3921.

[PS] G. Pólya and G. Szegö, *Isoperimetric inequalities in mathematical physics*, Princeton Univ. Press, 1951. MR0043486 (13:270d)

St.-Petersburg State University
E-mail address: al.il.nazarov@gmail.com
E-mail address: reznikov@ymail.com

Amer. Math. Soc. Transl.
(2) Vol. **229**, 2010

Non-divergence Elliptic Equations of Second Order with Unbounded Drift

M. V. Safonov

Dedicated to Nina N. Ural'tseva

ABSTRACT. We consider uniformly elliptic equations and inequalities of second order in the *nondivergence* form $Lu = a_{ij}D_{ij}u + b_iD_iu = 0$ $(\le 0,\ \ge 0)$, in a domain $\Omega \subset \mathbb{R}^n$. We derive an interior Harnack inequality, a boundary Harnack inequality and a comparison theorem for Lipschitz boundaries in the case $b_i \in L^n$, and a Hopf-Oleinik-type estimate near a flat boundary in the case when the normal component of the vector $(b_1, \dots, b_n)$ belongs to L^q, $q > n$. Our main tools are special Landis-type growth lemmas.

1. Introduction

In this paper, we derive interior and boundary pointwise estimates for solutions to the equation

$$Lu := a_{ij}D_{ij}u + b_iD_iu = 0 \quad \text{in a domain} \quad \Omega \subset \mathbb{R}^n, \tag{1.1}$$

and the inequalities $Lu \le 0$ or $Lu \ge 0$, under minimal assumptions on the functions a_{ij} and b_i. We use the summation convention over repeated indices and the notation $D_i := \partial/\partial x_i$, $D_{ij} := D_iD_j$. We assume that a_{ij} and b_i are measurable functions on $\mathbb{R}^n$, a_{ij} satisfy the *uniform ellipticity condition*

$$a_{ij}(x) = a_{ji}(x), \qquad \nu\,|\xi|^2 \le a_{ij}(x)\xi_i\xi_j \le \nu^{-1}|\xi|^2 \quad \forall \xi, x \in \mathbb{R}^n, \tag{1.2}$$

with a constant $\nu \in (0,1]$, and b_i, $f \in L^n(\Omega)$. Throughout the paper, the operator L in (1.1) is applied to functions u in the class $W(\Omega) := W^{2,n}_{loc}(\Omega) \cap C(\overline{\Omega})$, which implies, in particular, that u, D_iu, $D_{ij}u$ belong to the Lebesgue space $L^n(\Omega')$ for any bounded open set $\Omega' \subset \overline{\Omega'} \subset \Omega$.

The following estimate is crucial for our considerations. It can be considered as a particular case (when $m = n$) of the estimate (5.6) in the paper by A.D. Aleksandrov [**A63**] (see also [**GT83**], (9.14)).

2000 *Mathematics Subject Classification.* Primary 35J15, 35J67; Secondary 35B05.
Key words and phrases. Second-order elliptic equations, Harnack inequality, boundary Hopf lemma, measurable coefficients.

THEOREM 1.1. *Let Ω be a bounded open set in $\mathbb{R}^n$, and let u be a function in $W(\Omega)$ such that $Lu \geq f$ in Ω. Suppose that the coefficients a_{ij} satisfy (1.2) and that b_i, $f \in L^n(\Omega)$. Then*

$$\sup_{\Omega} u \leq \sup_{\partial\Omega} u + N \operatorname{diam} \Omega \cdot e^{NS} \, ||f_-||_{n,\Omega}, \tag{1.3}$$

where $f_- := \max\{-f, 0\}$,

$$S := S(\Omega) := \int_{\Omega} |\mathbf{b}|^n dx, \qquad \mathbf{b} := (b_1, \ldots, b_n), \tag{1.4}$$

$$||g||_{n,\Omega} := \left(\int_{\Omega} |g|^n dx \right)^{1/n} \quad \text{- the norm of} \quad g \quad \text{in } L^n(\Omega),$$

and N is a positive constant depending only on n and ν.

COROLLARY 1.2. *Let Ω be a bounded open set in $\mathbb{R}^n$, and let u be a function in $W(\Omega)$ such that $Lu \geq 0$ in Ω. Then the maximum of u on $\overline{\Omega}$ is attained on $\partial\Omega$.*

REMARK 1.3. Note that the equality $Lu = f$ in (1.1) and the inequality $Lu \geq f$ in Theorem 1.1 hold true almost everywhere in Ω with respect to the Lebesgue measure on $\mathbb{R}^n$. It is easy to see that the function $u(x) := 1 - |x|^2$ satisfies

$$\Delta u + b_i D_i u = 0 \qquad \text{in} \qquad B_1 := \{x \in \mathbb{R}^n : |x| < 1\},$$

where $b_i(x) := -nx_i \cdot |x|^{-2} \in L^{n-\varepsilon}(B_1)$ for arbitrarily small $\varepsilon > 0$, and the maximum of u on $\overline{B_1}$ is not attained on ∂B_1. Therefore, the assumption $b_i \in L^n$ is essential for the whole theory, which is based on the maximum principle.

REMARK 1.4. Note that the quantity $S(\Omega)$ in (1.4) is scaling invariant in the following sense. If $u = u(x)$ satisfies the equation $Lu = 0$ in Ω, then for any constant $k > 0$, the function $\tilde{u}(y) := u(ky)$ satisfies $D_i\tilde{u} = kD_iu$, $D_{ij}\tilde{u} = k^2 D_{ij}u$; hence

$$\tilde{L}\tilde{u} := \tilde{a}_{ij} D_{ij}\tilde{u} + \tilde{b}_i D_i \tilde{u} = 0 \quad \text{in} \quad \tilde{\Omega} := \{y \in \mathbb{R}^n : ky \in \Omega\},$$

where $\tilde{a}_{ij}(y) := a_{ij}(ky)$, $\tilde{b}_i(y) := kb_i(ky)$. Therefore, $|\tilde{\mathbf{b}}| := \left(\sum \tilde{b}_i^2\right)^{1/2}$ satisfies

$$\tilde{S}(\tilde{\Omega}) := \int_{\tilde{\Omega}} |\tilde{\mathbf{b}}(y)|^n dy = \int_{\tilde{\Omega}} |\mathbf{b}(ky)|^n k^n dy = \int_{\Omega} |\mathbf{b}(x)|^n dx =: S(\Omega).$$

Obviously, this argument also works for inequalities $Lu \leq 0$ or $Lu \geq 0$ in Ω.

In the next section, we will use the scaling invariance of $S(\Omega)$, together with the maximum principle, in order to prove some special *growth lemmas* (Lemmas 2.1 and 2.5) and a *doubling property* of solutions (Lemma 2.2) in the case $|\mathbf{b}| \in L^n$, i.e. $S(\Omega) < \infty$. The growth lemmas were first introduced by E.M. Landis [**La67, La71**]. Among other applications of these lemmas, Landis gave alternative proofs of results by De Giorgi and Moser on Hölder regularity and Harnack inequalities for solutions to second-order elliptic equations in *divergence* form. These results were extended to the equations in *nondivergence* form with $|\mathbf{b}| \in L^\infty$ in our joint work with N.V. Krylov [**KS80, S80**] (see also the book [**K85**]). For this purpose, we also used some variants of growth lemmas. Now this technique became standard; see the paper by H. Aimar, L. Forzani, and R. Toledano [**AFT01**], in which they treat Hölder and Harnack properties from an abstract point of view.

In Section 3, we use the growth lemmas and the doubling property of solutions in order to derive the interior and boundary Harnack inequalities and the Hölder estimates for solutions of equations (1.1) with $|\mathbf{b}| \in L^n$. Note that the Hölder regularity in the case $|\mathbf{b}| \in L^n$ was proved earlier by O.A. Ladyzhenskaya and N.N. Ural'tseva [**LU85**]. However, the Hölder constants in that paper depend on the modulus of continuity of $|\mathbf{b}|$ in L^n, or more precisely, on the constant $\rho > 0$ for which the norms of $|\mathbf{b}|$ in $L^n(\Omega \cap B_\rho(x))$ are small enough for all $x \in \Omega$. In our estimates, all the constants depend on $\mathbf{b}$ only through the quantity $S(\Omega)$ in (1.4).

Finally, in Section 4, we prove a Hopf-Oleinik-type estimate near the flat boundary $\{x_n = 0\}$ for positive solutions of the equation $a_{ij}D_{ij}u + b_iD_iu = 0$ with $|\mathbf{b}| \in L^n$, $b_n \in L^q$, $q > n$. A close result is contained in the paper by O.A. Ladyzhenskaya and N.N. Ural'tseva [**LU88**], Lemma 4.4, under a bit stronger assumption $|\mathbf{b}| \in L^q$, $q > n$. See also a recent work by A.I. Nazarov and N.N. Ural'tseva [**NU09**] for further results concerning elliptic and parabolic, divergence and non-divergence, equations with unbounded lower-order coefficients.

Notation. $x = (x_1, \dots, x_{n-1}, x_n) = (x', x_n)$ are vectors or points in $\mathbb{R}^n$, $(x, y) := x_iy_i$ is the scalar product of $x, y \in \mathbb{R}^n$ (the summation convention is implied), and $|x| := (x, x)^{1/2}$ is the length of $x \in \mathbb{R}^n$. The standard orthonormal basis in $\mathbb{R}^n$ consists of n vectors $\{\mathbf{e_1}, \dots, \mathbf{e_n}\}$, so that $x = x_i\mathbf{e_i}$ for all $x \in \mathbb{R}^n$. $\operatorname{tr} a := \sum_i a_{ii}$ is the trace of an $n \times n$ matrix $a = [a_{ij}]$. For $c \in \mathbb{R}^1$, $[c]$ denotes its integer part (the maximal integer $\le c$), $c_+ := \max\{c, 0\}$, $c_- := \max\{-c, 0\}$.

$B_r(x) := \{y \in \mathbb{R}^n : |y - x| < r\}$ is a ball of radius $r > 0$ centered at $x \in \mathbb{R}^n$.

$B'_r(x') := \{y' \in \mathbb{R}^{n-1} : |y' - x'| < r\}$ is a similar ball in $\mathbb{R}^{n-1}$ centered at $x' \in \mathbb{R}^{n-1}$. $B_r := B_r(0)$, $B'_r := B'_r(0)$.

$\partial\Omega$ is the boundary of an open set $\Omega \subset \mathbb{R}^n$; $|\Omega|$ is its Lebesgue measure.

The notation $A := B$, or $B =: A$, means "$A = B$ by definition". Throughout the paper, N, c (with indices or without) denote different constants depending only on the prescribed quantities, such as n, ν, etc. This dependence is indicated in parentheses: $N = N(n, \nu, \dots)$, $c = c(n, \nu, \dots)$.

2. Growth lemmas

The first growth lemma can be treated as a growth lemma for "narrow" domains.

LEMMA 2.1 (First growth lemma). *Let Ω be an open set in $\mathbb{R}^n$, and let $u \in W(\Omega)$, $x_0 \in \mathbb{R}^n$, and $r > 0$ be such that*

$$u \ge 0, \quad Lu \ge 0 \quad \text{in} \quad \Omega, \quad \text{and} \quad u = 0 \quad \text{on} \quad (\partial\Omega) \cap B_{2r}(x_0).$$

We claim that for an arbitrary constant $\beta_1 \in (0,1)$, there is a constant $\mu_1 \in (0,1)$, depending only on n, ν, and S, such that from the estimate for the Lebesgue measure

$$|\Omega \cap B_{2r}(x_0)| \le \mu_1 \cdot |B_{2r}| \tag{2.1}$$

it follows that

$$M_r := \sup_{\Omega \cap B_r(x_0)} u \le \beta_1 \cdot M_{2r}. \tag{2.2}$$

PROOF. Using translation in $\mathbb{R}^n$, and multiplying u by a constant, we can assume $x_0 = 0$, $M_r > 0$, and $M_{2r} = 1$. Moreover, since $S(\Omega)$ is invariant with respect to rescaling (Remark 1.4), we can choose any convenient value $r > 0$.

(a) We first take $r = 1$ and assume that $S := S(\Omega) \le 1$. By Corollary 1.2, the maximum of u on $\overline{\Omega \cap B_1}$ is attained at some point $y_0 \in \Omega \cap (\partial B_1)$. Consider the function $v(x) := u(x) - |x - y_0|^2$ on the set $\Omega' := \Omega \cap B_1(y_0)$. Since $v \le u = 0$ on $(\partial\Omega) \cap B_2$, and $v = u - 1 \le 0$ on $\overline{\Omega} \cap (\partial B_1(y_0))$, we have $v \le 0$ on $\partial\Omega'$. Moreover,

$$Lv = Lu - 2\,\mathrm{tr}\,a - 2\,(\mathbf{b}, x - y_0) \ge f := -2n\nu^{-1} - 2|\mathbf{b}| \quad \text{in} \quad \Omega'.$$

Applying Theorem 1.1 to the function v in Ω', and having in mind that $S(\Omega) \le 1$ and $|\Omega'| \le \mu_1 \cdot |\Omega \cap B_2| \le \mu_1 \cdot |B_2|$, we get

$$M_1 := \sup_{\Omega \cap B_1} u = u(y_0) = v(y_0) \le N_0 \cdot ||f||_{n,\Omega'} \le N_1 \cdot \left(\mu_1^{1/n} + S^{1/n}\right), \tag{2.3}$$

with constants $N_0,\ N_1 \ge 1$, depending only on n and ν.

Now fix an arbitrary constant $\beta_1 \in (0, 1)$, and set $\mu_0 := S_0 := (\beta_1/2N_1)^n$. If $S \le S_0$, we can take $\mu_1 = \mu_0$, and then the desired estimate $M_1 \le \beta_1$ follows from (2.3).

(b) In the remaining case $S > S_0$, we set $m := [S/S_0] + 1 > S/S_0 > 1$. It is convenient to take $r := 4m$. Then we divide $\Omega \cap (B_{8m} \setminus \overline{B_{4m}})$ into m disjoint sets

$$\Omega_k := \Omega \cap \left(B_{4m+4k} \setminus \overline{B_{4m+4k-4}}\right) \quad \text{for} \quad k = 1, 2, \dots, m.$$

We have

$$\sum_{k=1}^{m} \int_{\Omega_k} |\mathbf{b}|^n dx \le \int_{\Omega} |\mathbf{b}|^n dx =: S.$$

Since $S < mS_0$, at least one of the integrals on the left side $\le S_0$. Fix the corresponding index $k = k_0$ for which this is the case. Finally, we set $\mu_1 := (4m)^{-n}\mu_0$. By the maximum principle,

$$M_r := \sup_{\Omega \cap B_{4m}} u \le \sup_{\Omega \cap B_{4m+4k_0-2}} u = u(y) \tag{2.4}$$

for some $y \in \Omega \cap (\partial B_{4m+4k_0-2})$. Note that the set $\Omega \cap B_2(y)$ is contained in Ω_{k_0}; hence the integral of $|\mathbf{b}|^n$ over this set $\le S_0$, and moreover, from (2.1) it follows that

$$|\Omega \cap B_2(y)| \le |\Omega \cap B_{8m}| \le \mu_1 \cdot |B_{8m}| = \mu_0 \cdot |B_2|.$$

From the previous part (a) of the proof it follows that

$$u(y) \le \sup_{\Omega \cap B_1(y)} u \le \beta_1 \cdot \sup_{\Omega \cap B_2(y)} u \le \beta_1 \cdot M_{2r}.$$

Together with (2.4), this completes the proof. □

LEMMA 2.2 (Doubling property). *Let $x_0 \in \mathbb{R}^n$, $r > 0$, and a function v in $W(B_{4r}(x_0))$ be such that*

$$v \ge 0, \quad Lv \le 0 \quad \text{in} \quad B_{4r}(x_0), \quad \text{and} \quad v \ge 1 \quad \text{in} \quad B_r(x_0). \tag{2.5}$$

Then

$$v \ge \lambda = \lambda(n, \nu, S) > 0 \quad \text{in} \quad B_{2r}(x_0). \tag{2.6}$$

PROOF. Without loss of generality, we assume $x_0 = 0$, $r = 1$. In the part (a) below, we consider the case $n = 1$, part (b) deals with $n \ge 2$ and small S, and part (c) deals with general $S < \infty$. We need to treat the case $n = 1$ separately, because in our approach, we use the fact that the spheres $\partial B_r \subset \mathbb{R}^n$ are connected. This fails if $n = 1$, because in this case ∂B_r consists of two disjoint points $\pm r$.

(a) In the case $n = 1$, we have

$$v \geq 0, \quad av'' + bv' \leq 0 \quad \text{in} \quad (-4,4), \quad \text{and} \quad v \geq 1 \quad \text{on} \quad [-1,1],$$

where $0 < \nu \leq a \leq \nu^{-1}$ and $b \in L^1(-4,4)$. Using properties of weak derivatives (the *product formula* (7.18) and the *chain rule* in Lemma 7.5 of [**GT83**]), one can rewrite the above differential inequality as follows:

$$(2.7) \qquad (Av')' \leq 0, \quad \text{where} \quad A(x) := \exp\left[\int_0^x \frac{b(y)}{a(y)}dy\right].$$

The function $A(x)$ satisfies

$$(2.8) \qquad N_0^{-1} \leq A(x) \leq N_0 := \exp(\nu^{-1}S) \quad \text{in} \quad (-4,4), \qquad S := \int_{-4}^{4} |b(x)|dx.$$

If suffices to get the estimate $v \geq \lambda = \lambda(n,\nu,S) > 0$ on the interval $[-2,-1]$, because a similar estimate on the symmetric interval $[1,2]$ can be obtained by replacing x with $-x$. Denote $\kappa := \min\limits_{[-2,-1]} v$. We can assume $\kappa \leq 1/2$, because otherwise the estimate $v \geq \lambda$ holds true with $\lambda = 1/2$. By the maximum principle, i.e., Corollary 1.2 applied to the function $-v$, we have $\kappa = v(-2)$. Since v is continuous and $v(-2) = \kappa \leq 1/2 < 1 \leq v(-1)$, there is $h \in (0,1]$ such that

$$v(-2+h) = 1, \quad \text{and} \quad v(x) < 1 \quad \text{for all} \quad x \in [-2,-2+h).$$

Furthermore, for each $y \in (-4,-2+h)$ the minimum of $v(x)$ on $[y,-2+h]$ is attained at the point y. This means that the function $v(x)$ is nondecreasing on $(-4,-2+h]$, and therefore $v' \geq 0$ in $(-4,-2+h)$. From (2.7) it follows that the function Av' is nonincreasing. Together with (2.8), this implies that

$$\begin{aligned}\kappa = v(-2) &\geq \int_{-2-h}^{-2} v'dx \geq \frac{1}{N_0}\int_{-2-h}^{-2} Av'dx \geq \frac{1}{N_0}\int_{-2}^{-2+h} Av'dx \\ &\geq \frac{1}{N_0^2}\int_{-2}^{-2+h} v'dx = \frac{1}{N_0^2}[v(-2+h) - v(-2)] \geq \frac{1}{2N_0^2} =: \lambda.\end{aligned}$$

This proves the estimate (2.6) in the case $n = 1$.

(b) In the rest of the proof of this lemma, we assume $n \geq 2$. First consider the case when $S := S(\Omega) \leq S_0 = S_0(n,\nu)$, a small positive constant to be chosen below in (2.10). Note that

$$\begin{aligned}L_0\left(|x|^{-\gamma_0}\right) &:= a_{ij}D_{ij}\left(|x|^{-\gamma_0}\right) = \gamma_0 \cdot \left[\frac{(\gamma_0+2)a_{ij}x_ix_j}{|x|^2} - \operatorname{tr} a\right] \cdot |x|^{-\gamma_0-2} \\ &\geq \gamma_0 \cdot \left[(\gamma_0+2)\nu - n\nu^{-1}\right] \cdot |x|^{-\gamma_0-2} \geq 0 \quad \text{for} \quad x \neq 0,\end{aligned}$$

provided $\gamma_0 + 2 \geq n\nu^{-2}$. Fix such a $\gamma_0 = \gamma_0(n,\nu) > 0$, and consider the function

$$w(x) := \frac{|x|^{-\gamma_0} - 4^{-\gamma_0}}{1 - 4^{-\gamma_0}} \quad \text{in} \quad \Omega_1 := \{x \in \mathbb{R}^n : 1 < |x| < 4\}.$$

We have $L_0 w \geq 0$ in Ω_1, $w \leq v$ on $\partial\Omega_1$. Therefore,

$$L(w - v) \geq Lw = L_0 w + (b, Dw) \geq (b, Dw) \quad \text{in} \quad \Omega_1,$$

and by Theorem 1.1,

$$(2.9) \qquad \sup_{\Omega_1}(w - v) \leq N_0 \cdot ||(b, Dw)||_{n,\Omega_1} \leq N_1 \cdot \|\,|\mathbf{b}|\,\|_{n,\Omega} = N_1 \cdot S^{1/n},$$

with some constants N_0, $N_1 \geq 1$ depending only on n and ν. Next, fix the constants

(2.10) $$\lambda_0 := \frac{2^{-\gamma_0} - 4^{-\gamma_0}}{2 \cdot (1 - 4^{-\gamma_0})} \in \left(0, \frac{1}{2}\right) \quad \text{and} \quad S_0 := \left(\frac{\lambda_0}{N_1}\right)^n,$$

which also depend only on n and ν. Then from $S \leq S_0$ and (2.9) it follows that $w - v \leq \lambda_0$ in Ω_1. On the other hand, obviously $w \geq 2\lambda_0$ on the set $\{1 \leq |x| \leq 2\} \subset \Omega_1$. Therefore, $v(x) \geq \lambda_0$ for $1 \leq |x| \leq 2$. Since also $v(x) \geq 1$ for $|x| < 1$, the desired estimate (2.6) holds true with $\lambda = \lambda_0(n, \nu) > 0$, provided $S \leq S_0 = S_0(n, \nu)$.

(c) Now it remains to consider the case $n \geq 2$, $S > S_0$. As in the proof of Lemma 2.1, we set $m := [S/S_0] + 1 > S/S_0 > 1$. Choose $\rho_0 \in (0, 1/2)$ of order $1/2$ and $y_1, \ldots, y_m \in \partial B_{1/2}$ such that the balls $B_{\rho_0}(y_k)$ are disjoint for $k = 1, \ldots, m$. Then the "tubes"

$$T_k := B_4 \cap \Big(\bigcup_{t \geq 1} B_{\rho_0}(ty_k)\Big), \qquad k = 1, \ldots, m,$$

are also disjoint. Since $S < mS_0$, we can fix $k = k_1$ such that the integral of $|\mathbf{b}|^n$ over T_{k_1} does not exceed S_0. Further, divide $B_4 \setminus \overline{B_2}$ into m disjoint shells

$$\Omega_k := B_{r_k} \setminus \overline{B_{r_{k-1}}}, \quad \text{where} \quad r_k := 2 + \frac{2k}{m}, \quad k = 1, \ldots, m,$$

and choose $k = k_2$ such that the integral of $|\mathbf{b}|^n$ over Ω_{k_2} does not exceed S_0. Finally, set

$$R := 2 + \frac{2k_2 - 1}{m} \in \left(2 + \frac{1}{m}, 4 - \frac{1}{m}\right), \qquad \rho := \frac{1}{4} \min\left\{\rho_0, \frac{1}{m}\right\}.$$

Fix an arbitrary $y \in \partial B_R$. One can move the ball $B_{4\rho}(z)$ continuously inside T_{k_1} from $z = y_{k_1} \in \partial B_{1/2}$ to $z = 2R \cdot y_{k_1} \in \partial B_R$, and then inside Ω_{k_2} to $z = y$. Therefore, there is a sequence

$$z_0 := y_{k_1}, z_1, \ldots, z_{j_0}, z_{j_0+1}, \ldots, z_{m_0} := y,$$

such that

(2.11) $$B_{4\rho}(z_j) \subset T_{k_1} \quad \text{for} \quad j \leq j_0, \quad B_{4\rho}(z_j) \subset \Omega_{k_2} \quad \text{for} \quad j \geq j_0 + 1,$$

$|z_{j+1} - z_j| \leq \rho$ for all $j = 0, 1, \ldots, m_0 - 1$, and m_0 is of order m.

We claim that

(2.12) $$v \geq \lambda_0^j \quad \text{in} \quad B_\rho(z_j) \quad \text{for} \quad j = 0, 1, \ldots, m_0,$$

where $\lambda_0 = \lambda_0(n, \nu) > 0$ is a constant in part (b), i.e. in (2.10), which corresponds to $S = S_0$. For $j = 0$, this inequality is contained in (2.5), because $B_\rho(z_0) = B_\rho(y_{k_1}) \subset B_{1/2}(y_{k_1}) \subset B_1$. Moreover, if (2.12) is true for some j, then the function $v_j := \lambda_0^{-j} v$ satisfies (2.5) with $x_0 = z_j$ and $r = \rho$. By our construction of the sets T_{k_1} and Ω_{k_2}, from (2.11) it follows that $S(B_{4\rho}(z_j)) \leq S_0$ for all j. Therefore, we can apply the preceding part (b) of this proof, which yields $v_j \geq \lambda_0$ in $B_{2\rho}(z_j)$. Since $|z_{j+1} - z_j| \leq \rho$, we have $B_\rho(z_{j+1}) \subset B_{2\rho}(z_j)$; hence $v \geq \lambda_0^{j+1}$ on $B_\rho(z_{j+1})$. By induction, (2.12) holds true for all $j = 0, 1, \ldots, m_0$. Here m_0 does not exceed a constant m_1 depending only on n, ν, and S. Therefore,

$$v(y) = v(z_{m_0}) \geq \lambda_0^{m_0} \geq \lambda_0^{m_1} =: \lambda = \lambda(n, \nu, S) > 0.$$

Here y is an arbitrary point in ∂B_R. From the inequalities $Lv \leq 0$ in B_R and $v \geq \lambda$ on ∂B_R it follows that $v \geq \lambda$ on B_R. Since $R > 2 = 2r$, the desired estimate (2.6) follows. □

COROLLARY 2.3. *Let* $x_0 \in \mathbb{R}^n$, $r > 0$, *and a function* v *in* $W(B_r(x_0))$ *be such that*

$$v \geq 0, \quad Lv \leq 0 \quad \text{in} \quad B_r(x_0), \quad \text{and} \quad v \geq 1 \quad \text{in} \quad B_{\varepsilon r}(x_0), \tag{2.13}$$

where $0 < \varepsilon \leq 1/2$. *Then*

$$v \geq \varepsilon^\gamma \quad \text{in} \quad B_{r/2}(x_0), \tag{2.14}$$

where $\gamma = \gamma(n, \nu, S) = -\log_2 \lambda > 0$, *and* λ *is the constant in the previous lemma.*

PROOF. We assume $x_0 = 0$. Set $r_j := 2^{-j}r$ for $j = 1, 2, \ldots$, and choose a natural k such that $2^{-k-1} < \varepsilon \leq 2^{-k}$, so that $r_{k+1} < \varepsilon r \leq r_k$. By our assumptions, $v \geq 1$ on $B_{\varepsilon r} \supset B_{r_{k+1}}$. The previous lemma with $r = r_{k+1}$ yields $v \geq \lambda$ in B_{r_k}. Repeatedly using this lemma again, we get

$$v \geq \lambda^2 \quad \text{in} \quad B_{r_{k-1}}, \quad \ldots, \quad v \geq \lambda^k \quad \text{in} \quad B_{r_1} = B_{r/2}.$$

Since $\lambda^k = 2^{-k\gamma} \geq \varepsilon^\gamma$, the corollary is proved. □

The following technical lemma will help us to deal with functions defined on a ball B_r rather than on a general open set $\Omega \subset \mathbb{R}^n$.

LEMMA 2.4 (Extension lemma). *Let* Ω *be an open set in* $\mathbb{R}^n$, *and let* u *be a function in* $W(\Omega)$, *such that*

$$u \geq 0, \quad Lu \geq 0 \quad \text{in} \quad \Omega, \quad \text{and} \quad u = 0 \quad \text{in} \quad (\partial\Omega) \cap B_r,$$

where $B_r := B_r(x_0)$ *for some* $r > 0$ *and* $x_0 \in \mathbb{R}^n$. *We claim that there are functions* $u_\varepsilon \in W(B_r)$ *defined for each* $\varepsilon > 0$, *such that*

$$u_\varepsilon \geq 0, \quad Lu_\varepsilon \geq 0 \quad \text{in} \quad B_r, \qquad u_\varepsilon \equiv 0 \quad \text{in} \quad \overline{B_r} \setminus \Omega,$$

and $u_\varepsilon \to u$ *as* $\varepsilon \to 0^+$ *uniformly on* $\Omega \cap B_r$.

PROOF. We partially follow [**CS07**], pp. 6–7. Fix a standard function

$$0 \leq \eta \in C^\infty(\mathbb{R}^1), \quad \text{such that} \quad \eta(t) = 0 \quad \text{for} \quad |t| \geq 1, \quad \text{and} \quad \int_{\mathbb{R}^1} \eta(t)\, dt = 1,$$

and set $\eta_\varepsilon(t) := \varepsilon^{-1}\eta(\varepsilon^{-1}t - 2)$ for $\varepsilon > 0$, $t \in \mathbb{R}^1$. These are smooth functions vanishing on $\mathbb{R}^1 \setminus [\varepsilon, 3\varepsilon]$. Further, by repeated integration of η_ε, define the functions $G_\varepsilon \in C^\infty(\mathbb{R}^1)$ satisfying the properties

$$G_\varepsilon \equiv 0 \text{ on } (-\infty, \varepsilon], \quad G_\varepsilon'' \equiv \eta_\varepsilon \text{ on } \mathbb{R}^1.$$

Now we set

$$u_\varepsilon := G_\varepsilon(u) \text{ in } \Omega, \quad u_\varepsilon \equiv 0 \text{ on } \overline{B_r} \setminus \Omega.$$

From the properties of η_ε it follows that $G_\varepsilon' \geq 0$, $G_\varepsilon'' \geq 0$, and $(u - 3\varepsilon)_+ \leq u_\varepsilon \leq (u - \varepsilon)_+$. Since $u = 0$ on the set $(\partial\Omega) \cap B_r$, the functions u_ε vanish near this set. Hence we have $u_\varepsilon \in W(B_r)$, $u_\varepsilon \geq 0$, and $|u_\varepsilon - u| \leq 3\varepsilon$ on Ω. Finally,

$$Lu_\varepsilon = LG_\varepsilon(u) = G_\varepsilon'(u) \cdot Lu + G_\varepsilon''(u) \cdot a_{ij} D_i u\, D_j u \geq 0 \text{ in } \Omega.$$

The lemma is proved. □

In comparison with the first growth lemma (Lemma 2.1), the next lemma states that, roughly speaking, from (2.1) with $\mu_1 < 1$, (2.2) follows with $\beta_1 < 1$.

LEMMA 2.5 (Second growth lemma). *Let Ω be an open set in $\mathbb{R}^n$, and let $u \in W(\Omega)$, $x_0 \in \mathbb{R}^n$, and $r > 0$ be such that*

$$u \geq 0, \quad Lu \geq 0 \quad \text{in} \quad \Omega, \quad \text{and} \quad u = 0 \quad \text{on} \quad (\partial\Omega) \cap B_{2r}(x_0).$$

We claim that for arbitrary $\mu_2 \in (0,1)$, there is a constant $\beta_2 = \beta_2(n, \nu, S, \mu_2) \in (0,1)$, such that from

$$(2.15) \qquad |\Omega \cap B_r(x_0)| \leq \mu_2 \cdot |B_r|$$

it follows that

$$(2.16) \qquad M_r := \sup_{\Omega \cap B_r(x_0)} u \leq \beta_2 \cdot M_{2r}.$$

Replacing Ω by $\Omega \cap B_{2r}(x_0)$, and u by const $\cdot\, u$ we can assume that $0 \leq u \leq M_{2r} = 1$ in Ω. Taking

$$v := 1 - u, \quad \mu := 1 - \mu_2 \in (0,1), \quad \text{and} \quad \beta := 1 - \beta_2 \in (0,1),$$

we see that this lemma follows from the following one.

LEMMA 2.6. *Let Ω be an open set in $\mathbb{R}^n$, and let $v \in W(\Omega)$, $x_0 \in \mathbb{R}^n$, and $r > 0$ be such that*

$$v \geq 0, \quad Lv \leq 0 \quad \text{in} \quad \Omega, \quad \text{and} \quad v \geq 1 \quad \text{on} \quad (\partial\Omega) \cap B_{2r}(x_0).$$

We claim that for arbitrary $\mu \in (0,1)$, there is a constant $\beta = \beta(n, \nu, S, \mu) \in (0,1)$ such that from $|B_r(x_0) \setminus \Omega| \geq \mu \cdot |B_r|$ it follows that $v \geq \beta$ on $\Omega \cap B_r(x_0)$.

PROOF. We follow the lines of the proof of Lemma 2.3 in [**S80**], though some details are different. As before, we assume $x_0 = 0$. Moreover, applying the extension lemma (Lemma 2.4) to the function $u := 1 - v$, we can also assume that the function v belongs to $W(B_{2r})$ and satisfies $v \geq 0$, $Lv \leq 0$ in B_{2r}. One needs to show that from $|B_r \cap \{v \geq 1\}| \geq \mu \cdot |B_r|$ with $0 < \mu < 1$ it follows $v \geq \beta = \beta(n, \nu, S, \mu) \in (0,1)$ on B_r.

First consider the case

$$(2.17) \qquad |B_r \cap \{v \geq 1\}| \geq \mu_0 \cdot |B_r|, \quad \text{where} \quad \mu_0 := 1 - \mu_1(n, \nu, S, 1/2),$$

i.e. μ_1 is the constant in Lemma 2.1 corresponding to $\beta_1 = 1/2$. Then

$$\begin{aligned} |B_r \cap \{u > 0\}| = |B_r \cap \{v < 1\}| &= |B_r| - |B_r \cap \{v \geq 1\}| \\ &\leq (1 - \mu_0) \cdot |B_r| = \mu_1 \cdot |B_r|. \end{aligned}$$

By Lemma 2.1 applied to $\Omega' := \Omega \cap B_r \cap \{u > 0\}$ and $r/2$ in place of r, we get $u \leq 1/2$, and $v = 1 - u \geq 1/2$ on $B_{r/2}$. Now by the doubling property (Lemma 2.2), $v \geq \beta_0 = \beta_0(n, \nu, S) := \lambda/2 > 0$ on B_r. We have proved that from (2.17) it follows that $v \geq \beta_0 = \beta_0(n, \nu, S) > 0$ on B_r, so that the estimate $v \geq \beta > 0$ holds true for $\mu \geq \mu_0$ with $\beta = \beta_0$.

Now consider the remaining case when the set $\Gamma := B_r \cap \{v \geq 1\}$ satisfies $\mu \cdot |B_r| \leq |\Gamma| < \mu_0 \cdot |B_r|$. Almost every point x in the set Γ is its density point, which implies that $B_\rho(x) \subset B_r$ and $|\Gamma \cap B_\rho(x)| > \mu_0 \cdot |B_\rho|$ for small $\rho > 0$. One can include $B_\rho(x)$ into a monotone continuous family of balls B^θ, $0 \leq \theta \leq 1$, such that $B^0 = B_\rho(x)$ and $B^1 = B_r$. Then $\varphi(\theta) := |\Gamma \cap B^\theta| / |B^\theta|$ is a continuous function on $[0,1]$ with the boundary values $\varphi(0) > \mu_0 > \varphi(1)$. Therefore, for some intermediate value $\theta_0 \in (0,1)$, we have $\varphi(\theta_0) = \mu_0$, and the corresponding ball $B := B^{\theta_0}$ satisfies

$$(2.18) \qquad B \subset B_r, \qquad |B \cap \Gamma| = |B \cap \{v \geq 1\}| = \mu_0 \cdot |B|.$$

Denote by Γ_1 the union of all the balls B satisfying (2.18). By the previous argument, $v \geq \beta_0 > 0$ on Γ_1. Obviously $|\Gamma \setminus \Gamma_1| = 0$, because Γ_1 contains all the density points of Γ. Further, by the *simple Vitali lemma*, there is a finite collection of *disjoint* balls $B^{(1)}, B^{(2)}, \ldots, B^{(m)}$, satisfying (2.18), such that

$$\sum_{j=1}^{m} |B^{(j)}| \geq c_0 \cdot |\Gamma_1|, \quad \text{where} \quad c_0 = c_0(n) > 0.$$

This implies that

$$\begin{aligned} |\Gamma_1 \setminus \Gamma| &\geq \sum_{j=1}^{m} |B^{(j)} \cap (\Gamma_1 \setminus \Gamma)| = \sum_{j=1}^{m} |B^{(j)} \setminus \Gamma| \\ &= (1-\mu_0) \sum_{j=1}^{m} |B^{(j)}| \geq (1-\mu_0)c_0 \cdot |\Gamma_1| \geq (1-\mu_0)c_0 \cdot |\Gamma| \end{aligned}$$

and

$$(2.19) \qquad |B_r \cap \{v \geq \beta_0\}| \geq |\Gamma_1| \geq c_1 \cdot |\Gamma| = c_1 \cdot |B_r \cap \{v \geq 1\}| \geq c_1\mu \cdot |B_r|,$$

where $c_1 = c_1(n, \nu, S) := 1 + (1-\mu_0)c_0 > 1$. If the left side is $< \mu_0 \cdot |B_r|$, then we can apply this estimate again with the function $\beta_0^{-1}v$ in place of v, then to $\beta_0^{-2}v$, etc., which gives us

$$|B_r \cap \{v \geq \beta_0^k\}| \geq c_1^k \mu \cdot |B_r| \quad \text{for} \quad k = 1, 2, \ldots.$$

This iteration stops when the left side becomes $\geq \mu_0 \cdot |B_r|$. Since $c_1^k \mu \leq 1$, we must have $k \leq k_0 = k_0(n, \nu, S, \mu) := [-\ln \mu / \ln c_1]$. Finally, since the function $\beta_0^{-k_0}v$ satisfies (2.17), it follows that $\beta_0^{-k_0}v \geq \beta_0 > 0$ on B_r, and the estimate $v \geq \beta > 0$ on B_r holds true with $\beta = \beta(n, \nu, S, \mu) := \beta_0^{k_0+1} > 1$. The lemma is proved. □

The Hölder regularity of solutions to the nonhomogeneous equations $Lu = f$ with $f \in L^n$ follows from Theorem 1.1 and Lemma 2.5 in the same way as Theorem IV.2.5 in [**K85**], or Theorem 4.1 in [**S80**], are derived from the corresponding statements. The next theorem is quite similar to these results; therefore we formulate it without proof. We only note that the proof uses approximation of u by solutions of equations with regular coefficients, so that some auxiliary boundary value problems have solutions. This part is provided by the approximation Lemma 4.2 below.

THEOREM 2.7. *Let u be a function in $W(B_{2r})$, $r > 0$, such that $Lu = f$ in B_{2r}, where $f \in L^n(B_{2r})$. Then there are constants $\alpha \in (0,1)$ and $N > 0$, depending only on n, ν, and S, such that*

$$(2.20) \qquad \sup_{x,y \in B_r} \frac{|u(x) - u(y)|}{|x-y|^\alpha} \leq \frac{N}{r^\alpha} \cdot \Big(\sup_{B_{2r}} |u| + r \cdot ||f||_{n, B_{2r}} \Big).$$

3. Interior and boundary Harnack inequalities

THEOREM 3.1 (Interior Harnack inequality). *Let u be a function in $W(B_{8r})$ satisfying $u > 0$, $Lu := a_{ij}D_{ij}u + b_iD_iu = 0$ in B_{8r} for some $r > 0$. Then*

$$(3.1) \qquad \sup_{B_r} u \leq N_1 \cdot \inf_{B_r} u, \quad \text{where} \quad N_1 = N_1(n, \nu, S) \geq 1, \quad S := \int_{B_{8r}} |\mathbf{b}|^n dx.$$

PROOF. We partially follow the proof of Theorem 3.1 in [**S80**].

Without loss of generality, we assume $r = 1$. Let $\gamma = \gamma(n, \nu, S) > 0$ be the constant in Corollary 2.3. Since $2 - |x| \geq 1$ in the ball $B_1 := B_1(0)$, we have

$$\sup_{B_1} u \leq M := \sup_{B_2} (2 - |x|)^\gamma u = (2 - |x_0|)^\gamma u(x_0) \tag{3.2}$$

for some $x_0 \in B_2$. Further, consider the function

$$u_0 := u - \frac{u(x_0)}{2} \quad \text{in the ball} \quad B_\rho(x_0), \quad \text{where} \quad \rho := \frac{2 - |x_0|}{2}.$$

Since $2 - |x| \geq \rho$ in $B_\rho(x_0)$, we also have

$$\sup_{B_\rho(x_0)} u_0 < \sup_{B_\rho(x_0)} u \leq \sup_{B_\rho(x_0)} \left(\frac{2 - |x|}{\rho} \right)^\gamma u \leq \rho^{-\gamma} M = 2^\gamma u(x_0) = 2^{\gamma+1} u_0(x_0)$$

and

$$\sup_{B_{\rho/2}(x_0)} u_0 \geq u_0(x_0) > \beta_1 \cdot \sup_{B_\rho(x_0)} u_0, \quad \text{where} \quad \beta_1 = \beta_1(n, \nu, S) := 2^{-\gamma-1} > 0.$$

Now we can use Lemma 2.1 in an equivalent form: "if (2.2) fails, then (2.1) fails", with $\Omega := B_8 \cap \{u_0 > 0\}$, $r := \rho/2$, and u_0 in place of u. By this lemma, there is a constant $\mu_1 > 0$ depending only on n, ν, and S, such that

$$|B_\rho(x_0) \cap \{u_0 > 0\}| > \mu_1 \cdot |B_\rho|. \tag{3.3}$$

Next, the function $v := u/u_0(x_0)$ satisfies

$$v > 0, \quad Lv = 0 \quad \text{in} \quad \Omega' := B_8 \cap \{v < 1\} = B_8 \setminus \overline{\Omega},$$

and $v = 1$ on $(\partial\Omega') \cap B_8$. Moreover, by virtue of (3.3),

$$|B_\rho(x_0) \setminus \Omega'| = |B_\rho(x_0) \cap \{v \geq 1\}| = |B_\rho(x_0) \cap \{u_0 \geq 0\}| > \mu_1 \cdot |B_\rho|.$$

Applying Lemma 2.6 to the function v in Ω', with $r := \rho$, we obtain the estimate

$$v \geq \beta = \beta(n, \nu, S, \mu_1) > 0 \quad \text{in} \quad B_\rho(x_0).$$

By the choice of x_0 and ρ,

$$u = u(x_0) \cdot v \geq c_1 := \beta \cdot (2\rho)^{-\gamma} M \quad \text{in} \quad B_\rho(x_0).$$

Finally, we apply Corollary 2.3 to the function $c_1^{-1} u$ with $r = 6$ and $\varepsilon := \rho/6 \in (0, 1/6)$. In our case, $B_3(x_0) = B_{r/2}(x_0) \subset B_6(x_0) \subset B_8$, so that the estimate (2.14) implies $c_1^{-1} u \geq \varepsilon^\gamma$ in the ball $B_3(x_0)$, which contains $B_1 := B_1(0)$. Therefore,

$$\inf_{B_1} u \geq \inf_{B_3(x_0)} u \geq \varepsilon^\gamma c_1 = \left(\frac{\rho}{6} \right)^\gamma \cdot \beta \cdot (2\rho)^{-\gamma} M = \frac{\beta M}{12^\gamma}.$$

This estimate together with (3.20 imply the Harnack inequality (3.1) with $N_1 = N_1(n, \nu, S) := 12^\gamma \cdot \beta^{-1} \geq 1$. $\square$

As a standard consequence of the interior Harnack inequality, we have

THEOREM 3.2 (Liouville). *Let u be a bounded from above or from below function in $W(\mathbb{R}^n) := W^{2,n}(\mathbb{R}^n) \cap C(\mathbb{R}^n)$ such that $Lu := a_{ij} D_{ij} u + b_i D_i u = 0$ in the entire space $\mathbb{R}^n$, with a_{ij} satisfying (1.2) and $|\mathbf{b}| \in L^n(\mathbb{R}^n)$. Then $u = \text{const}$ in $\mathbb{R}^n$.*

Indeed, replacing u by const $\pm u$ is necessary; we can reduce the proof to the case $\inf\limits_{\mathbb{R}^n} u = 0$. In this case, taking the limit in (3.1) as $r \to \infty$, we get $u \equiv 0$ in $\mathbb{R}^n$.

The following theorem is a more general form of the interior Harnack inequality, which is convenient for applications.

THEOREM 3.3. *Let Ω be a bounded open set in $\mathbb{R}^n$, and let u be a function in $W(\Omega)$ satisfying $u > 0$, $Lu = 0$ in Ω. Then for arbitrary constant $\delta > 0$, such that the set $\Omega^\delta := \{x \in \Omega : \operatorname{dist}(x, \partial\Omega) > \delta\}$ is nonempty and* connected, *we have*

$$\sup_{\Omega^\delta} u \le N_2 \cdot \inf_{\Omega^\delta} u, \quad \text{where} \quad N_2 = N_2(n, \nu, S, \delta/\operatorname{diam}\Omega), \quad S := \int_\Omega |\mathbf{b}|^n dx. \tag{3.4}$$

PROOF. Fix points $x, y \in \Omega^\delta$. One can choose a sequence $x^{(0)}, x^{(1)}, \dots, x^{(m)}$ in Ω^δ, such that

$$x^{(0)} = x, \quad x^{(m)} = y, \quad \text{and} \quad |x^{(k)} - x^{(k-1)}| < \delta/8 \quad \text{for all} \quad k = 1, \dots, m,$$

and m does not exceed a number m_0 which depends only on n and $\delta/\operatorname{diam}\Omega$. Applying Theorem 3.1 to the balls $B_r(x^{(k)})$, $k = 1, \dots, m \le m_0$, we obtain

$$u(x) = u(x^{(0)}) \le N_1 u(x^{(1)}) \le \cdots \le N_1^m u(x^{(m)}) = N_1^m u(y) \le N_1^{m_0} u(y),$$

where $N_1 = N_1(n, \nu, S)$ is the constant in (3.1). Since the points $x, y \in \Omega^\delta$ can be selected in an arbitrary way, the inequality (3.4) follows with $N_2 := N_1^{m_0}$. □

We now proceed to the boundary estimates for positive solutions vanishing at the "bottom" of a Lipschitz cylinder. Here ψ is a Lipschitz function on $\mathbb{R}^{n-1}$ with a Lipschitz constant $K \ge 0$, i.e.

$$|\psi(x') - \psi(y')| \le K \cdot |x' - y'| \quad \text{for all} \quad x', y' \in \mathbb{R}^{n-1}. \tag{3.5}$$

For $r > 0$, denote

$$\begin{aligned} Q_r &:= \{x = (x', x_n) \in \mathbb{R}^n : |x'| < r,\ 0 < x_n - \psi(x') < r\}, \\ \Gamma_r &:= \{x = (x', x_n) \in \mathbb{R}^n : |x'| \le r,\ x_n = \psi(x')\} \subset \partial Q_r. \end{aligned} \tag{3.6}$$

The following theorem contains a Carleson-type estimate for equations (1.1) with $|\mathbf{b}| \in L^n$.

THEOREM 3.4 (Boundary Harnack inequality). *Let ψ be a function on $\mathbb{R}^{n-1}$ satisfying the Lipschitz condition (3.5), $\psi(0) = 0$, and let u be a function in $W(Q_{2r})$ for some $r > 0$, such that*

$$u > 0, \quad Lu = 0 \quad \text{in} \quad Q_{2r}, \quad \text{and} \quad u = 0 \quad \text{on} \quad \Gamma_{2r}, \tag{3.7}$$

where Q_{2r} and Γ_{2r} are defined according to (3.6). Then

$$\sup_{Q_r} u \le N_3 u(0, r), \quad \text{where} \quad N_3 = N_3(n, \nu, S, K) \ge 1. \tag{3.8}$$

PROOF. Since all the conditions here are invariant with respect to rescaling (see Remark 1.4), we can assume that $r = 1$.

(a) First we show that there is a constant $\gamma = \gamma(n, \nu, S, K) > 0$, such that

$$M := \sup_{Q_2} d^\gamma u \le N u(P_1), \quad \text{where} \quad d = d(x) := \operatorname{dist}(x, \partial Q_2), \tag{3.9}$$

the constant $N = N(n,\nu,S,K) \ge 1$, and $P_1 := (0,1)$. $N_1 = N_1(n,\nu,s) \ge 1$ is the constant in Theorem 3.1, and $P_1 := (0,1)$.

Fix $x \in Q_2$. The ball of radius $d(x)$ centered at x touches ∂Q_2 at some point y. A simple geometrical consideration shows that there is a smooth curve parameterized by the arc length $C := \{z = z(s),\ 0 \le s \le s_0\} \subset \overline{Q_2}$, connecting the points $y = z(0)$ and $P_1 = z(s_0)$, passing through the point x, i.e. $z(s_1) = x$ for some $s \in (0, s_0]$, and such that

$$cs \le d(z(s)) \le s \quad \text{for all} \quad s \in [0, s_0], \tag{3.10}$$

with a constant $c \in (0,1)$ depending only on K. (This property means that Q_2 is a *John domain.*)

Set $\theta := 1 - c/8$. Then for arbitrary $s \in (0, s_0]$ and $t \in [\theta s, s]$, we have

$$|z(t) - z(s)| \le |t - s| \le (1-\theta)s = \frac{cs}{8} \le \frac{d(z(s))}{8}.$$

From Theorem 3.1, with $r = d(z(s))/8$, it follows that

$$N_1 u(z(s)) \ge u(z(t)), \qquad 0 < \theta s \le t \le s \le s_0. \tag{3.11}$$

Fix a constant $\gamma = \gamma(n,\nu,S,K) > 0$ such that $1 \ge \theta^\gamma N_1$. Then

$$\varphi(s) := s^\gamma u(z(s)) \ge N_1(\theta s)^\gamma u(z(s)) \ge (\theta s)^\gamma u(z(\theta s)) = \varphi(\theta s), \qquad 0 < s \le s_0.$$

For each $s \in (0, s_0]$, there is an integer $k \ge 0$, such that $\theta s_0 < \theta^{-k} s \le s_0$. Using the previous inequalities, including (3.11) with $s = s_0$, we get

$$\varphi(s) \le \varphi(\theta^{-1}s) \le \cdots \le \varphi(\theta^{-k}s) \le s_0^\gamma N_1 u(z(s_0)) = s_0^\gamma N_1 u(P_1).$$

Note that $s_0 \le c^{-1} d(z(s_0)) \le c^{-1}$. Therefore, at the point $x = z(s_1)$,

$$d^\gamma u(x) = d^\gamma u(z(s_1)) \le s_1^\gamma u(z(s_1)) = \varphi(s_1) \le N u(P_1),$$

where $N := c^{-\gamma} N_1$. Since $x \in Q_2$ can be chosen in an arbitrary way, the estimate (3.9) follows.

(b) Our next step is to prove the estimate

$$M_0 := \sup_{Q_2} d_0^\gamma u \le N_0 M, \quad \text{where} \quad d_0 = d_0(x) := \operatorname{dist}(x, (\partial Q_2) \setminus \Gamma_2), \tag{3.12}$$

with a constant $N_0 = N_0(n,\nu,S,K) \ge 1$. Note that $d_0^\gamma u = 0$ on ∂Q_2; hence the *supremum* in (3.12) is attained at some point $x_0 \in Q_2$, i.e. $d_0^\gamma u(x_0) = M_0$. We claim that

$$d(x_0) \ge \varepsilon_0 d_0(x_0) \quad \text{with} \quad \varepsilon_0 = \varepsilon_0(n,\nu,S,K) \in (0, 1/4]. \tag{3.13}$$

The constant ε_0 will be specified below. Suppose (3.13) fails, i.e.

$$\rho := d(x_0) < \varepsilon_0 \rho_0, \quad \text{where} \quad \rho_0 := d_0(x_0), \quad 0 < \varepsilon_0 \le 1/4.$$

Since $4\rho < 4\varepsilon_0 d_0(x_0) < d_0(x_0)$, the intersection $(\partial Q_2) \cap B_{4\rho}(x_0)$ lies in Γ_2, so that $u = 0$ on this set. Further, the ball $B_\rho(x_0)$ touches ∂Q_2 at some point $y_0 \in \Gamma_2$. By (3.6), Γ_2 is the graph of a Lipschitz function $x_n = \psi(x')$ restricted to $|x'| \le 2$. It is easy to see that the measure

$$|B_{2\rho}(x_0) \setminus Q_2| \ge |B_\rho(y_0) \setminus Q_2| \ge \mu \rho^n \quad \text{with} \quad \mu = \mu(n,K) \in (0,1).$$

Now we can apply Lemma 2.5 with $r := 2\rho$ and $\mu_2 := 1 - 2^{-n}\mu$. By this lemma,

$$u(x_0) \le \sup_{Q_2 \cap B_{2\rho}(x_0)} u \le \beta \cdot \sup_{Q_2 \cap B_{4\rho}(x_0)} u, \quad \text{where} \quad \beta = \beta(n, \nu, S, K) \in (0, 1).$$

By the triangle inequality,

$$d_0(x) \ge \rho_0 - 4\rho > (1 - 4\varepsilon_0)\rho_0 \quad \text{on} \quad Q_2 \cap B_{4\rho}(x_0).$$

Combining these inequalities, we obtain

$$M = \rho_0^\gamma u(x_0) \le (1 - 4\varepsilon_0)^{-\gamma}\beta \cdot \sup_{Q_2 \cap B_{4\rho}(x_0)} d_0^\gamma u \le (1 - 4\varepsilon_0)^\gamma \beta \cdot M.$$

For small enough $\varepsilon_0 = \varepsilon_0(n, \nu, S, K) \in (0, 1)$, the right side is strictly less than M, and we get the desired contradiction.

The above argument proves the estimate (3.13), which in turn implies (3.12) with $N_0 := \varepsilon_0^{-\gamma} \ge 1$ as follows:

$$M_0 = d_0^\gamma u(x_0) \le \varepsilon_0^{-\gamma} d^\gamma u(x_0) \le \varepsilon_0^{-\gamma} M.$$

(c) Both the "top" and "bottom" portions of ∂Q_2 are graphs of Lipschitz functions $x_n = \psi(x') + c$, with $c = 0$ or 2. An elementary geometric reasoning shows that $d_0(x) \ge (1 + K^2)^{-1/2}$ on Q_1. Hence

$$\sup_{Q_1} u \le (1 + K^2)^{\gamma/2} \sup_{Q_1} d_0^\gamma u \le (1 + K^2)^{\gamma/2} M_0.$$

This estimate together with (3.12) and (3.9) yields the desired estimate (3.4). The lemma is proved. □

In the following Theorem 3.6, which is preceded with a technical Lemma 3.5, we deal with ratios u_1/u_2 of positive solutions. Note that only the numerator u_1 vanishes on Γ_{2r}, while u_2 is just a positive solution. In particular, in the case $u_2 \equiv 1$, Theorem 3.6 is reduced to Theorem 3.4.

Let $\psi = \psi(x')$ be a function on $\mathbb{R}^{n-1}$ satisfying the Lipschitz condition (3.5), and $\psi(0) = 0$. For $r > 0$ and $h \ge 0$, denote

$$\begin{aligned} Q_{r,h} &:= \{x = (x', x_n) \in \mathbb{R}^n : |x'| < r, \quad 0 < x_n - \psi(x') < h\}, \\ Q^+_{r,h} &:= \{x = (x', x_n) \in \mathbb{R}^n : |x'| < r, \quad h/2 < x_n - \psi(x') < h\}, \\ \Gamma_{r,h} &= \{x = (x', x_n) \in \mathbb{R}^n : |x'| \le r, \quad x_n = \psi(x') + h\}, \\ S_{r,h} &:= \{x = (x', x_n) \in \mathbb{R}^n : |x'| = r, \quad 0 < x_n - \psi(x') < h\}. \end{aligned} \tag{3.14}$$

Comparing this notation with (3.6), we see that $Q_r = Q_{r,r}$, $\Gamma_r = \Gamma_{r,0}$. For $r, h > 0$, the boundary $\partial Q_{r,h}$ of the "cylinder" $Q_{r,h}$ is the union of three disjoint sets: the "top" $\Gamma_{r,h}$, the "bottom" Γ_r, and the "lateral side" $S_{r,h}$. If $\psi \equiv 0$, this terminology is understood in the usual sense.

LEMMA 3.5. *Let w be a function in $W(Q_{r,h})$ for some $0 < h \le r$, such that*

$$Lw = 0 \quad \text{in} \quad Q_{r,h}, \qquad w \ge 0 \quad \text{on} \quad \Gamma_r, \tag{3.15}$$

and

$$\inf_{\Gamma_{r,h}} w \ge \sup_{S_{r,h}} (-w)_+. \tag{3.16}$$

We claim that there is a constant $\varepsilon_1 = \varepsilon_1(n, \nu, S, K) \in (0, 1/4]$ such that from $h \le \varepsilon_1 r$ it follows that

$$w(0, x_n) \ge 0 \quad \text{for} \quad 0 \le x_n \le h. \tag{3.17}$$

PROOF. Without loss of generality, we assume $r = 1$, and the left side of (3.16) is equal to 1. Then

$$(3.18) \qquad w \geq 1 \quad \text{on} \quad \Gamma_{1,h}, \qquad w \geq 0 \quad \text{on} \quad \Gamma_1, \qquad \text{and} \quad w \geq -1 \quad \text{on} \quad S_{1,h}.$$

(a) Consider the function

$$u := -w \quad \text{on the set} \quad \Omega := Q_{1,h} \cap \{w < 0\}.$$

Obviously, it satisfies

$$0 < u \leq 1, \quad Lu = 0 \quad \text{in} \quad \Omega; \qquad u = 0 \quad \text{on} \quad (\partial\Omega) \cap Q_{1,h}.$$

For each $x_0 \in \Omega \cap S_{3/4,h}$, the measure

$$|\Omega \cap B_{1/4}(x_0)| \leq |Q_{1,h}| \leq N_0 \cdot h \quad \text{with} \quad N_0 = N_0(n) > 0,$$

so that we can apply Lemma 2.1 with $x_0 \in \Omega \cap S_{3/4,h}$ and $r = 1/8$ to the function $u := -w$. Since also $u \leq 0$ on the remaining part of $\partial Q_{3/4,h}$, we obtain the estimate

$$(3.19) \qquad \sup_{Q_{3/4,h}} (-w)_+ = \sup_{S_{3/4,h}} (-w)_+ \leq \beta_1 = \beta_1(n, \nu, S, h) \to 0^+ \quad \text{as} \quad h \to 0^+.$$

(b) Next, set $w_1 := w + \beta_1$. By the maximum principle, (3.18), and (3.19), it follows that

$$w_1 \geq 0, \quad Lw_1 = 0 \quad \text{in} \quad Q_{3/4,h}; \qquad w_1 \geq 0 \quad \text{on} \quad \Gamma_{3/4}, \quad w_1 \geq 1 \quad \text{on} \quad \Gamma_{3/4,h}.$$

For an arbitrary $x_0 = (x_0', x_{0n}) \in \Gamma_{1/2,h}$ and $\rho := (1 + K^2)^{-1/2}h/2$, the intersection $(\partial Q_{3/4,h}) \cap B_{2\rho}(x_0)$ lies in the set $\Gamma_{3/4,h}$, which is the graph of a Lipschitz function. Therefore, the measure $|B_\rho(x_0) \setminus Q_{3/4,h}| \geq \mu \cdot |B_\rho|$ with a constant $\mu = \mu(n, K) \in (0, 1)$. Using Lemma 2.6 with $Q_{3/4,h}$, w_1, ρ in place of Ω, v, r, respectively, we get the estimate

$$w_1 \geq \beta = \beta(n, \nu, S, K) > 0 \quad \text{on} \quad Q_{3/4,h} \cap B_\rho(x_0).$$

Further, by Theorem 3.3, applied to the function w_1 on the intersection $Q_{3/4,h} \cap \{|x' - x_0'| < \rho\}$, it follows that

$$w_1(x_0 - t\mathbf{e_n}) \geq \beta_0 = \beta_0(n, \nu, S, K) > 0 \quad \text{for} \quad 0 \leq t \leq \frac{h}{2}.$$

Since x_0 is an arbitrary point in $\Gamma_{1/2,h}$, we get the estimate

$$(3.20) \qquad w_1 \geq \beta_0 > 0 \quad \text{in} \quad Q^+_{1/2,h}.$$

(c) It is important that the constant β_0 in (3.20) does not depend on h. By virtue of (3.19), one can choose the constant $\varepsilon_1 = \varepsilon_1(n, \nu, s, K) \in (0, 1/4]$ in such a way that from $h \in (0, \varepsilon_1]$ it follows that $2\beta_1 \leq \beta_0$. Then the estimates (3.20) and (3.19) imply that

$$(3.21) \qquad \inf_{Q^+_{1/2,h}} w = \inf_{Q^+_{1/2,h}} w_1 - \beta_1 \geq \beta_1 \geq \sup_{Q_{1/2,h}} (-w)_+.$$

Note that $\partial Q^+_{1/2,h}$ contains the set $\Gamma_{1/2,h/2} = \Gamma_{r/2,h/2}$, and $Q_{1/2,h}$ contains $S_{1/2,h/2} = S_{r/2,h/2}$. Therefore

$$\inf_{\Gamma_{r/2,h/2}} w \geq \sup_{S_{r/2,h/2}} (-w)_+.$$

This simply means that the inequality (3.16) remains true with r, h being replaced by $r/2, h/2$. Iterating this procedure, we can replace r, h by $2^{-k}r, 2^{-k}h$ for $k = 1, 2, \dots$. Correspondingly, the first inequality in (3.21) implies that

$$w \geq 0 \quad \text{on} \quad Q^+_{2^{-k-1}r, 2^{-k}h} \quad \text{for} \quad k = 0, 1, 2, \dots,$$

and (3.17) follows by continuity of w. The lemma is proved. □

THEOREM 3.6 (Comparison theorem). *Let ψ be a function on $\mathbb{R}^{n-1}$ satisfying the Lipschitz condition* (3.5), *$\psi(0) = 0$, and let u_1 and u_2 be functions in $W(Q_{3r})$, $r > 0$, such that*

$$u_{1,2} > 0, \quad Lu_{1,2} = 0 \quad \text{in} \quad Q_{3r}, \quad \text{and} \quad u_1 = 0 \quad \text{on} \quad \Gamma_{3r}. \tag{3.22}$$

Then

$$\sup_{Q_r} \frac{u_1}{u_2} \leq N_4 \cdot \frac{u_1(0,r)}{u_2(0,r)}, \quad \text{where} \quad N_4 = N_4(n, \nu, S, K) \geq 1. \tag{3.23}$$

PROOF. Multiplying $u_{1,2}$ by appropriate constants if necessary, we can assume that $u_1(0,r) = u_2(0,r) = 1$. By Theorem 3.4,

$$u_1 \leq N_3 \quad \text{in} \quad Q_r, \tag{3.24}$$

where $N_3 = N_3(n, \nu, S, K) \geq 1$ is the constant in this theorem. Then set $h := \varepsilon_1 r$, where $\varepsilon_1 = \varepsilon_1(n, \nu, S, K) \in (0,1)$ is the constant in the previous lemma. Applying the Harnack inequality, Theorem 3.3, to the function u_2, we obtain

$$u_2 \geq c_0 = c_0(n, \nu, S, K) > 0 \quad \text{on} \quad Q_r \setminus Q_{r,h}. \tag{3.25}$$

Finally, we set

$$N_4 := 2N_3/c_0 \qquad \text{and} \qquad w := N_4 u_2 - u_1.$$

Then

$$w \geq N_4 c_0 - N_3 \geq N_3 \quad \text{on} \quad Q_r \setminus Q_{r,h} \qquad \text{and} \qquad -w \leq u_1 \leq N_3 \quad \text{in} \quad Q_r.$$

Therefore, the function w satisfies all the assumptions of the previous lemma, which implies that $w(0, x_n) \geq 0$ for $0 \leq x_n \leq r$.

This construction remains valid if we move the origin $0 \in \mathbb{R}^n$ to any point $y = (y', y_n) \in \Gamma_r$. Under this translation, the set Q_{2r} will be replaced by

$$Q_{2r}(y) := \{x = (x', x_n) \in \mathbb{R}^n : \ |x' - y'| < 2r, \ \ 0 < x_n - \psi(x') < 2r\},$$

which is a subset of Q_{3r}. As a result, we have $w \geq 0$, or equivalently, $u_1/u_2 \leq N_4$, on the whole set Q_1. The theorem is proved. □

COROLLARY 3.7. *Let u_1 and u_2 be functions in $W(Q_{3r})$, $r > 0$, satisfying* (3.22), *and in addition, $u_2 = 0$ on Γ_{3r}. Then*

$$\sup_{Q_r} \frac{u_1}{u_2} \leq N_4^2 \cdot \inf_{Q_r} \frac{u_1}{u_2}, \tag{3.26}$$

where $N_4 = N_4(n, \nu, S, K) \geq 1$ is the constant in (3.23).

PROOF. Since both functions u_1 and u_2 vanish on Γ_{3r}, we can interchange u_1 and u_2 in (3.23), so that

$$\left(\inf_{Q_r} \frac{u_1}{u_2} \right)^{-1} = \sup_{Q_r} \frac{u_2}{u_1} \leq N_4 \cdot \frac{u_2(0,r)}{u_1(0,r)}.$$

Multiplying both sides of this inequality by the corresponding sides of (3.23), we get the desired estimate (3.26). □

4. Boundary Hopf-Oleinik estimates

In a particular case $\psi \equiv 0$ and $b \equiv 0$, the function $u_2(x) = u_2(x', x_n) := x_n$ satisfies $Lu = 0$ and vanishes on the boundary $\{x_n = 0\}$. In this case, $Q_r := \{|x'| < r, 0 < x_n < r\}$, and the estimate (3.26) provides both upper and lower bounds for the ratio $u_1(x)/x_n$ near the origin in $\mathbb{R}^n$. In 1952, the lower bounds of such a kind for solutions of uniformly elliptic equations $Lu = 0$ with $|\mathbf{b}| \in L^\infty$ were independently obtained by E. Hopf [**H52**] and O.A. Oleinik [**O52**]. They considered domains Ω satisfying the *interior sphere condition* at a point $y_0 \in \partial\Omega$; i.e., there exists a ball $B_r(x_0) \subset \Omega$ such that $(\partial\Omega) \cap \partial B_r(x_0) = \{y_0\}$. The Hopf-Oleinik estimates state that for any function $u \in C^2(\Omega) \cap C(\overline{\Omega})$ satisfying $u > 0$, $Lu \le 0$ in Ω, and $u(y_0) = 0$, and any vector $\mathbf{v} \in \mathbb{R}^n$ such that $(\mathbf{v}, x_0 - y_0) > 0$, we have $t^{-1}u(y_0 + t\mathbf{v}) \ge \text{const} > 0$ for small $t > 0$. See the books by M.H. Protter and H.F. Weinberger [**PW67**], and by D. Gilbarg and N.S. Trudinger [**GT83**], for further references on this subject. Here we only mention that in a majority of sources, such kinds of estimates are obtained by means of a more or less standard *barrier technique*, which requires the boundedness (at least locally) of the coefficients.

Our approach uses special iterations based on Theorem 1.1 and Lemma 2.1. It allows us to derive a Hopf-Oleinik-type estimate in the case $|\mathbf{b}| \in L^q$, $q > n$. On the other hand, this estimate fails if $|\mathbf{b}| \in L^n$, as the following example shows.

EXAMPLE 4.1. Consider the functions

$$u_1(x) := \frac{x_n}{\big|\ln|x|\big|} \quad \text{and} \quad u_2(x) := x_n \cdot \big|\ln|x|\big|$$

in the cylinder $Q := \{x = (x', x_n) \in \mathbb{R}^n : |x'| < 1/2,\ 0 < x_n < 1/2\}$, extended as $u_1 = u_2 = 0$ on $(\partial Q) \cap \{x_n = 0\}$. Each of these two functions can be considered as a positive solution to the equation

$$\Delta u + (b, Du) := \Delta u + b_i D_i u = 0 \quad \text{in} \quad Q,$$

where the vector function $b := -\Delta u \cdot |Du|^{-2} Du$ satisfies

$$|\mathbf{b}| = \frac{|\Delta u|}{|Du|} \le \frac{\text{const}}{|x| \cdot \big|\ln|x|\big|} \in L^n(Q) \quad \textit{for} \quad n \ge 2.$$

However, in any neighborhood of $0 \in \mathbb{R}^n$, $\inf(u_1/x_n) = 0$ and $\sup(u_2/x_n) = \infty$.

In fact, if $|\mathbf{b}| \in L^q$ with $q > n$, then any positive solution of the equation $Lu = 0$ in Q_{2r} vanishing on Γ_{2r} (we follow the notation (3.6) with $\psi \equiv 0$) satisfies a two-sided estimate

$$0 < \inf_{Q_r} \frac{u}{x_n} \le \sup_{Q_r} \frac{u}{x_n} < \infty. \tag{4.1}$$

Here both the upper estimate $\sup(u/x_n) < \infty$ and the lower estimate $\inf(u/x_n) > 0$ are contained in the paper by O.A. Ladyzhenskaya and N.N. Ural'tseva ([**LU88**], Lemmas 2.3 and 4.4, respectively). In Theorem 4.3 below, the lower bound is extended to a bit more general case $|\mathbf{b}| \in L^n$, $b_n \in L^q$ with $q > n$. The upper estimate can be obtained similarly, with some simplifications. The following lemma helps us to reduce the proofs of such kinds of results to the case when the coefficients of L are smooth.

LEMMA 4.2 (Approximation lemma). *Let Ω be a bounded open set in $\mathbb{R}^n$ satisfying an interior cone condition; i.e., there are constants $K > 0$ and $r_0 > 0$ such that for each $y \in \partial\Omega$, there is a Cartesian coordinate system centered at the point $y = 0$, such that the cone*

$$(4.2) \qquad C := \{x = (x', x_n) \in \mathbb{R}^n : \; K \cdot |x'| < x_n < K r_0\} \subset \Omega^c := \mathbb{R}^n \setminus \Omega.$$

Let u be a function in $W^{2,n}(\Omega) \cap C(\overline{\Omega})$ satisfying the inequality

$$(4.3) \qquad Lu := a_{ij} D_{ij} u + b_i D_i u \le f \quad \text{in} \quad \Omega,$$

where the a_{ij} satisfy (1.2), $|\mathbf{b}| \in L^n(\Omega)$, $f \in L^n(\Omega)$.

We claim that there are approximations of functions a_{ij}, b_i, f by functions $a_{ij}^\varepsilon, b_i^\varepsilon, f^\varepsilon \in C^\infty(\overline{\Omega})$, $\varepsilon > 0$, such that the a_{ij}^ε satisfy (1.2),

$$(4.4) \qquad a_{ij}^\varepsilon \to a_{ij} \quad \text{a.e. in} \quad \Omega; \qquad b_i^\varepsilon \to b_i, \quad f^\varepsilon \to f \quad \text{in} \quad L^n(\Omega)$$

as $\varepsilon \to 0^+$, and the solutions $u^\varepsilon \in C^\infty(\Omega) \cap C(\overline{\Omega})$ of the Dirichlet problems

$$(4.5) \qquad L^\varepsilon u^\varepsilon := a_{ij}^\varepsilon D_{ij} u^\varepsilon + b_i^\varepsilon D_i^\varepsilon u = f^\varepsilon \quad \text{in} \quad \Omega, \qquad u^\varepsilon = u \quad \text{on} \quad \partial\Omega,$$

satisfy

$$(4.6) \qquad \sup_\Omega (u^\varepsilon - u) \to 0 \quad \text{as} \quad \varepsilon \to 0^+.$$

The solvability of the problems (4.5) in Lipschitz domains for equations with smooth, or even Hölder or just continuous, coefficients is well known; see e.g. [**M67**], Theorem 3.

PROOF. **(a)** We first consider the case $|\mathbf{b}| \in L^\infty(\Omega)$. The coefficients a_{ij} are defined on the whole space $\mathbb{R}^n$, and also b_i, f can be extended from Ω to $\mathbb{R}^n$ by setting $b_i, f \equiv 0$ on $\mathbb{R}^n \setminus \Omega$. One can approximate a_{ij}, b_i, f by convolutions $a_{ij}^\varepsilon := a_{ij} * \eta^\varepsilon$, $b_i^\varepsilon := b_i * \eta^\varepsilon$, $f^\varepsilon := f * \eta^\varepsilon$ with standard kernels η^ε, $\varepsilon > 0$, satisfying

$$0 \le \eta^\varepsilon \in C^\infty(\mathbb{R}^n), \quad \eta^\varepsilon \equiv 0 \quad \text{on} \quad \mathbb{R}^n \setminus B_\varepsilon, \quad \text{and} \quad \int_{\mathbb{R}^n} \eta^\varepsilon(x)\, dx = 1.$$

Then $a_{ij}^\varepsilon, b_i^\varepsilon, f^\varepsilon \in C^\infty(\mathbb{R}^n)$, the a_{ij}^ε satisfy (1.2), and

$$(4.7) \qquad a_{ij}^\varepsilon \to a_{ij}, \quad b_i^\varepsilon \to b_i \quad \text{as} \quad \varepsilon \to 0^+ \quad \text{a.e. in} \quad \Omega.$$

This convergence follows from the properties of the Lebesgue sets (see [**St70**], Sec. I.1.8). Moreover, $f^\varepsilon \to f$ in $L^n(\mathbb{R}^n)$ as $\varepsilon \to 0^+$ (see [**GT83**], Sec. 7.2). Having in mind the application of Theorem 1.1 to $u^\varepsilon - u$, we need to estimate from below the functions

$$L^\varepsilon(u^\varepsilon - u) = L^\varepsilon u^\varepsilon - Lu + (L - L^\varepsilon)u \ge F^\varepsilon := f^\varepsilon - f + (L - L^\varepsilon)u.$$

We have

$$(L - L^\varepsilon)u := (a_{ij} - a_{ij}^\varepsilon) D_{ij} u + (b_i - b_i^\varepsilon) D_i u \to 0 \quad \text{in} \quad L^n(\Omega) \quad \text{as} \quad \varepsilon \to 0^+.$$

(Here we use our additional assumption $b_i \in L^\infty$.) By Theorem 1.1, from the property (4.7) it follows that

$$\sup_\Omega (u^\varepsilon - u) \le N \cdot ||F^\varepsilon||_{n,\Omega} \to 0 \quad \text{as} \quad \varepsilon \to 0^+.$$

(b) In the general case $|\mathbf{b}| \in L^n(\Omega)$, fix a small constant $\delta > 0$, and choose an open subset $\Omega_\delta \subset \Omega$, such that $|\mathbf{b}| \in L^\infty(\Omega \setminus \overline{\Omega_\delta})$, and the norms in $L^n(\Omega_\delta)$,

$$||a_{ij} D_{ij} u||_{n,\Omega_\delta}, \quad ||b_i||_{n,\Omega_\delta}, \quad ||f||_{n,\Omega_\delta} \le \delta.$$

Define the functions $\overline{b_i}, \overline{f}$ as follows:

$$\overline{b_i} \equiv 0, \quad \overline{f} := a_{ij} D_{ij} u \quad \text{on} \quad \Omega_\delta, \qquad \overline{b_i} \equiv b_i, \quad \overline{f} \equiv f \quad \text{on} \quad \Omega \setminus \Omega_\delta.$$

The inequality (4.3) remains valid with the modified b_i and f:

$$\overline{L}u := a_{ij} D_{ij} u + \overline{b_i} D_i u \leq \overline{f}.$$

By the choice of Ω_δ, we also have

$$(4.8) \qquad ||\overline{b_i} - b_i||_{n,\Omega} = ||b_i||_{n,\Omega_\delta} \leq \delta, \qquad ||\overline{f} - f||_{n,\Omega} = ||a_{ij} D_{ij} u - f||_{n,\Omega_\delta} \leq 2\delta.$$

Since $\overline{b_i} \in L^\infty$, we can apply the argument in (a) correspondingly with

$$b_i^{\delta,\varepsilon} := \overline{b_i} * \eta^\varepsilon, \; f^{\delta,\varepsilon} := \overline{f} * \eta^\varepsilon, \; u^{\delta,\varepsilon}, \quad \text{in place of} \quad b_i^\varepsilon, \;, f^\varepsilon, \; u^\varepsilon.$$

One can go through this construction for each $\delta_k := 2^{-k}$, $k = 1, 2, \ldots$. Therefore, there is a sequence $\varepsilon_k \searrow 0$ such that

$$||b_i^{\delta_k,\varepsilon} - \overline{b_i}||_{n,\Omega} \leq \delta_k, \quad ||f^{\delta_k,\varepsilon} - \overline{f}||_{n,\Omega} \leq \delta_k, \quad \text{and} \quad \sup_\Omega (u^{\delta_k,\varepsilon} - u) \leq \delta_k$$

for all $\varepsilon \in (0, \varepsilon_k]$. Here $\overline{b_i}$ and $\overline{f}$ are defined above for $\delta = \delta_k$. Together with (4.8), the first two of these inequalities imply that

$$||b_i^{\delta_k,\varepsilon} - b_i||_{n,\Omega} \leq 2\delta_k, \quad ||f^{\delta_k,\varepsilon} - f||_{n,\Omega} \leq 3\delta_k.$$

Finally, for each $\varepsilon > 0$, we take $a_{ij}^\varepsilon := a_{ij} * \eta^\varepsilon$, as in part (a), and define $b_i^\varepsilon, f^\varepsilon$ in two steps: (i) find k from the relations $\varepsilon_{k+1} < \varepsilon \leq \varepsilon_k$; then (ii) take $b_i^\varepsilon := b_i^{\delta_k,\varepsilon}$, $f^\varepsilon := f^{\delta_k,\varepsilon}$. This choice of $a_{ij}^\varepsilon, b_i^\varepsilon, f^\varepsilon$ satisfies all the requirements of our lemma. The lemma is proved. □

THEOREM 4.3. *Let Ω be an open set in $\mathbb{R}^n$, and let u be a function in $W(\Omega)$ satisfying $u > 0$ and $Lu = 0$ in Ω. Let $y_0 \in \partial\Omega$ and $r > 0$ be such that in a Cartesian coordinate system centered at $y_0 = 0$, the cylinder*

$$(4.9) \qquad Q_{2r} := B'_{2r} \times (0, 2r) = \{x = (x', x_n) \in \mathbb{R}^n : \; |x'| < 2r, \;\; 0 < x_n < 2r\}$$

is contained in Ω, and $u(0) = 0$. Suppose that

$$(4.10) \qquad S := \int_{Q_{2r}} |\mathbf{b}|^n dx < \infty \qquad \text{and} \quad S_1 := r^{q-n} \int_{Q_{2r}} |b_n|^q dx < \infty,$$

where $q = \text{const} > n$. Then

$$(4.11) \qquad x_n^{-1} u(0, x_n) \geq c_1 \cdot r^{-1} u(0, r) \quad \text{for} \quad 0 < x_n \leq r,$$

where $c_1 = c_1(n, \nu, S, S_1, q) \in (0, 1]$.

REMARK 4.4. This theorem provides a Hopf-Oleinik-type estimate for equations with $|\mathbf{b}| \in L^q$, $q > n$. One can modify this formulation in different directions.

(i) One can replace the condition $Q_{2r} \subset \Omega$ with $y_0 \in (\partial Q_{2r}) \cap (\partial\Omega)$ by the interior sphere condition $B_r \subset \Omega$ with $y_0 \in (\partial B_r) \cap (\partial\Omega)$; in both cases $u(y_0) = 0$. One case is reduced to another by an appropriate C^2-transformation of coordinates.

(ii) Using Lemma 4.2, and the comparison principle, one can replace the equality $Lu = 0$ in Ω by the inequality $Lu \leq 0$ in Ω.

(iii) By the interior Harnack inequality, from (4.11) it follows that for any vector $\mathbf{v} = (v_1, \ldots, v_n) \in \mathbb{R}^n$ such that $v_n > 0$, we have $t^{-1} u(t\mathbf{v}) \geq \text{const} > 0$ for small $t > 0$.

PROOF OF THEOREM 4.3. All the quantities in the above formulation are invariant with respect to transformations $x \to \text{const} \cdot x$ and $u \to \text{const} \cdot u$. Without loss of generality, we assume $r = 1$ and $u(0, 1) = 1$. By the previous lemma, $u \geq \limsup_{\varepsilon \to 0^+} u^\varepsilon$, where u^ε are solutions to the Dirichlet problem (4.5) with coefficients $a^\varepsilon_{ij}, b^\varepsilon_i \in C^\infty(\overline{\Omega})$, and $f^\varepsilon \equiv 0$. Therefore, we can also assume $a_{ij}, b_i \in C^\infty(\overline{\Omega})$.

Set $\delta := 1 - n/q \in (0, 1)$, $\alpha := \delta/2$, and for $k = 1, 2, \dots$, $r_k := 2^{-k}$, $h_k := r_k^{1+\alpha}$,

$$\Omega_k := Q_{r_k} := B'_{r_k} \times (0, r_k), \quad \Omega_k^+ := \Omega_k \cap \{x_n > h_k\}, \quad \Omega_k^- := \Omega_k \cap \{x_n < h_k\}.$$

By the Harnack inequality, the quantities

$$m_k := \inf_{\Omega_k^+} x_n^{-1} u > 0, \quad k = 1, 2, \dots, \tag{4.12}$$

are estimated from below by positive constants depending on the prescribed quantities and k. Our goal is to eliminate the dependence on k. From the definition of m_k it follows that

$$v_k := m_k x_n - u \leq 0 \quad \text{on} \quad \overline{\Omega_k} \cap \big(\{x_n = 0\} \cup \{x_n = h_k\}\big).$$

We can "split" v_k into two functions: $v_k = w_k + z_k$ on Ω_k^-, where w_k and z_k are solutions of the problems

$$\begin{aligned} Lw_k = Lv_k = m_k b_n \quad &\text{in} \quad \Omega_k^-, \qquad w_k = 0 \quad \text{on} \quad \partial\Omega_k^-; \\ Lz_k = 0 \quad &\text{in} \quad \Omega_k^-, \qquad z_k = v_k \quad \text{on} \quad \partial\Omega_k^-. \end{aligned} \tag{4.13}$$

These problems have classical solutions because of our smoothness assumptions on the coefficients a_{ij}, b_i. By Theorem 1.1,

$$\sup_{\Omega_k^-} w_k \leq N r_k m_k \cdot ||b_n||_{n, \Omega_k^-} \quad \text{with} \quad N = N(n, \nu, S).$$

The last factor is estimated by Hölder's inequality ($1/n = 1/q + 1/p$):

$$||b_n||_{n, \Omega_k^-} \leq ||b_n||_{q, \Omega_k^-} \cdot ||\, 1 \,||_{p, \Omega_k^-} \leq N S_1^{1/q} r_k^{n/p}, \qquad N = N(n).$$

Since $n/p = 1 - n/q = \delta > 0$, we get

$$w_k \leq N r_k^{1+\delta} m_k \quad \text{in} \quad \Omega_k^-. \tag{4.14}$$

Here and in the rest of the proof, N denotes different constants depending only on n, ν, S, S_1, q. Our next step is to evaluate, for $0 < \rho \leq r_k$ and fixed k,

$$M(\rho) := \sup_{D(\rho)} (z_k)_+, \quad \text{where} \quad D(\rho) := B'_\rho \times (0, h_k).$$

Consider the case $M(\rho) > 0$ and $0 < \rho \leq r_k - h_k$. Since $z_k = v_k \leq 0$ on $\{x_n = 0\} \cup \{x_n = h_k\}$, by the maximum principle, the value $M(\rho)$ is attained by z_k on the lateral side of $D(\rho)$:

$$M(\rho) = z_k(x(\rho)) = z_k(x'(\rho), x_n(\rho)), \quad \text{where} \quad |x'(\rho)| = \rho, \quad 0 \leq x_n(\rho) \leq h_k.$$

In a subcase $0 \leq x_n(\rho) \leq h_k/2$, one can apply Lemma 2.5 to the function z_k with $\Omega := \Omega_k^- \cap \{z_k > 0\}$, $x_0 := (x'(\rho), 0)$, and $r := h_k/2$. Obviously, the measure condition (2.15) holds true with $\mu_2 = 1/2$. From this lemma and the maximum principle, it follows that

$$M(\rho) = z_k(x(\rho)) \leq \sup_{\Omega \cap B_{h_k/2}(x_0)} z_k \leq \beta \cdot \sup_{\Omega \cap B_{h_k}(x_0)} z_k \leq \beta \cdot M(\rho + h_k),$$

where $\beta = \beta(n, \nu, S) \in (0, 1)$. If $h_k/2 \le x_n \le h_k$, then these inequalities remain valid with $x_0 := (x'(\rho), h_k)$.

Let k be large enough, so that the integer part $j := [r_{k+1}/h_k] = [2^{k\alpha-1}] \ge 1$. Then $r_{k+1} + jh_k \le 2r_{k+1} = r_k$, and iterating the previous estimate, we get

$$M(r_{k+1}) \le \beta \cdot M(r_{k+1} + h_k) \le \cdots \le \beta^j \cdot M(r_{k+1} + jh_k) \le \beta^j M(r_k) \le \beta^j m_k h_k.$$

Here

$$\beta^j \le \beta^{-1} \cdot \beta^{2^{k\alpha-1}} = \beta^{-1} \cdot \exp\big(-cr_k^{-\alpha}\big), \quad \text{where} \quad c := -(\ln\beta)/2 > 0.$$

This estimate together with (4.14) yields the estimate for $m_k x_n - u =: v_k = w_k + z_k$:

$$m_k x_n - u \le N \cdot \Big(r_k^{1+\delta} + \exp\big(-cr_k^{-\alpha}\big)\Big) \cdot m_k \quad \text{in} \quad D(r_{k+1}). \tag{4.15}$$

From the definition (4.12) of m_k it follows that if $m_k > m_{k+1}$, then

$$m_{k+1} = \inf_{\Delta_k} x_n^{-1} u, \quad \text{where} \quad \Delta_k := \big(\Omega_{k+1}^+ \setminus \overline{\Omega_k^+}\big) \subset D(r_{k+1}). \tag{4.16}$$

Since $x_n \ge h_{k+1} = r_{k+1}^{1+\alpha} = 2^{-1-\alpha} r_k^{1+\alpha}$, and $\delta = 2\alpha$, the properties (4.15) and (4.16) imply that

$$m_k - m_{k+1} \le N \cdot \Big(r_k^{\alpha} + r_k^{-1-\alpha} \exp\big(-cr_k^{-\alpha}\big)\Big) \cdot m_k.$$

We proved the last inequality in the case $m_k > m_{k+1}$. In the opposite case $m_k \le m_{k+1}$ it is obvious, so this inequality holds true without any restrictions.

Since the exponential term converges to 0 as $k \to \infty$ faster than any positive power of r_k, there are large natural numbers k_0 and N_0, depending only on n, ν, S, S_1, q, such that

$$m_k - m_{k+1} \le N_0 r_k^{\alpha} m_k, \quad \text{and} \quad N_0 r_k^{\alpha} \le 1/2 \quad \text{for all} \quad k \ge k_0.$$

From $m_{k+1} \ge \big(1 - N_0 r_k^{\alpha}\big) m_k$, together with an elementary inequality $\ln(1-t) \ge -2t$ for $0 \le t \le 1/2$, we obtain

$$\ln m_{k+1} - \ln m_k \ge \ln\big(1 - N_0 r_k^{\alpha}\big) \ge -2N_0 r_k^{\alpha}, \qquad k \ge k_0.$$

For any integer $k_1 > k_0$,

$$\ln m_{k_1} - \ln m_{k_0} = \sum_{k=k_0}^{k_1-1} \big(\ln m_{k+1} - \ln m_k\big) \ge -2N_0 \sum_{k=k_0}^{\infty} r_k^{\alpha} =: N_1.$$

This implies the estimate

$$m_k \ge c_0 = c_0(n, \nu, S, S_1, q) > 0 \quad \text{for all} \quad k > k_0.$$

In particular, $x_n^{-1} u(0, x_n) \ge c_0$ for all $x_n \in (0, r_{k_0+1})$. By the Harnack inequality, this estimate holds true for $x_n \ge r_{k_0+1}$ as well, possibly with a different constant, because k_0 depends only on the prescribed quantities. This proves the desired estimate (4.11). □

REMARK 4.5. In our recent paper [**S08**], the Oleinik-Hopf-type estimate (4.11) was extended to the case when Q_r is represented in the form (3.6) with

$$\psi(x') := \psi_0(|x'|), \quad \text{where} \quad \psi_0 \ge 0,\ \psi_0' \ge 0, \quad \text{and} \quad \int_0^1 t^{-2} \psi(t)\, dt < \infty. \tag{4.17}$$

This condition is sharp in a sense that if the integral in (4.17) diverges, then (4.11) fails. However, in [**S08**] we assumed $b_i \equiv 0$. In the present paper, we impose the "complementary" conditions $\psi \equiv 0$, and $|\mathbf{b}| \in L^q$, $q > n$. In our forthcoming work, we plan to combine the results and techniques of these two papers in order to cover the general case: ψ satisfying (4.17), and $|\mathbf{b}| \in L^q$, $q > n$.

Acknowledgements. The author is thankful to Nicolai V. Krylov for stimulating conversations. A part of this paper was completed in June 2009, when the author took part in the INdAM intensive period in Milan, Italy, organized by Ugo Gianazza, Sandro Salsa, and Vincento Vespri. The author thanks the organizers for the invitation and useful discussions. Special thanks are due to Nina N. Ural'tseva, who raised some open problems which motivated the entire research related to this paper. Finally, the author wishes to thank the referee for corrections and valuable suggestions.

References

[AFT01] H. Aimar, L. Forzani, and R. Toledano, *Hölder regularity of solutions of PDE's: A geometrical view*, Comm. Partial Differential Equations **26** (2001), no. 7–8, 1145–1173. MR1855275 (2002j:35053)

[A63] A.D. Aleksandrov, *Uniqueness conditions and estimates for the solution of the Dirichlet problem*, Vestnik Leningrad Univ. **18** (1963), no. 3, 5–29 (in Russian). English transl. in Amer. Math. Soc. Transl. (2) **68** (1968), 89–119. MR0164135 (29:1434)

[CS07] S. Cho and M.V. Safonov, *Hölder regularity of solutions to second order elliptic equations in non-smooth domains,* Journal of Boundary Value Problems, Special Volume 2007, 24 pp.

[GT83] D. Gilbarg and N.S. Trudinger, *Elliptic Partial Differential Equations of Second Order*, 2nd ed., Springer-Verlag, Berlin–Heidelberg–New York–Tokyo, 1983. MR737190 (86c:35035)

[H52] E. Hopf, *A remark on linear elliptic differential equations of second order,* Proc. Amer. Math. Soc., **3** (1952), 791–793. MR0050126 (14:280b)

[K85] N.V. Krylov, *Nonlinear Elliptic and Parabolic Equations of the Second Order*, Nauka, Moscow, 1985 (in Russian). English transl.: D. Reidel, Dordrecht, 1987. MR815513 (87h:35002); MR0901759 (88d:35005)

[KS80] N.V. Krylov and M.V. Safonov, *A certain property of solutions of parabolic equations with measurable coefficients*, Izvestia Akad. Nauk SSSR, ser. Matem. **44**(1980), 161–175 (in Russian). English transl. in Math. USSR Izvestija, **16** (1981), 151–164. MR563790 (83c:35059)

[La67] E.M. Landis, *A new proof of E. De Giorgi's theorem*, Trudy Moskov. Mat. Obshch. **16** (1967), 319–328 (in Russian). English transl. in Trudy Moscow Math. Soc. **1967** (1968). MR0224975 (37:574)

[La71] E.M. Landis, *Second Order Equations of Elliptic and Parabolic Type,* "Nauka", Moscow, 1971 (in Russian). English transl.: Translations of Mathematical Monographs, Vol. 17, Amer. Math. Soc., Providence, RI, 1997. MR0320507 (47:9044); MR1487894 (98k:35034)

[LU85] O.A. Ladyzhenskaya and N.N. Ural'tseva, *Estimates of the Hölder constant for functions satisfying a uniformly elliptic or a uniformly parabolic quasilinear inequality with unbounded coefficients*, Zap. Nauchn. Sem. Leningrad. Otdel. Mat. Inst. Steklov (LOMI) **147** (1985), 72–94 (in Russian). English transl. in J. Soviet Math. **37** (1987), 837–851. MR821476 (87e:35090)

[LU88] ———, *Estimates on the boundary of a domain for the first derivatives of functions satisfying an elliptic or parabolic inequality*, Trudy Mat. Inst. Steklov **179** (1988), 102–125 (in Russian). English transl. in Proc. Steklov Inst. Math. **179** (1989), 109–135. MR0964915 (89k:35083)

[M67] K. Miller, *Barriers on cones for uniformly elliptic operators*, Ann. Mat. Pura Appl. (4) **76** (1967), 93–105. MR0221087 (36:4139)

[NU09] A.I. Nazarov and N.N. Ural'tseva, *Qualitative properties of solutions to elliptic and parabolic equations with unbounded lower-order coefficients*, Preprints of the St. Petersburg Math. Soc. 2009-05, 9 pp.

[O52] O.A. Oleinik, *On properties of solutions of certain boundary problems for equations of elliptic type,* Mat. Sb. (N. S.) **30** (1952), 695–702 (in Russian). MR0050125 (14:280a)

[PW67] M.H. Protter and H.F. Weinberger, *Maximim Principles in Differential Equations,* Prentice-Hall, Inc., Englewood Cliffs, N.J., 1967. MR0219861 (36:2935)

[S80] M.V. Safonov, *Harnack's inequality for elliptic equations and the Hölder property of their solutions,* Zap. Nauchn. Sem. Leningrad. Otdel. Mat. Inst. Steklov (LOMI) **96** (1980), 272–287 (in Russian). English transl. in J. Soviet Math. **21**, no. 5 (1983), 851–863. MR579490 (82b:35045)

[S08] ———, *Boundary estimates for positive solutions to second order elliptic equations,* arXiv:0810.0522v2 [math.AP], 20 pp.

[St70] E. Stein, *Singular Integrals and Differentiability Properties of Functions.* Princeton University Press, Princeton, NJ, 1970. MR0290095 (44:7280)

School of Mathematics, University of Minnesota, Minneapolis, Minnesota 55455
E-mail address: `safonov@math.umn.edu`

Amer. Math. Soc. Transl.
(2) Vol. **229**, 2010

Global solvability of Navier–Stokes equations for a nonhomogeneous non-Newtonian fluid

V. V. Zhikov and S. E. Pastukhova

This paper is dedicated to Nina Nikolaevna Uraltseva

ABSTRACT. We propose a new approach to prove the existence of weak solutions to generalized or modified Navier–Stokes equations for nonhomogeneous viscous incompressible fluids.

1. Setting of the problem

The model Navier–Stokes system for an incompressible fluid is

$$\frac{\partial u}{\partial t} - \operatorname{div}(A - u \otimes u) + \nabla\pi = f,$$

$$\operatorname{div} u = 0, \tag{1.1}$$

$$A = A(Du) = |Du|^{p-2}Du, \quad p > 1.$$

The unknowns are the velocity $u = (u_1, ..., u_d)$ and the pressure π, $\nabla = \nabla_x$, $\operatorname{div} = \operatorname{div}_x$, $x \in \mathbb{R}^d$, $u \otimes u = \{u_i u_j\}_{i,j=1}^d$, and Du is the symmetric part of the gradient ∇u. The fluid density is assumed to be a constant, $\rho = 1$. For $p = 2$, when the viscous stress tensor $A(Du) = Du$ is linear, we obtain the classical Navier–Stokes equations governing a Newtonian fluid. In the case of $p \neq 2$, the tensor $A(Du)$ is nonlinear, which corresponds to a non-Newtonian fluid.

Consider the following generalization of system (1.1) for an incompressible fluid with a variable density ρ, which is an unknown in the problem along with the velocity u and the pressure π:

$$\frac{\partial}{\partial t}\rho + \operatorname{div}(\rho u) = 0,$$

$$\frac{\partial}{\partial t}(\rho u) - \operatorname{div}(A - \rho u \otimes u) + \nabla\pi = \rho f, \tag{1.2}$$

$$\operatorname{div} u = 0.$$

2000 *Mathematics Subject Classification.* Primary 54C40, 14E20; Secondary 46E25, 20C20.
Key words and phrases. Continuity equation, passing to the limit in the viscous term.
The authors are supported by grant RFFI (08-01-00616 and 09-01-12157).

In contrast to (1.1), this system includes continuity equation $(1.2)_1$, and the basic equation is more complicated because it involves the unknown ρ.

Let Ω be a bounded domain in $\mathbb{R}^d$ ($d \geq 2$) with a Lipschitz boundary, and let $Q_T = \Omega \times (0, T)$, where $T > 0$. We study the system (1.2) in the cylinder Q_T, supplementing it with the conditions

$$(1.3) \qquad u|_{\partial\Omega\times(0,T)} = 0, \quad \rho|_{t=0} = \rho_0 \geq 0, \quad \rho u|_{t=0} = m_0.$$

The data of the problem are assumed to be such that

$$f \in L^\infty(Q_T), \quad \rho_0 \in L^1(\Omega), \quad m_0 = 0 \text{ a.e. on the set } \{\rho_0 = 0\}, \quad m_0 \in (L^1(\Omega))^d.$$

The case of $p < 2$, which corresponds to so-called pseudoplastic flows, is of special interest in applications.

We deal with weak solutions of the initial-boundary value problem.

Define the space of smooth solenoidal vectors

$$C^\infty_{0,sol}(\Omega) = \{\varphi \in (C^\infty_0(\Omega))^d : \ \mathrm{div}\varphi = 0\},$$

and let V be the closure of the set $C^\infty_{0,sol}(\Omega)$ in $(W^{1,p}(\Omega))^d$.

DEFINITION 1.1. The (weak) solution to the problem (1.2), (1.3) is a pair of functions $\rho \geq 0$, u such that
(i) $\rho \in L^\infty(0, T, L^1(\Omega))$, $u \in L^p(0, T, V)$, $\rho u^2 \in L^1(Q_T)$;
(ii) for arbitrary $\varphi \in C^\infty_0(\Omega)$, $\psi \in C^\infty_{0,sol}(\Omega)$ and t_0, $t_1 \in [0, T]$, the integral identities

$$\int_\Omega \rho\varphi dx|_{t_0}^{t_1} - \int_{t_0}^{t_1} \int_\Omega \rho u \cdot \nabla\varphi dx dt = 0,$$

$$\int_\Omega \rho u \cdot \psi dx|_{t_0}^{t_1} - \int_{t_0}^{t_1} \int_\Omega ((\rho u \otimes u - A) \cdot \nabla\psi - \rho f \cdot \psi) dx dt = 0$$

hold and the limiting relations

$$(1.4) \qquad \lim_{t\to 0} \int_\Omega \rho\varphi dx = \int_\Omega \rho_0\varphi dx, \quad \lim_{t\to 0} \int_\Omega \rho u \cdot \psi dx = \int_\Omega m_0 \cdot \psi dx$$

are satisfied.

Actually, it is required in Definition 1.1 that the scalar functions of t, $\int_\Omega \rho\varphi dx$ and $\int_\Omega \rho u \cdot \psi dx$, are absolutely continuous on $[0, T]$. That is why there is no problem with the existence of these integrals at any moment $t \in [0, T]$. Other integrals in the identities of Definition 1.1 are also correctly defined, due to the condition (i).

Note that the limiting relation $(1.4)_2$ deals not with ρu but with its projection onto the corresponding space of solenoidal vectors. The replacement of the test function $\psi \in C^\infty_{0,\mathrm{sol}}(\Omega)$ in $(1.4)_2$ by $\psi \in (C^\infty_0(\Omega))^d$ has to be analysed separately.

We give here the most simple definition, in which test functions are independant of t. For the continuity equation, another variant of the weak setting will also be used. We base our definition on the following assertion.

LEMMA 1.2. *Assume that $v \in L^1(Q_T)$, $K \in (L^1(Q_T))^d$. Then the following conditions are equivalent:*

1) $\frac{\partial v}{\partial t} - \mathrm{div}K = 0$ in the sense of distributons in Q_T; i.e.,

$$(1.5) \qquad \int_{Q_T} (-v\frac{\partial\varphi}{\partial t} + K \cdot \nabla\varphi) dx dt \quad \forall\varphi \in C^\infty_0(Q_T);$$

2) for arbitrary $\eta \in C_0^\infty(\Omega)$, the scalar function $g(t) = \int_\Omega v\eta dx$ is absolutely continuous on $(0,T)$ and

$$\frac{d}{dt}\int_\Omega v\eta dx + \int_\Omega K\cdot\nabla\eta dx = 0 \text{ for a.e. } t\in[0,T]. \tag{1.6}$$

Thus, the continuity equation holds in the sense of distributions in $\Omega\times(0,T)$. Under some conditions, it holds even in the sense of distributions in $\mathbb{R}^d\times(0,T)$; see Section 8.

2. Existence theorems

Define the lower bounds for admissble values of p:

$$p_1 = \frac{3d}{d+2}, \quad p_0 = \max\left\{\frac{d+\sqrt{3d^2+4d}}{d+2},\ p_1\right\}. \tag{2.1}$$

We have $p_0 = \frac{d+\sqrt{3d^2+4d}}{d+2} < 2$ for $d\le 3$; $p_0 = p_1 = 2$ for $d=4$, and $p_0 = p_1 > 2$ for $d>4$.

The case of a bounded initial density ρ_0 is considered first.

THEOREM 2.1. *Let $\rho_0\in L^\infty(\Omega)$, $\frac{m_0^2}{\rho_0}\in L^1(\Omega)$. Then, for $p\ge p_0$, problem (1.2), (1.3) has a weak solution such that*

$$\rho\in L^\infty(Q_T), \quad \rho u^2\in L^\infty(0,T,L^1(\Omega)),$$

and the energy inequality

$$\frac{1}{2}\int_\Omega \rho u^2 dx|_0^t + \int_0^t\int_\Omega A(Du)\cdot Du dx ds \le \int_0^t\int_\Omega \rho f\cdot u dx ds, \quad t\in[0,T] \tag{2.2}$$

holds. Moreover, if $\rho_0\equiv 1$, then $\rho(\cdot,t)\equiv 1$ for any $t\in[0,T]$.

The weak solvability of the initial-boundary value problem for system (1.1) with $p\ge p_0$ was established in [**1**], [**2**]. The definition of a weak solution to this problem is naturally obtained from Definition 1.1 in view of the fact that $\rho\equiv 1$.

Now, consider the case of an unbounded density such that $\rho_0\in L^\gamma(\Omega)$ for some $\gamma>1$, which depends on p. Let

$$p_1(\gamma) = \frac{3d}{d+2-d\gamma^{-1}}$$

and $p_2(\gamma)$ be a positive root of the quadratic equation

$$p^2(d+2-d\gamma^{-1}) - 2dp - 2d = 0. \tag{2.3}$$

Define

$$p_0(\gamma) = \begin{cases} \gamma' = \frac{\gamma}{\gamma-1}, & \text{if } \gamma\in(1,\frac{d}{d-1}],\\ \max\{p_1(\gamma),p_2(\gamma),\gamma'\}, & \text{if } \gamma\in[\frac{d}{d-1},\infty). \end{cases} \tag{2.4}$$

It is easy to see that $p_0(\gamma)$ is continuous, since $\max\{p_1(d'),p_2(d'),d\} = d$.

THEOREM 2.2. *Assume that $\rho_0\in L^\gamma(\Omega)$, $\frac{m_0^2}{\rho_0}\in L^1(\Omega)$ and $p\ge p_0(\gamma)$ with the strict inequality if $\gamma=\frac{d}{d-1}$. Then problem (1.2), (1.3) has a weak solution that satisfies the energy inequality (2.2). Moreover,*

$$\begin{gathered}\rho\in L^\infty(0,T,L^\gamma(\Omega)),\\ \rho u^2\in L^\infty(0,T,L^1(\Omega)) \quad \rho u\in L^\infty(0,T,(L^k(\Omega))^d),\ k=\frac{2\gamma}{\gamma+1}.\end{gathered} \tag{2.5}$$

For fixed d and $\gamma \to \infty$, (2.3) becomes $p^2(1+\frac{2}{d}) - 2p - 2 = 0$, whose positive root is equal to $\frac{d+\sqrt{3d^2+4d}}{d+2}$, so that $p_0(\gamma)$ in (2.4) and p_0 in (2.1) are related by the convergence $p_0(\gamma) \to p_0$. For $d = 3$, we have $p_0 < 1.86$, whence $p_0(\gamma) < 2$ for sufficiently large γ. Thus, Theorems 2.1 and 2.2 partially cover the case of pseudoplastic flows.

Let us comment on the definition of $p_0(\gamma)$. In our considerations, it is important that the continuity equation $(1.2)_1$, with a fixed vector $u \in L^p(0,T,V)$ in it, is uniquely solvable. This is guaranteed by the condition $p \geq \gamma'$, which is contained in the requirement $p \geq p_0(\gamma)$. The other constraints on p in this requirement are associated with the convective term $\operatorname{div}(\rho u \times u)$ in the framework of the natural energy space defined by the viscous term and the derivative $\frac{\partial}{\partial t}(\rho u)$, and that is a distinctive feature of the method in question. We have the inequality

$$\|\rho u^2\|_{L^1(0,T,L^{p'}(\Omega))} \leq c_0\{\|\nabla u\|^2_{L^p(Q_T)} + \sup_{0\leq t\leq T}\int_\Omega \rho u^2 dx\}, \tag{2.6}$$

which means the indicated "layerwise" subordination of the convective term. Complete subordination would mean a similar inequality with $\|\rho u^2\|_{L^1(0,T,L^{p'}(\Omega))}$ on the left-hand side replaced by $\|\rho u^2\|_{L^{p'}(Q_T)}$. This requires more restrictive conditions on p. Specifically, for $\rho_0 \equiv 1$ (i.e., for system (1.1)) we need the condition $p \geq \frac{3d+2}{d+2}$, known from [**3**],[**4**].

The definition of the weak solution implies for the density ρ the property of weak continuity over the variable t. Actually, the density is continuous over t in a strong sense.

THEOREM 2.3. *The following assertions are valid:*

$$\rho \in C(0,T,L^\alpha(\Omega))\ \forall\alpha \geq 1 \quad \textit{under the conditions of Theorem 2.1,}$$

$$\rho \in C(0,T,L^\beta(\Omega)),\quad \beta = \frac{\gamma p}{\gamma + p}, \quad \textit{under the conditions of Theorem 2.2.}$$

This phenomenon is determined exclusively by the nature of the continuity equation

$$\frac{\partial}{\partial t}\rho + \operatorname{div}(\rho u) = 0$$

with given vector $u \in L^p(0,T,V)$ in it and the initial condition $\rho|_{t=0} = \rho_0 \in L^\gamma(\Omega)$, $\frac{1}{\gamma}+\frac{1}{p} \leq 1$. The facts of such kind are well known [**5**]. At the end of the paper, we give an Appendix, where the corresponding properties of the continuity equation are expounded.

To bring this to a conclusion, note that global solvability of the Navier–Stokes equations for a nonhomogeneous Newtonian fluid (see (1.2) with $p = 2$) was first proved by A. Kazhikhov in [**6**] (see also [**7**]) under the assumption that ρ_0, $\rho_0^{-1} \in L^\infty(\Omega)$. The result was extended by P.-L. Lions to an arbitrary bounded density ρ_0 [**8**].

3. Galerkin approximations

3.1. Galerkin system and its solvability. The solution of the problem (1.2) is constructed as a limit of Galerkin approximations. We describe these approximations in the simplest situation, when $u_0 = \rho_0^{-1} m_0 \in H$, where H is a closure of the set $C^\infty_{0,sol}(\Omega)$ in $(L^2(\Omega))^d$.

Let $\{\omega_i(x)\}_{i=1}^{\infty}$ be an orthonomal in $(L^2(\Omega))^d$ system of functions from $C^{\infty}_{0,sol}(\Omega)$, whose linear combinations are dense in V, and let P_n be the orthogonal projector on the linear span $[\omega_1,\dots,\omega_n]$. Actually, the choice of the basis is subordinate to the additional condition, which will be clarified later at the very moment when it is demanded (see Section 7).

Approximate the Cauchy data (1.3) with functions

$$\rho_{0n}\in C^{\infty}(\overline{\Omega}),\ u_{0n}=P_nu_0,\ n\in\mathbb{N}, \tag{3.1}$$

where $\frac{1}{n}\le\rho_{0n}$, $\rho_{0n}\to\rho_0$ in $L^{\gamma}(\Omega)$.

Consider the system

$$\begin{aligned}&\frac{\partial}{\partial t}\rho_n+u_n\cdot\nabla\rho_n=0,\\ &\rho_n\frac{\partial}{\partial t}u_n+\rho_n(u_n\cdot\nabla)u_n-\operatorname{div}A(Du_n)=\rho_nf,\end{aligned} \tag{3.2}$$

with Cauchy data from (3.1). The solution of the system is a pair of functions

$$\rho_n(t,x),\quad u_n(t,x)=\sum_{i=1}^{n}a_{ni}(t)\omega_i(x),\quad a_n(t)=\{a_{ni}(t)\}_{i=1}^{n}\in(C[0,T])^n, \tag{3.3}$$

which satisfy the first equation in the classical sense and the second equation in the sense of an integral identity with test functions from $[\omega_i,\dots,\omega_n]$.

Begin with a priori estimates for solutions ρ_n, u_n. Firstly, thanks to the choice of an initial function and from properties of the continuity equation (see Section 8)

$$c_0^{-1}\le\rho_n(x,t)\le c_0,\quad (x,t)\in Q_T, \tag{3.4}$$

where the constant $c_0>0$ depends on n. Secondly, the energy equality

$$\frac{1}{2}\int_{\Omega}\rho_nu_n^2dx|_0^t+\int_0^t\int_{\Omega}A(Du_n)\cdot Du_ndxds=\int_0^t\int_{\Omega}\rho_nf\cdot u_ndxds,\quad t\in[0,T] \tag{3.5}$$

holds. For its derivation take the following integral equalities ensueing from (3.2):

$$\int_0^t\int_{\Omega}[\rho_nu_n\cdot u_n'+\rho_nu_n\cdot\nabla(\frac{u_n^2}{2})]dxds=\int_{\Omega}\rho_n\frac{u_n^2}{2}dx|_0^t,\quad u_n'=\frac{\partial u_n}{\partial t},$$

$$\int_0^t\int_{\Omega}[\rho_nu_n\cdot u_n'+(\rho_nu_n\times u_n)\cdot\nabla u_n-A(Du_n)\cdot Du_n+\rho_nf\cdot u_n]dxds=\int_{\Omega}\rho_nu_n^2dx|_0^t$$

and substract one from another taking into account the relation $(u_n\times u_n)\cdot\nabla u_n=u_n\cdot\nabla\frac{u_n^2}{2}$.

Due to (3.5) and the Korn inequality

$$\int_{\Omega}|\nabla u|^pdx\le c_K\int_{\Omega}|Du|^pdx,\quad u\in(W_0^{1,p}(\Omega))^d,$$

we have the following uniform-in-n estimate for u_n:

$$\sup_{t\in(0,T)}\int_{\Omega}\rho_nu_n^2dx+\int_0^T\int_{\Omega}|\nabla u_n|^pdxdt\le c_1, \tag{3.6}$$

and, consequently, because of the orthonormality of $\{\omega_i\}$, the estimate for the vector $a_n(t)$ (see (3.3)) is

$$|a_n(t)|^2=\int_{\Omega}|u_n(t)|^2dx\le(\sup\frac{1}{\rho_n})\int_{\Omega}\rho_nu_n^2dx\le c_1c_0,\quad t\in(0,T). \tag{3.7}$$

We pass to the solvability of (3.1), (3.2). The proof is divided into steps.

1° Temporarily we deal with the system

$$(3.8)\qquad \begin{gathered}\frac{\partial}{\partial t}\rho + v\cdot\nabla\rho = 0, \quad \rho|_{t=0} = \rho_{0n},\\ \rho\frac{\partial}{\partial t}u + \rho(v\cdot\nabla)u - \operatorname{div} A(Du) = \rho f, \quad u|_{t=0} = u_{0n},\end{gathered}$$

where $v = \sum_{i=1}^n b_i(t)\omega_i(x)$ and $b(t) = \{b_i(t)\}_{i=1}^n \in (C[0,T])^n$ is a fixed vector. The solution is considered in the same sense as for (3.1), (3.2). The first equation defines uniquely the density $\rho(t,x)$. Then, for given ρ, v the second equation defines uniquely the vector $u(t,x) = \sum_{i=1}^n a_i(t)\omega_i(x)$. In fact, an appropriate integral identity on the test functions from $\psi = \omega_i$, $i = 1,...,n$ yields for the unknown vector a,

$$(3.9)\qquad G\frac{da}{dt} = F(t,a), \quad a|_{t=0} = a_0.$$

Here $G = \{g_{ij}\}_{i,j=1}^n$, $g_{ij} = \int_\Omega \rho\omega_i\cdot\omega_j dx$, is a nondegenerate matrix of Gram,

$$(3.10)\qquad \begin{gathered}a_0 = \{a_{0i}\}_{i=1}^n, \quad a_{0i} = (u_0,\omega_i)_{L^2(\Omega)^d}, \quad F = \{F_i\}_{i=1}^n,\\ F_i(t,a) = -b_s(t)a_j(t)\int_\Omega \rho(\omega_s\cdot\nabla)\omega_j\cdot\omega_i dx\\ + \int_\Omega A(Du)\cdot D\omega_i dx - \int_\Omega \rho f\cdot\omega_i dx.\end{gathered}$$

We assume summation over repeated indices from 1 to n.

By the Carathéodory theorem (see [**10**]), the Cauchy problem (3.9) is solvable on some interval $(0,t_0)$. We may state that $t_0 = T$ in view of the a priori estimates

$$c_0^{-1} \le \sup_{Q_T}\rho(t,x) \le c_0, \qquad \sup_{t\in(0,T)}|a(t)|^2 \le c_1c_0,$$

which are derived in the same way as (3.4)-(3.7).

2° Let us show the uniqueness of the solution constructed just now. Similarly as before (see the derivation of (3.5)), we prove for the difference of two solutions $w = u^{(1)} - u^{(2)}$ the equalities

$$\int_0^t\int_\Omega(\rho w\cdot w' + \rho v\cdot\nabla\frac{w^2}{2})dxds = \int_\Omega \rho\frac{w^2}{2}dx\,|_0^t,$$

$$\int_0^t\int_\Omega[\rho w\cdot w' + (\rho v\times w)\cdot\nabla w - (A^{(1)} - A^{(2)})\cdot\nabla w]dxds = \int_\Omega \rho w^2 dx|_0^t,$$

where $A^{(i)} = A(Du^{(i)})$, $i = 1,2$. Hence, by the relations

$$(v\times w)\cdot\nabla w = v\cdot\nabla\frac{w^2}{2}, \quad (A(\xi) - A(\eta))\cdot(\xi-\eta) \ge 0, \quad w|_{t=0} = 0,$$

it follows that

$$\int_\Omega \rho w^2(t,x)dx \le 0 \quad \Rightarrow w(\cdot,t) = 0 \quad \Rightarrow u^{(1)} = u^{(2)},$$

and the solution u is unique.

3° Take the set $K_r = \{a(t) \in (C[0,T])^n : \|a\|_{(C[0,T])^n} \le r, \quad a|_{t=0} = a_0\}$, where $a_0 \in \mathbb{R}^n$ is a fixed vector from (3.9), (3.10) and $r^2 = c_1c_0^{-1}$, and consider the mapping $v \to u$ given by the system (3.8). The corresponding operator $S : b(t) \to$

$a(t)$ is continuous in $(C[0,T])^n$ and maps the convex closed set K_r on its compact part. We omit here the proof of these facts.

Thus, the operator S satisfies the Schauder fixed point theorem, according to which there exists a function $u = Sv = v$. Evidently, the solution of (3.1), (3.2) is obtained.

3.2. Passing to the limit in the Galerkin system. The constructed functions ρ_n, u_n satisfy the energy equality (3.5) and are uniform with respect to n estimates

$$\begin{gathered} \|\rho_n\|_{L^\infty(0,T,L^\gamma(\Omega))} \le c, \ \|\rho_n u_n^2\|_{L^\infty(0,T,L^1(\Omega))} \le c, \\ \|\nabla u_n\|_{L^p(Q_T)} \le c, \ \|\rho_n u_n\|_{L^\infty(0,T,L^k(\Omega))} \le c, \ k = 2\gamma/(\gamma+1). \end{gathered} \tag{3.11}$$

The first of them is determined by the properties of the continuity equation (see Section 8) and the choice of the initial function ρ_{0n} in (3.1). The other two were already mentioned ealier (see (3.6)), and the last one stems from the first two estimates due to Lemma 6.2. The uniform estimates ensure the convergence (maybe after selecting the subsequence)

$$\begin{gathered} \rho_n \overset{*}{\rightharpoonup} \rho \ \text{ in } \ L^\infty(0,T,L^\gamma(\Omega)), \\ u_n \rightharpoonup u \text{ in } \ L^p(0,T,W^{1,p}(\Omega)), \quad A(Du_n) \rightharpoonup z \text{ in } \ L^{p'}(Q_T). \end{gathered} \tag{3.12}$$

Here and hereafter the symbols $\rightharpoonup$, $\overset{*}{\rightharpoonup}$, $\to$ denote the weak, the $*$- weak and the strong convergence, respectively.

We write down the integral identities for ρ_n and u_n:

$$\begin{gathered} \int_\Omega \rho_n \varphi dx|_{t_0}^{t_1} + \int_{t_0}^{t_1}\int_\Omega \rho_n u_n \cdot \nabla\varphi dxdt = 0, \quad \varphi \in C_0^\infty(\Omega), \\ \int_\Omega \rho_n u_n \cdot \psi dx|_{t_0}^{t_1} + \int_{t_0}^{t_1}\int_\Omega [A(Du_n) - \rho_n u_n \otimes u_n]\cdot\nabla\psi dxdt \\ - \int_{t_0}^{t_1}\int_\Omega \rho_n f \cdot \psi dxdt \end{gathered} \tag{3.13}$$

with a test function $\psi \in [\omega_1, \dots, \omega_{n_0}]$, where n_0 is fixed, $n \ge n_0$, t_0, $t_1 \in [0,T]$. It is required to pass to the limit in these identities as $n \to \infty$. To that end we use the folowing result, which is proved in Sections 7, 8.

LEMMA 3.1. *At least on a subsequence*

$$\rho_n u_n \overset{*}{\rightharpoonup} \rho u \ \ in \ \ L^\infty(0,T,(L^k(Q_T))^d), \quad k = \frac{2\gamma}{\gamma+1}, \tag{3.14}$$

$$\lim_{n\to\infty}\int_\Omega \rho_n u_n \cdot \psi dx = \int_\Omega \rho u\cdot\psi dx, \quad t\in[0,T], \quad \psi \in C_{0,sol}^\infty(\Omega), \tag{3.15}$$

$$\rho_n \to \rho \ in \ C(0,T,L^\beta(\Omega)), \quad \beta = \frac{p\gamma}{\gamma+p}, \tag{3.16}$$

$$\sqrt{\rho_n}u_n \to \sqrt{\rho}u \ \ in \ (L^2(Q_T))^d. \tag{3.17}$$

All the terms in (3.13), except for one, contain the density ρ_n. One can find their limits with the help of Lemma 3.1. For example, (3.17) leads to convergence $\rho_n u_n \otimes u_n \to \rho u \otimes u$ in $L^1(Q_T)^{d\times d}$, which allows us to take the limit of the convective term.

As concerns the integral with $A(Du_n)$, we have, according to (3.12),

$$\lim_{n\to\infty}\int_{t_0}^{t_1}\int_\Omega A_n\cdot\nabla\psi dxdt=\int_{t_0}^{t_1}\int_\Omega z\cdot\nabla\psi dxdt.$$

It remains to show that the weak limit z of $A_n = A(Du_n)$ is of natural structure, i.e.

$$z = A = A(Du). \tag{3.18}$$

This item constitutes the central part in our proof, and the whole Section 4 is devoted to it.

Once (3.18) is proved, we should obtain for ρ, u equalities just similar to (3.13) with test functions $\psi \in [\omega_1, \dots, \omega_{n_0}]$. Hence, by density arguments, we come to the integral identities in Definition 1.1. As a result, the pair ρ, u turns out to be a weak solution to (1.2), (1.3). Moreover, the properties (2.5) are valid. Along with this, the convergence (3.16) ensures for the function ρ the higher continuity property stated in Theorem 2.3.

We proceed by passing to the limit in the energy equality (3.5). By (3.17) and (3.14),

$$\lim_{n\to\infty}\int_\Omega \rho_n u_n^2 dx=\int_\Omega \rho u^2 dx \text{ for a.e. } t\in[0,T], \tag{3.19}$$

$$\lim_{n\to\infty}\int_0^t\int_\Omega \rho_n f\cdot u_n dxd\tau=\int_0^t\int_\Omega \rho f\cdot u dxd\tau.$$

Besides, $A(Du)\cdot Du = |Du|^p$, and thanks to the weak lower semicontinuity of the L^p-norm,

$$\liminf_{n\to\infty}\int_0^t\int_\Omega A(Du_n)\cdot Du_n dxd\tau \geq \int_0^t\int_\Omega |Du|^p dxd\tau.$$

Therefore, (3.5) leads to (2.2).

4. Passing to the limit in the viscous term

In establishing the relation (3.18), a key role is played by the properties of the sequence $A_n\cdot\nabla u_n$, which is always bounded but is not necessarily weakly compact in $L^1(Q_T)$. However, without loss of generality, we assume the weak convergence

$$A_n\cdot\nabla u_n dxdt \rightharpoonup d\mu. \tag{4.1}$$

Here, μ is a finite Borel measure on the compact set $\bar Q_T = \bar\Omega\times[0,T]$, and

$$d\mu = adxdt + \mu^s$$

is its decomposition into absolutely continuous (with respect to the Lebesgue measure in Q_T) and singular components. Convergence (4.1) means (in view of the boundedness $A_n\cdot\nabla u_n$ in $L^1(Q_T)$) that

$$\lim_{n\to\infty}\int_{Q_T}\varphi A_n\cdot\nabla u_n dxdt=\int_{\bar Q_T}\varphi d\mu,\quad \varphi\in C^\infty(\bar Q_T).$$

In the sequel, we use the well-known general properties of the weak convergence of nonnegative measures (see [**11**]):

$$\liminf_{n\to\infty}\int_M A_n\cdot\nabla u_n dxdt \geq \int_M d\mu \text{ for any open set } M\subset \bar{Q}_T,$$
$$\limsup_{n\to\infty}\int_M A_n\cdot\nabla u_n dxdt \leq \int_M d\mu \text{ for any closed set } M\subset \bar{Q}_T, \tag{4.2}$$

and, also, established in [**12**] the property of the limit measure μ (more exactly, of its absolutely continuous component a) related to the monotonicity of the function $A(\xi)$.

LEMMA 4.1. *(i) $a \geq z\cdot\nabla u$ a.e. in Q_T;*
(ii) if $a = z\cdot\nabla u$, then $z = A$.

Clearly, to prove (3.18) it suffices to show that

$$\int_\Omega adx|_{t=t_0} \leq \int_\Omega z\cdot\nabla udx|_{t=t_0} \text{ for a.e. } t_0\in[0,T], \tag{4.3}$$

in which case the general lower semicontinuity property (i) implies $a = z\cdot\nabla u$ and we can apply the property (ii).

Proof of inequality (4.3). Recalling that $A_n\cdot\nabla u_n$ is the density for the viscous energy of Galerkin approximations, we shall use the corresponding energy equality and integral identities. Hereafter, for the simplification of formulae, let $f = 0$.

From (3.5) and $(3.13)_2$ by subtraction, we find

$$I_n = \int_{t_0}^{t_1}\int_\Omega A_n\cdot\nabla u_n dxdt$$

$$= \int_{t_0}^{t_1}\int_\Omega (A_n - \rho_n u_n\otimes u_n)\cdot\nabla\psi dxdt - \int_\Omega (\frac{1}{2}\rho_n u_n^2 - \rho_n u_n\cdot\psi)dx|_{t_0}^{t_1} = I_{n,1} + I_{n,2}.$$

After transforming we estimate from above the term $I_{n,2}$:

$$2I_{n,2} = -\int_\Omega \rho_n(u_n - \psi)^2dx|_{t_0}^{t_1} + \int_\Omega \rho_n\psi^2dx|_{t_0}^{t_1}$$
$$\leq \int_\Omega \rho_n(u_n-\psi)^2dx|_{t=t_0} + \int_{t_0}^{t_1}\int_\Omega \rho_n u_n\cdot\nabla(\psi^2)dxdt;$$

thereby,

$$I_n \leq \int_{t_0}^{t}\int_\Omega [(A_n - \rho_n u_n\otimes u_n)\cdot\nabla\psi + \rho_n u_n\cdot\nabla\frac{\psi^2}{2}]dxdt + \frac{1}{2}\int_\Omega \rho_n(u_n-\psi)^2dx|_{t=t_0}.$$

Hence, by $(4.2)_1$, $(3.12)_3$ and limit relations from Lemma 3.1,

$$\int_{\bar{\Omega}\times(t_0,t_1)} d\mu \leq \int_{t_0}^{t_1}\int_\Omega [(z-\rho u\otimes u)\cdot\nabla\psi + \rho u\cdot\nabla\frac{\psi^2}{2}]dxdt + \frac{1}{2}\int_\Omega \rho(u-\psi)^2dx|_{t=t_0}, \tag{4.4}$$

since the set $\bar{\Omega}\times(t_0,t_1)$ is open in $\bar{Q}_T$. Here, t_0 is selected in such a way that the convergence (3.19) holds at t_0 and $u(\cdot,t_0)\in(W_0^{1,p}(\Omega))^d$.

We check that we can set $\psi(x) = u(x,t_0)$ in (4.4) by taking the approximation

$$\psi_N\to\psi \quad \text{in} \quad (W_0^{1,p}(\Omega))^d, \quad \psi_N\in C_{0,\mathrm{sol}}^\infty(\Omega), \quad \psi(x) = u(x,t_0). \tag{4.5}$$

It follows from (2.6) that $\int_{t_0}^{t_1} \rho u \otimes u dt \in L^{p'}(\Omega)$; therefore,

$$\int_{t_0}^{t_1} \int_\Omega \rho u \otimes u \cdot \nabla \psi_N dx dt = \int_\Omega (\int_{t_0}^{t_1} \rho u \otimes u dt) \cdot \nabla \psi_N dx \to \int_\Omega (\int_{t_0}^{t_1} \rho u \otimes u dt) \cdot \nabla \psi dx.$$

By a similar reason and due to Lemma 6.3,

$$\begin{aligned} &\lim_{N\to\infty} \int_{t_0}^{t_1} \int_\Omega z \cdot \nabla \psi_N dx dt = \int_\Omega \int_{t_0}^{t_1} z \cdot \nabla \psi dx dt, \\ &\lim_{N\to\infty} \int_{t_0}^{t_1} \int_\Omega \rho u \cdot \nabla \frac{\psi_N^2}{2} dx dt = \int_\Omega \int_{t_0}^{t_1} \rho u \cdot \nabla \frac{\psi^2}{2} dx dt. \end{aligned} \tag{4.6}$$

By the continuity of the embedding $W_0^{1,p}(\Omega) \subset L^{p^*}(\Omega)$, where p^* is the Sobolev conjugate exponent, we have $\psi_N \to \psi$ in $(L^{p^*}(\Omega))^d$, $\psi(x) = u(x, t_0)$. Together with the estimate

$$\|\rho(u - \psi_N)^2\|_{L^1(\Omega)} \le \|\rho\|_{L^\gamma(\Omega)} \|u - \psi_N\|^2_{L^{p^*}(\Omega)}, \text{ where } \frac{1}{\gamma} + \frac{2}{p^*} < 1,$$

this yields

$$\lim_{N\to\infty} \int_\Omega \rho(u - \psi_N)^2 dx|_{t=t_0} = 0.$$

The necessary constraint for the exponents is fulfilled under the condition (2.4). In fact, for example, if $p < d$, then $p^* = \frac{dp}{d-p}$ and $\frac{1}{\gamma} + \frac{2}{p^*} < 1 \Longleftrightarrow \frac{1}{\gamma} + \frac{2}{p} < 1 + \frac{2}{d}$. Note that (2.4) implies that

$$p \ge p_1(\gamma) = \frac{3d}{d + 2 - d\gamma^{-1}} \Longleftrightarrow \frac{1}{\gamma} + \frac{3}{p} < 1 + \frac{2}{d}.$$

Thus, using the approximation (4.5) in the inequality (4.4) we obtain after passage to the limit,

$$\int_{\bar\Omega \times (t_0, t_1)} d\mu \le \int_{t_0}^{t_1} \int_\Omega [(z - \rho u \otimes u) \cdot \nabla \psi + \rho u \cdot \nabla \frac{\psi^2}{2}] dx dt, \quad \psi(x) = u(x, t_0). \tag{4.7}$$

Setting in (4.7) $t_1 = t_0 + h$, where $h > 0$, we deduce

$$\begin{aligned} &\int_\Omega \frac{1}{h} \int_{t_0}^{t_0+h} a\, dt dx \le \frac{1}{h} \int_{\bar\Omega \times (t_0, t_0+h)} d\mu \\ &\le \int_\Omega \frac{1}{h} \int_{t_0}^{t_0+h} (z - \rho u \otimes u) dt \cdot \nabla \psi dx + \int_\Omega \frac{1}{h} \int_{t_0}^{t_0+h} \rho u dt \cdot \nabla \frac{\psi^2}{2} dx, \quad \psi = u(\cdot, t_0), \end{aligned} \tag{4.8}$$

since $a dx dt \le d\mu$. Assume that t_0 is a Lebesgue point for $z - \rho u \otimes u$ as a function from $L^1(0, T, L^{p'}(\Omega))$, for ρu as a function from $L^1(0, T, L^k(\Omega))$ and for a as a function from $L^1(0, T, L^1(\Omega))$. Then, by the properties of Lebesgue points, it follows from (4.8) as $h \to 0$ that

$$\int_\Omega a dx|_{t=t_0} \le \int_\Omega [(z - \rho u \otimes u) \cdot \nabla u + \rho u \cdot \nabla \frac{u^2}{2}] dx|_{t=t_0} = \int_\Omega z \cdot \nabla u dx|_{t=t_0}, \tag{4.9}$$

where the last simplification is due to the identity $(u \otimes u) \cdot \nabla u = u \cdot \nabla \frac{u^2}{2}$. Finally, we prove the property (4.3) and, consequently (see Lemma 4.1), the required relation (3.18). The latter one ensures the passage to the limit in the viscous term (see

Section 3) and, thus, the existence theorem. By the way, we get the following decomposition of the limit measure in (4.1):

$$d\mu = |Du|^p dxdt + d\mu^s.$$

5. Energy equalities

We continue the analysis of the limit measure μ started in Section 4. As before, for notational brevity, consider $f = 0$, in which case the energy equality for the Galerkin approximation is written as

$$\frac{1}{2}\int_\Omega \rho_n u_n^2 dx|_0^t + \int_0^t \int_\Omega A_n \cdot \nabla u_n dxd\tau = 0, \quad 0 \le t \le T. \tag{5.1}$$

Earlier it was shown that passing to the limit does not preserve, generally, the equality in (5.1). One can state only that the left-hand side is nonpositive for a.e. $t \in [0,T]$. However, it is easy to restore the energy balance by means of the limit measure, and thereafter"energy" equality is obtained in the form

$$\frac{1}{2}\int_\Omega \rho u^2 dx|_0^t + \int_0^t \int_\Omega |Du|^p dxd\tau + \int_{\bar\Omega\times[0,t)} d\mu^s = 0 \quad \text{for a.e. } t \in [0,T]. \tag{5.2}$$

In fact, because of (3.19), the integrals over the sections in (5.1) converge; i.e.,

$$\lim_{n\to\infty} \int_\Omega \rho_n u_n^2 dx|_0^t = \int_\Omega \rho u^2 dx|_0^t \quad \text{for a.e. } t \in [0,T].$$

Clearly, for the sections $\bar\Omega_t = \bar\Omega \times \{t\}$, the equality $\mu(\bar\Omega_t) = 0$ holds, except for a countable set $t \in [0,T]$ at most. Hence, in view of (4.2),

$$\lim_{n\to\infty} \int_0^t \int_\Omega A_n \cdot \nabla u_n dxd\tau = \int_{\bar\Omega\times[0,t)} d\mu,$$

where t is taken out of the exceptional set. Since the representation

$$d\mu = |Du|^p dxdt + d\mu^s \tag{5.3}$$

is already established (see Section 4), the equality (5.2) is also proved.

Consider the monotone function $F(t) = \int_{\bar\Omega\times[0,t)} d\mu$, that is, the projection of the measure μ on the time axis. From (4.8), (4.9), (5.3) and the equalities $a = |Du|^p$, $z = |Du|^{p-2}Du$, it follows that

$$F'(t) = \int_{\Omega_t} |Du|^p dx \text{ for a.e. } t \in [0,T],$$

whence

$$\frac{d}{dt}\int_{\bar\Omega\times[0,t)} d\mu^s = 0 \quad \text{for a.e. } t \in [0,T].$$

This means that *the projection of the singular component μ^s on the time axis is itself a singular measure.* In this case, differentiation of (5.2) implies the equality

$$\frac{d}{dt}\int_{\Omega_t} \frac{\rho u^2}{2} dx + \int_{\Omega_t} |Du|^p dx = 0 \text{ for a.e. } t \in [0,T].$$

We sum up the result.

THEOREM 5.1. *Under the assumptions of Theorems 2.1 and 2.2, for the solution to problem (1.2), (1.3), we have the following energy equality in differential form:*

$$\frac{d}{dt}\int_{\Omega_t}\frac{\rho u^2}{2}dx+\int_{\Omega_t}|Du|^p dx=\int_{\Omega_t}\rho f\cdot u dx \quad \text{for a.e.} \quad t\in[0,T]. \tag{5.4}$$

Note that the function $t\to\int_{\Omega_t}\frac{\rho u^2}{2}dx$ is not necessarily absolutely continuous. Therefore, an equality in (2.2) cannot be achieved by integrating (5.4) with respect to t.

6. Estimate of the convective term

In what follows, $\|\cdot\|_q=\|\cdot\|_{L^q(\Omega)}$, and p^* denotes the Sobolev conjugate exponent, i.e.

$$p^*=\begin{cases}\frac{pd}{d-p}, \text{ if } p<d,\\ \infty, \text{ if } p>d,\\ \text{any finite value, if } p=d.\end{cases}$$

LEMMA 6.1. *Let* $\rho\in L^\infty(0,T,L^\gamma(\Omega))$, $\rho u^2\in L^\infty(0,T,L^1(\Omega))$ $u,\in L^p(0,T,(W_0^{1,p}(\Omega))^d$, *and let* p, γ *be such that*

$$\frac{1}{\gamma}+\frac{3}{p}\le 1+\frac{2}{d} \quad \text{and} \quad p\ge p_2(\gamma) \text{ if } p<d, \tag{6.1}$$

$$p>\gamma' \text{ if } p=d, \quad p\ge\gamma' \text{ if } p>d,$$

where $p_2(\gamma)$ *is the positive root of the quadratic equation (2.3).*

Then $\rho u^2\in L^1(0,T,L^{p'}(\Omega))$ *for some* p' *and an estimate of the type (2.6) is valid.*

PROOF. We consider, here, only the case $p<d$. Denote $s=p'$. By the Hölder and Sobolev ineqalities, since $\rho^s u^{2s}=(\rho u^2)^\delta\rho^{s-\delta}u^{2(s-\delta)}$, we deduce

$$\|\rho^s u^{2s}\|_1\le\|\rho u^2\|_1^\delta\|\rho\|_\gamma^{s-\delta}\|u\|_{p^*}^{2(s-\delta)}\le C\|\nabla u\|_p^{2(s-\delta)}, \tag{6.2}$$

where $\delta+\frac{s-\delta}{\gamma}+2\frac{s-\delta}{p^*}=1$ and $0\le\delta\le1$, i.e.

$$\delta=\frac{1-s(\frac{1}{\gamma}+\frac{2}{p^*})}{1-(\frac{1}{\gamma}+\frac{2}{p^*})}\in[0,1]. \tag{6.3}$$

Since $s>1$, then (6.3) is true whenever $\frac{1}{s}\ge\frac{1}{\gamma}+\frac{2}{p^*}$, which agrees with the constraint $(6.1)_1$.

By (6.2), $\|\rho u^2\|_s\le C\|\nabla u\|_p^{\frac{2(s-\delta)}{s}}$; therefore,

$$\int_0^T\|\rho u^2\|_s dt<\infty \text{ whenever } \frac{2(s-\delta)}{s}\le p, \text{ i.e. } \delta\ge\frac{2-p}{2}s,$$

and by account of (6.3),

$$\frac{1-s\alpha}{1-\alpha}\ge\frac{2-p}{2}s, \quad \alpha=\frac{1}{\gamma}+\frac{2}{p^*}, \quad s=p'.$$

Hence, elementary calculations yield the quadratic inequality $p^2(1-\frac{1}{\gamma}+\frac{2}{d})-2p-2\ge0$. The only positive root of the corresponding quadratic equation was denoted earlier by $p_2(\gamma)$. Thus, under suppositions (6.1) we have $\rho u^2\in L^1(0,T,L^{p'}(\Omega))$, the estimate of the type (2.6) being valid. □

Remark. One can readily verify that the constraints on the exponents p, γ in Lemma 6.1 are fulfilled under suppositions of Theorem 2.2. Besides, if $\gamma = \infty$, then the constraints on the exponent p in Lemma 6.1 are reduced to the inequality $p \geq p_0$, where p_0 is defined in (2.1).

We give, here, another boundedness lemma.

LEMMA 6.2. *Suppose that* $\rho \in L^\infty(0,T,L^\gamma(\Omega))$, $\gamma > 1$, *and* $\rho u^2 \in L^\infty(0,T, L^1(\Omega))$. *Then* $\rho u \in L^\infty(0,T,L^k(\Omega))$, $k = \frac{2\gamma}{\gamma+1}$, *the corresponding estimate being valid.*

PROOF. Starting with $(\rho u)^{\frac{2\gamma}{\gamma+1}} = (\rho u^2)^{\frac{\gamma}{\gamma+1}} \rho^{\frac{\gamma}{\gamma+1}}$, we deduce, by the Hölder inequality, $\|\rho u\|_k \leq \|\rho u^2\|_1^{\frac{\gamma}{\gamma+1}} \|\rho\|_\gamma^{\frac{1}{\gamma+1}}$, whence the result follows.

The following assertion was used in Section 4 for the justification of (4.7) via the approximations (4.5). $\square$

LEMMA 6.3. *Let* $d > p \geq p_1(\gamma)$ *and consider* $\Phi(v) = (\rho u \otimes v) \cdot \nabla v$, *functions* $\rho \in L^\infty(0,T,L^\gamma(\Omega))$ *and* $u \in L^1(0,T,(W_0^{1,p}(\Omega))^d)$ *being fixed. Then* $\Phi(v)$: $(W_0^{1,p}(\Omega))^d \to L^1(\Omega\times(0,T))$ *is a continuous operator.*

PROOF. Let $v_N \to v$ in $W_0^{1,p}(\Omega)^d$. We have to prove that $\Phi(v_N) \to \Phi(v)$ in $L^1(Q_T)$. From the representation

$$\begin{aligned}\Phi(v_N) - \Phi(v) &= (\rho u \otimes v_N) \cdot \nabla v_N - (\rho u \otimes v) \cdot \nabla v \\ &= (\rho u \otimes v_N) \cdot \nabla(v_N - v) + (\rho u \otimes (v_N - v)) \cdot \nabla v\end{aligned}$$

and the Hölder and Sobolev inequalities, it follows that

$$\begin{aligned}\int_\Omega |\Phi(v_N) - \Phi(v)| dxdt &\leq \|\rho\|_\gamma \|u\|_{p^*} (\|v_N\|_{p^*} \|\nabla(v_N - v)\|_p + \|v_N - v\|_{p^*} \|\nabla v\|_p) \\ &\leq C_{\rho,v} \|\nabla u\|_p \|\nabla(v_N - v)\|_p, \quad \text{whenever } \frac{1}{\gamma} + \frac{2}{p^*} + \frac{1}{p} \leq 1.\end{aligned}$$

Therefore,

$$\int_0^T \int_\Omega |\Phi(v_N) - \Phi(v)| dxdt \leq C_{\rho,v} \|\nabla(v_N - v)\|_p \int_0^T \|\nabla u\|_p dt,$$

which gives the continuity of the operator $\Phi(v)$. The condition on exponents just means that $p \geq p_1(\gamma)$. $\square$

Remark. For $p \geq d$ the operator $\Phi(v)$ possesses the same continuity property provided $p \geq \gamma'$ with strict inequality in the case $p = d$. Thus, under the suppositions of Theorem 2.2 the operator $\Phi(v)$ is always continuous.

7. Compactness lemmas

Our aim is to prove relations (3.14), (3.15), and (3.17). We shall do this in several steps.

Step 1. First, recall the abstract version of the Arzelà-Ascoli theorem.

Let M be a compact metric space with a distance function d_M. Denote by $C(0,T,M)$ the space of continuous functions on the segment $[0,T]$ with values in M. The sequence $\{v_n\}_{n=1}^\infty$ from $C(0,T,M)$ is called equi-continuous if

$$\forall \varepsilon > 0 \quad \exists \delta : d_M(v_n(t), v_n(t')) \leq \varepsilon, \quad n = 1, 2, ..., \text{ whenever } |t - t'| < \delta.$$

Arzelà-Ascoli Theorem [13, Ch. XII]. *Assume that the sequence* $\{v_n\}_{n=1}^{\infty}$ *from* $C(0,T,M)$ *be equi-continuous. Then there exists* $v \in C(0,T,M)$ *such that* $v_n \to v$ *in* $C(0,T,M)$*, at least on a subsequence.*

Let X be a separable Banach space and X^*_{weak} its dual endowed with $*$-weak topology; let $\langle x^*, x\rangle$ be a value of the functional $x^* \in X^*$ at $x \in X$. It is known that the $*$-weak topology is metrizable on the bounded sets in X^*. According to the construction of the corresponding metric, the general Arzelà-Ascoli theorem, in connection with X^*_{weak}, may be formulated as the following.

Corollary. *Let* $\{v_n\}_{n=1}^{\infty}$ *from* $C(0,T,X^*_{weak})$ *be uniformly bounded:* $\|v_n(t)\|_{X^*} \le c$, $t \in [0,T]$, $n = 1,2,...$, *and let the sequence of scalar functions* $\langle v_n(t), \varphi\rangle$, $t \in [0,T]$, *be equi-continuous for any fixed* φ *belonging to a dense subspace in the space* X. *Then there exists* $v \in C(0,T,X^*_{weak})$ *such that* $v_n \to v$ *in* $C(0,T,X^*_{weak})$*, at least on a subsequence.*

Step 2. We pass to study the continuity equation.

LEMMA 7.1. *Assume that* ρ_n, u_n, $n = 1,2,...$ *solve the continuity equation in the sense of distributions in* $\Omega \times (0,T)$ *and that*

$$\rho_n \overset{*}{\rightharpoonup} \rho \text{ in } L^\infty(0,T,L^\gamma(\Omega)), \quad u_n \rightharpoonup u \text{ in } L^p(0,T,(W^{1,p}(\Omega))^d), \tag{7.1}$$

$$\gamma > (p^*)', \tag{7.2}$$

$$\rho_n u_n \text{ are bounded in } L^\infty(0,T,(L^k(\Omega))^d) \tag{7.3}$$

for a certain $k > 1$. *Then, at least on a subsequence,*

$$\rho_n u_n \overset{*}{\rightharpoonup} \rho u \text{ in } L^\infty(0,T,(L^k(\Omega))^d), \tag{7.4}$$

and ρ, u *solve the continuity equation in the sense of distributions in* $\Omega \times (0,T)$.

PROOF. We have to verify (7.4); all the rest follows as a corollary. The sequence $\rho_n \in L^\infty(0,T,L^\gamma(\Omega))$ is bounded (see (7.1)) and weakly equi-continuous due to the continuity equation. In fact, for any $\varphi \in C_0^\infty(\Omega)$,

$$|\int_\Omega \rho_n \varphi dx|_{t_0}^{t_1} \le |\int_{t_0}^{t_1}\int_\Omega \rho_n u_n \cdot \nabla\varphi dx dt| \le C_\varphi \|\rho_n u_n\|_{L^\infty(0,T,L^k(M))}(t_1 - t_0).$$

By the Arzelà-Ascoli theorem and compact embedding, $L^\gamma(\Omega) \subset W^{-1,p'}(\Omega) \equiv (W_0^{1,p}(\Omega))^*$, which stems from (7.2), we have

$$\rho_n \to \rho \text{ in } C(0,T,L^\gamma_{weak}(\Omega)), \quad \rho_n \to \rho \text{ in } C(0,T,W^{-1,p'}(\Omega)), \tag{7.5}$$

at least on a subsequence. From $(7.5)_2$ and $(7.1)_2$ we can derive (7.4). □

Step 3. Consider the Galerkin approximations ρ_n, u_n, constructed before. We have relations $(3.12)_1$, $(3.12)_2$ and the boundedness property $(3.11)_4$ with $k = \frac{2\gamma}{\gamma+1} > 1$. So, by Lemma 7.1, the convergence (3.14) is derived. It remains to take note of (7.2). In the case $p < d$,

$$\gamma > (p^*)' \Leftrightarrow \frac{1}{\gamma'} > \frac{1}{p^*} \Leftrightarrow \frac{1}{p} < 1 + \frac{1}{d} - \frac{1}{\gamma} \Leftrightarrow p > \frac{d}{d+1-d\gamma^{-1}}, \tag{7.6}$$

which is ensured by (2.4). Actually, the inequality $p \ge p_1(\gamma) = \frac{3d}{d+2-d\gamma^{-1}}$, contained in (2.4), is even stronger than (7.6). In the case $p \ge d$, the condition (7.2) is trivial.

Step 4. Suppose that $p \geq p_0(\gamma)$, then $p \geq \gamma'$ (see (2.4)). Then therefore, by virtue of Lemma 8.4, the functions $\sqrt{\rho_n}$, u_n solve the continuity equation, and Lemma 7.1 may be applied to $\sqrt{\rho_n}$. By the way, the supposition (7.3) means, here, the boundedness of $\sqrt{\rho_n}u_n$ in $L^\infty(0,T,(L^2(\Omega))^d)$, which is equivalent to $(3.11)_2$. As a result,

$$\sqrt{\rho_n}u_n \overset{*}{\rightharpoonup} \sqrt{\rho}u \text{ in } L^\infty(0,T,(L^2(\Omega))^d). \tag{7.7}$$

It is important that the weak limit of $\sqrt{\rho_n}$ is equal to $\sqrt{\rho}$. This rather fine property is ensured by the strong convergence (3.16), proved in Lemma 8.5.

Step 5. Consider $g_n = P_n(\rho_n u_n)$, where P_n is the orthogonal projector on the linear span $[\omega_1, \ldots, \omega_n]$ introduced in the beginning of Section 3.1. Additionally, suppose the property $\|P_n\|_{W^{1,p}\to W^{1,p}} \leq 1$, which is possible by the special choice of the basis $\{\omega_i(x)\}_{i=1}^\infty$; see **[4, Ch.1, § 6]**, **[14, Appendix]**.

By definition of the Galerkin system, the vector $\rho_n u_n$ satisfies the identity $(3.13)_2$ on test functions from $[\omega_1, \ldots, \omega_n]$, or in short,

$$\int_\Omega g_n \cdot \psi dx|_{t_0}^{t_1} = \int_{t_0}^{t_1}\int_\Omega B_n \cdot \nabla P_n \psi dxdt, \quad \psi \in C^\infty_{0,sol}(\Omega), \tag{7.8}$$

where B_n is bounded in $L^1(0,T,(L^{p'}(\Omega))^{d\times d})$ thanks to estimates (3.11) and Lemma 6.1. From (7.8) the weak equi-continuity of g_n follows. Then, by the Arzelà-Ascoli theorem,

$$g_n \to P(\rho u) \text{ in } C(0,T,(L^k_{weak}(\Omega)^d)), \quad g_n \to P(\rho u) \text{ in } C(0,T,W^{-1,p'}(\Omega)^d), \tag{7.9}$$

where P is a projector on the space of solenoidal vectors. In particular, (3.15) is proved.

Since u_n is solenoidal by construction, the limit vector u is also solenoidal. Accordingly, from $(7.9)_2$, $(7.1)_2$ we deduce

$$\int_0^T\int_\Omega \rho_n u_n \cdot u_n dxdt = \int_0^T\int_\Omega g_n \cdot u_n dxdt = \int_0^T \langle g_n, u_n\rangle_{W^{-1,p'},W_0^{1,p}} dt \to$$
$$\int_0^T \langle P\rho u, u\rangle_{W^{-1,p'},W_0^{1,p}} dt = \int_0^T\int_\Omega P\rho u\cdot udxdt = \int_0^T\int_\Omega \rho u \cdot udxdt,$$

i.e. $\lim_{n\to\infty} \|\rho_n u_n^2\|_{L^1(Q_T)} = \|\rho u^2\|_{L^1(Q_T)}$. In combination with (7.7) this leads to the strong convergence (3.17).

It remains to explain the conclusion $(7.9)_2$. Here, we use the compact embedding $L^k(\Omega) \subset W^{-1,p'}(\Omega)$, $k = \frac{2\gamma}{\gamma+1}$, due to the proper correlation between the exponents k and p. For example, in the case $p < d$ we need $k > (p^*)' \Leftrightarrow \frac{1}{k'} > \frac{1}{p^*} \Leftrightarrow \frac{1}{2} - \frac{1}{2\gamma} > \frac{1}{p} - \frac{1}{d} \Leftrightarrow \frac{2}{p} + \frac{1}{\gamma} < 1 + \frac{2}{d}$. The last inequality is fulfilled under the condition $p \geq p_0(\gamma)$ implying (see(2.4)) $p \geq p_1(\gamma) = \frac{3d}{d+2-d\gamma^{-1}} \Leftrightarrow \frac{3}{p} + \frac{1}{\gamma} \leq 1 + \frac{2}{d}$.

8. Appendix: On the continuity equation

In this Appendix, for the reader's convenience we collect data on the continuity equation which were used in our constructions and proofs. Many results of the corresponding theory developed in [**5**], [**8**], [**9**] should be adapted to the case of a non-Newtonian fluid.

8.1. Conservation laws. Consider the equation for the unknown density ρ:

$$(8.1)\qquad \frac{\partial}{\partial t}\rho + \operatorname{div}(\rho u) = 0 \text{ in the sense of distributions in } \Omega \times (0,T).$$

For the time being, we don't assume that the vector u is solenoidal here.

LEMMA 8.1. *Let $u \in L^p(0,T,(W_0^{1,p}(\Omega))^d)$, $\rho \in L^{p'}(0,T,L^{p'}(\Omega))$, $p \geq 1$, solve (8.1), where Ω is a bounded Lipschitz domain. Then the functions u, ρ solve the continuity equation in the sense of distributions in $R^d \times (0,T)$ provided they were extended to be zero outside $\Omega \times (0,T)$.*

As a consequence, *the total mass conservation* holds: $\int_\Omega \rho(x,t)dx$ is independant of $t \in [0,T]$.

Lemma 8.1 shows when it is possible, in the integral identity for (8.1), to take test functions not vanishing at $\partial\Omega \times (0,T)$. The proof of Lemma 8.1, based on the Hardy inequality

$$\|\frac{u(\cdot)}{\operatorname{dist}(\cdot,\partial\Omega)}\|_{L^p(\Omega)} > c(p,\Omega)\|u\|_{W_0^{1,p}(\Omega)}, \quad p \geq 1,$$

is given for $p = 2$ in [**9**], and it sustains the general case $p > 1$ without particular changes.

Hereafter, as for (8.1), we assume

$$(8.2)\quad \rho \in L^\infty(0,T,L^\gamma(\Omega)), \quad u \in L^p(0,T,V), \quad \frac{1}{\gamma} + \frac{1}{p} \leq 1,\ p \in (1,\infty), \gamma \in (1,\infty].$$

Note that the vector u is already solenoidal. This property entails for the density *the invariance of the distribution function.*

LEMMA 8.2. *Meas $\{x \in \Omega : \alpha \leq \rho(x,t) \leq \beta\}$ is independent of $t \in [0,T]$, for any $\alpha \leq \beta$.*

As evident consequences from Lemma 8.2, one obtains the following properties of the density:

a) $\rho(x,0) \geq 0 \Longrightarrow \rho(x,t) \geq 0 \quad \forall t \in [0,T]$;

b) $\|\rho(\cdot,t)\|_{L^1(\Omega)} = \|\rho(\cdot,0)\|_{L^1(\Omega)}$;

c) $\|\rho(\cdot,t)\|_{L^\infty(\Omega)} = \|\rho(\cdot,0)\|_{L^\infty(\Omega)}$;

d) $\frac{\partial}{\partial t}\rho + \operatorname{div}(\rho u) = 0, \rho(\cdot,0) = 0 \Longrightarrow \rho \equiv 0$ (uniqueness of the solution to the Cauchy problem).

By the Risz-Torin interpolation theorem [**15**], properties b) and c) imply:

e) $\|\rho(\cdot,t)\|_{L^q(\Omega)} = \|\rho(\cdot,0)\|_{L^q(\Omega)}, \quad 1 \leq q \leq \infty$ (conservation of any Lebesgue norm).

Lemma 8.2 is proved via the regularization method, which is often used for the derivation of various properties of the continuity equation [**5**],[**8**],[**9**].

8.2. Regularization and its consequences. Let $\omega_\varepsilon(\cdot) = \varepsilon^{-d}\omega(\frac{\cdot}{\varepsilon})$ be a smoothing kernel, $\varepsilon \in (0,1)$, where $\omega \in C_0^\infty(\mathbb{R}^d)$, $\int_{\mathbb{R}^d} \omega dx = 1, \omega \geq 0, \operatorname{supp}\omega < \{|x| < 1\}$. As usual, smoothing (or regularization) is introduced by means of convolution $\rho^\varepsilon = [\rho]^\varepsilon = \rho * \omega_\varepsilon$.

The following "lemma about commutator with smoothing" is well known [**5**], [**8**], [**9**].

LEMMA 8.3. *(i) Let* $\rho \in L^\gamma(\mathbb{R}^d)$, $u \in (W^{1,p}(\mathbb{R}^d))^d$, $1 \le \gamma \le \infty$, $1 \le p<\infty$, $\frac{1}{\gamma}+\frac{1}{p} \le 1$. *Define* $r_\varepsilon = [\mathrm{div}(u\rho)]^\varepsilon - \mathrm{div}(u[\rho]^\varepsilon)$. *Then*

$$\|r_\varepsilon\|_{L^\beta(\mathbb{R}^d)} \le C\|u\|_{W^{1,p}(\mathbb{R}^d)}\|\rho\|_{L^\gamma(\mathbb{R}^d)}, \quad \beta = \frac{p\gamma}{p+\gamma}, \tag{8.3}$$

with the constant C *independant of* ε, u, ρ. *Moreover,* $r_\varepsilon \to 0$ *in* $L^\beta(\mathbb{R}^d)$ *as* $\varepsilon \to 0$.
(ii) Let $\rho \in L^\infty(0,T,L^\gamma(\mathbb{R}^d))$, $u \in L^1(0,T,(W^{1,p}(\Omega))^d)$. *Then*

$$r_\varepsilon \to 0 \text{ in } L^1(0,T,L^\beta(\mathbb{R}^d)) \text{ as } \varepsilon \to 0, \tag{8.4}$$

with r_ε *and* β *as above.*

As a corollary we get

LEMMA 8.4. *Let* $\rho \ge 0$ *solve (8.1) under supposition (8.2). Let* $B \in C^1([0,\infty),\mathbb{R})$, *and* $B(t)$, $B'(t)$ *be bounded. Then*

$$\frac{\partial}{\partial t}B(\rho) + \mathrm{div}(uB(\rho)) = 0 \text{ in the sense of distributions in } \Omega \times (0,T), \tag{8.5}$$

$$\int_\Omega B(\rho(x,t))dx = const. \tag{8.6}$$

The assertion holds true also for the function $B(t) = \sqrt{t}$.

PROOF. 1°. We make regularization of both parts of (8.1). As a result,

$$\frac{\partial}{\partial t}\rho^\varepsilon + \mathrm{div}(u\rho^\varepsilon) = r_\varepsilon \text{ a.e. in } \Omega \times (0,T), \tag{8.7}$$

with r_ε defined in Lemma 8.3. Multiplying (8.7) by $B'(\rho^\varepsilon)$ yields the equation

$$\frac{\partial}{\partial t}B(\rho^\varepsilon) + \mathrm{div}(uB(\rho^\varepsilon)) = B'(\rho^\varepsilon)r_\varepsilon \text{ a.e. in } \Omega \times (0,T). \tag{8.8}$$

Using (8.4) in view of (8.2) and, also, the boundedness properties of $B(t)$, $B'(t)$, we deduce

$$B(\rho^\varepsilon) \to B(\rho), \quad uB(\rho^\varepsilon) \to uB(\rho), \quad B'(\rho^\varepsilon)r_\varepsilon \to 0 \text{ in } L^1(\Omega \times (0,T)). \tag{8.9}$$

So, passing to the limit in (8.8) leads to (8.5) and, by Lemma 8.1, the relation (8.6) is true.

2°. Now let $B(t) = \sqrt{t}$ and approximate $\sqrt{t}$ on $[0,\infty)$ with Lipschitz functions $B_\delta(t) = (\sqrt{t}-\delta)_+$, $\delta > 0$. The preceding arguments are feasible for the function B_δ. For justification, write down the analogue of (8.8),

$$\frac{\partial}{\partial t}B_\delta(\rho^\varepsilon) + \mathrm{div}(uB_\delta(\rho^\varepsilon)) = B_\delta'(\rho^\varepsilon)r_\varepsilon \quad \text{a.e. in} \quad \Omega \times (0,T), \tag{8.10}$$

for which the limit relations of the type (8.9) hold with B_δ instead of B and any fixed δ. In fact, $|B_\delta'(t)| < c_\delta$ a.e., $B_\delta(t) \le \sqrt{t} < t+1$ and condition (8.2) is assumed.

We pass to the limit in (8.10), first, over ε, and, then, over δ, according to the Lebesgue dominated convergence theorem. Thus, we consequently obtain

$$\frac{\partial}{\partial t}B_\delta(\rho) + \mathrm{div}(uB_\delta(\rho)) = 0, \quad \frac{\partial}{\partial t}\sqrt{\rho} + \mathrm{div}(u\sqrt{\rho}) = 0$$

in the sense of distributions in $\Omega \times (0,T)$, which is required. The proof of the lemma is complete.

With the help of Lemma 8.4, we prove Lemma 8.2. To this end, fix an arbitrary segment $[\alpha,\beta] \subset \mathbb{R}$ and approximate its characteristic function $\chi = \chi_{[\alpha,\beta]}$ with smooth functions χ_n, such that $0 \le \chi_n \le 1$, $\chi = 0$ outside $[\alpha, \beta]$, $\chi_n = 1$ on $(\alpha + \frac{1}{n}, \beta - \frac{1}{n})$, where $n \in \mathbb{N}$ is so large, that $\frac{2}{n} < \beta - \alpha$. Write down (8.6) for $B = \chi_n$ and pass to the limit in it as $n \to \infty$:

$$\frac{d}{dt} \int_\Omega \chi_n(\rho) dx = 0 \Longrightarrow \frac{d}{dt} \int_\Omega \chi(\rho) dx = 0,$$

and the lemma is proven. □

8.3. Continuity properties. By means of regularization, we study the continuity of the solution to (8.1) considered as a function of t with values in Lebesgue spaces.

LEMMA 8.5. *(i) Let ρ, u solve (8.1) under condition (8.2) and $\rho|_{t=0} = \rho_0 \in L^\gamma(\Omega)$. Then*

$$\rho \in C(0, T, L^\beta(\Omega)), \quad \beta = \frac{\gamma p}{\gamma + p} \in [1, \gamma). \tag{8.11}$$

(ii) Let ρ_n, u_n ($n \in \mathbb{N}$) satisfy the conditions of part (i), and

$$\|\rho_n\|_{L^\infty(0,T,L^\gamma(\Omega))} \le c, \quad \|u_n\|_{L^p(0,T,V)} \le c, \tag{8.12}$$

$$\rho_n|_{t=0} = \rho_{n_0} \to \rho_0 \text{ in } L^\beta(\Omega), \quad u_n \rightharpoonup u \text{ in } L^p(0, T, V). \tag{8.13}$$

Let ρ be the solution of (8.1) with the vector u from (8.13), $\rho|_{t=0} = \rho_0$. Then

$$\rho_n \to \rho \text{ in } C(0, T, L^\beta(\Omega)).$$

PROOF. We start with (8.8), where $\rho^\varepsilon = \rho * \omega_\varepsilon$ and $B(t) = t^\beta$. Integrating it yields

$$\int_\Omega (\rho^\varepsilon)^\beta dx|_0^\tau = \int_0^\tau \int_\Omega \beta r_\varepsilon (\rho^\varepsilon)^{\beta-1} dx dt, \quad \tau \in [0, T].$$

Hence, by the Hölder and Young inequalities,

$$A_\varepsilon^\beta \equiv \sup_{t \in [0,T]} \|\rho^\varepsilon(\cdot, t)\|_{L^\beta(\Omega)}^\beta \le \|\rho_0 * \omega_\varepsilon\|_{L^\beta(\Omega)}^\beta + \beta A_\varepsilon^{\beta - 1} \int_0^T \|r_\varepsilon\|_{L^\beta(\Omega)} dx,$$

$$A_\varepsilon \le c_0(\|\rho_0 * \omega_\varepsilon\|_{L^\beta(\Omega)} + \|r_\varepsilon\|_{L^1(0,T,L^\beta(\Omega))}) \le c_1, \tag{8.14}$$

due to Lemma 8.3. Thus, ρ^ε is bounded in $C(0, T, L^\beta(\Omega))$.

In the same way we deduce the estimate, similar to (8.14), for the difference of two smoothings:

$$\|\rho^{\varepsilon_1} - \rho^{\varepsilon_2}\|_{C(0,T,L^\beta(\Omega))} \le c_0(\|\rho_0 * \omega_{\varepsilon_1} - \rho_0 * \omega_{\varepsilon_2}\|_{L^\beta(\Omega)} + \|r_{\varepsilon_1} - r_{\varepsilon_2}\|_{L^1(0,T,L^\beta(\Omega))}),$$

whence, by (8.4) and properties of smoothing, $\rho^\varepsilon \to \rho$ in $C(0, T, L^\beta(\Omega))$, which proves (8.11).

In order to prove part *(ii)*, we use the preceding arguments towards the smoothings $\rho^\varepsilon = \rho * \omega_\varepsilon$, $\rho_n^\varepsilon = \rho_n * \omega_\varepsilon$ and their difference, taking into account relations (8.12) and $(8.13)_1$. The details are omitted. □

8.4. Proof of Lemma 1.2. Here several equivalent versions of the weak setting of (8.1) are given, including (1.5) and (1.6). To this end, Steklov smoothing is used. Recall its definition,

$$f_h(x,t) = \frac{1}{h}\int_t^{t+h} f(x,s)\,ds, h>0, \quad f \in L^1(Q_T), \quad (x,t) \in Q_{T-h}$$

and properties.

LEMMA 8.6. *Let $f \in L^\gamma(Q_T)$, $\gamma \geq 1$. Then $f_h \to f$ in $L^\gamma(Q_{T-\delta})$ for any $\delta > 0$ and $f_h(\cdot,t) \to f(\cdot,t)$ in $L^\gamma(\Omega)$ for a.e. $t \in [0,T]$ as $h \to 0$.*

Pass to Lemma 1.2. Taking in (1.5) $\varphi = \eta\chi$, $\eta \in C_0^\infty(\Omega)$, $\chi \in C_0^\infty(0,T)$, we obtain

$$\int_0^T \chi'(t)\int_\Omega v\eta\,dxdt = \int_0^T \chi(t)\int_\Omega K\cdot\nabla\eta\,dxdt \text{ or } \int_0^T \chi'(t)g(t)\,dt = \int_0^T \chi(t)G(t)\,dt.$$

Since $g, G \in L^1(0,T)$, then $g \in W^{1,1}(0,T)$ and property *2)* is obvious.

Now we prove the opposite assertion. Integrating (1.6) leads to

$$\int_\Omega v\eta\,dxdt\Big|_{t_1}^{t_2} + \int_{t_1}^{t_2}\int_\Omega K\cdot\nabla\eta\,dxdt = 0. \tag{8.15}$$

Fixing $t_1 = t, t_2 = t+h < T$, where $h > 0$, we rewrite (8.15) in terms of the Steklov smoothings

$$\int_{\Omega\times\{t\}} \left(\frac{\partial v_h}{\partial t}\eta + K_h\cdot\nabla\eta\right)dx = 0.$$

Here the test function η may be dependent on t, and after integrating over t we arrive at

$$\int_{t_1}^{t_2}\int_\Omega \left(\frac{\partial v_h}{\partial t}\varphi + K_h\cdot\nabla\varphi\right)dxdt = 0 \quad \forall\varphi \in C^\infty(\overline{Q}_T), \quad \varphi|_{\partial\Omega\times[0,T]} = 0. \tag{8.16}$$

Setting in (8.16) $t_1 = 0$, $t_2 = T$, $\varphi \in C_0^\infty(Q_T)$, via convergences (see Lemma 8.6)

$$\int_{Q_T} \frac{\partial v_h}{\partial t}\varphi\,dxdt = -\int_{Q_T} v_h\frac{\partial\varphi}{\partial t}\,dxdt \to -\int_{Q_T} v\frac{\partial\varphi}{\partial t}\,dxdt,$$

$$\int_{Q_T} K_h\cdot\nabla\varphi\,dxdt \to \int_{Q_T} K\cdot\nabla\varphi\,dxdt$$

we come to (1.5). The proof of the lemma is complete.

References

[1] V.V. Zhikov, *Solvability of generalized Navier–Stokes equations,* Doklady Akad. Nauk **423:5** (2008), 597–602; English transl., Doklady Mathematics, **78:3** (2008), 896–900. MR2498572

[2] ———, *An approach to solvability of generalized Navier–Stokes equations,* Funktsional′nyi Analiz i Ego Prilozheniya, **43:3** (2009) (in Russian). MR2498572

[3] O. A. Ladyzhenskaya, *New equations for the description of the viscous incompressible fluids and solvability in the large of the boundary problems for them,* Trudy Mat. Instituta AN SSSR, **102** (1967), 85-104; English transl.: Boundary value problems of mathematical physics, Amer. Math. Soc., Providence, RI, 1970. MR0226907 (37:2493); MR0266725 (42:1628)

[4] J.-L. Lions, *Quelques méthodes de résolution des problèmes aux limites non linéaires,* Dunod, Paris, 1969. MR0259693 (41:4326)

[5] R. J. Diperna, P.-L Lions, *Ordinary differential equations, transport theory and Sobolev spaces,* Invent. Math. **98** (1989), 511–547. MR1022305 (90j:34004)

[6] A. V. Kazhikhov, *Resolution for boundary value problems for nonhomogeneous viscous fluids,* Dokl. Akad. Nauk SSSR, **216** (1974), 1008–1010 (in Russian).

[7] S. N. Antontsev, A. V. Kazhikhov, and V. N. Monakhov, *Boundary Value Problems in Nonhomogeneous Fluid Mechanics,* Nauka, Novosibirsk, 1983; English transl., North-Holland, Amsterdam, 1990. MR1035212 (91d:76018)

[8] P.-L. Lions, *Mathematical topics in fluid mechanics.* Vol. I. *Incompressible Models,* Oxford Sci. Publ., 1998.

[9] E. Feireisl, *Dynamics of viscous compressible fluids,* Oxford Univ. Press, 2004. MR2040667 (2005i:76092)

[10] E. A. Coddington and N. Levinson, *Theory of ordinary differential equations,* McGraw-Hill, New York, Toronto, London, 1955. MR0069338 (16:1022b)

[11] P. Billingsley. *Convergence of probability measures*, Wiley, New York–London–Sydney–Toronto, 1968. MR0233396 (38:1718)

[12] V. V. Zhikov, *On the technique for passing to the limit in nonlinear elliptic equations,* Funktsional′nyi Analiz i Ego Prilozheniya,**43:2** (2009), 19–38; English transl.: Functional Analysis and Its Applications, **43:2** (2009), 96–112. MR2542272

[13] J.L. Kelley, *General topology,* Van Nostrand, Princeton, NJ, 1957. MR0070144 (16:1136c)

[14] J. Málek, J. Nečas, M. Rokyta, and M. Růžička, *Weak and measure-valued solution to evolutionary PDE*, Chapman & Hall, London, 1996. MR1409366 (97g:35002)

[15] N. Dunford and J. Schwartz, *Linear Operators. General Theory*, Interscience, New York-London, 1958. MR0117523 (22:8302)

Vladimir State Humanities University, 600024, Vladimir, prospect Stroiteley, 11
E-mail address: zhikov@vladggu.ru

Moscow State Institute of Radioengineering, Electronics and Automatics (Technical University), 119454, Moscow, prospect Vernadskogo, 78
E-mail address: pas-se@yandex.ru

Titles in This Series

198 **R. A. Minlos, Senya Shlosman, and Yu. M. Suhov, Editors,** On Dobrushin's Way. From Probability Theory to Statistical Physics

197 **Vladimir Arnold, Maxim Kontsevich, and Anton Zorich, Editors,** Pseudoperiodic Topology

196 **Ya. Eliashberg, D. Fuchs, T. Ratiu, and A. Weinstein, Editors,** Northern California Symplectic Geometry Seminar

195 **O. V. Belegradek et al.,** Model Theory and Applications

194 **A. Astashkevich and S. Tabachnikov, Editors,** Differential Topology, Infinite-Dimensional Lie Algebras, and Applications (D. B. Fuchs' 60th Anniversary Collection)

193 **N. N. Uraltseva, Editor,** Proceedings of the St. Petersburg Mathematical Society, Volume V

192 **Lev Beklemishev, Mati Pentus, and Nikolai Vereshchagin,** Provability, Complexity, Grammars

191 **A. Yu. Morozov and M. A. Olshanetsky, Editors,** Moscow Seminar in Mathematical Physics

190 **S. Tabachnikov, Editor,** Differential and Symplectic Topology of Knots and Curves

189 **V. Buslaev, M. Solomyak, and D. Yafaev, Editors,** Differential Operators and Spectral Theory (M. Sh. Birman's 70th Anniversary Collection)

188 **O. A. Ladyzhenskaya, Editor,** Proceedings of the St. Petersburg Mathematical Society, Volume IV

187 **M. V. Karasev, Editor,** Coherent Transform, Quantization, and Poisson Geometry

186 **A. Khovanskiĭ, A. Varchenko, and V. Vassiliev, Editors,** Geometry of Differential Equations

185 **B. Feigin and V. Vassiliev, Editors,** Topics in Quantum Groups and Finite-Type Invariants (Mathematics at the Independent University of Moscow)

184 **Peter Kuchment and Vladimir Lin, Editors,** Voronezh Winter Mathematical Schools (Dedicated to Selim Krein)

183 **K. Nomizu, Editor,** Selected Papers on Harmonic Analysis, Groups, and Invariants

182 **V. E. Zakharov, Editor,** Nonlinear Waves and Weak Turbulence

181 **G. I. Olshanski, Editor,** Kirillov's Seminar on Representation Theory

180 **A. Khovanskiĭ, A. Varchenko, and V. Vassiliev, Editors,** Topics in Singularity Theory

179 **V. M. Buchstaber and S. P. Novikov, Editors,** Solitons, Geometry, and Topology: On the Crossroad

178 **V. Kreinovich and G. Mints, Editors,** Problems of Reducing the Exhaustive Search

177 **R. L. Dobrushin, R. A. Minlos, M. A. Shubin, and A. M. Vershik, Editors,** Topics in Statistical and Theoretical Physics (F. A. Berezin Memorial Volume)

176 **E. V. Shikin, Editor,** Some Questions of Differential Geometry in the Large

175 **R. L. Dobrushin, R. A. Minlos, M. A. Shubin, and A. M. Vershik, Editors,** Contemporary Mathematical Physics (F. A. Berezin Memorial Volume)

174 **A. A. Bolibruch, A. S. Merkur′ev, and N. Yu. Netsvetaev, Editors,** Mathematics in St. Petersburg

173 **V. Kharlamov, A. Korchagin, G. Polotovskiĭ, and O. Viro, Editors,** Topology of Real Algebraic Varieties and Related Topics

172 **K. Nomizu, Editor,** Selected Papers on Number Theory and Algebraic Geometry

171 **L. A. Bunimovich, B. M. Gurevich, and Ya. B. Pesin, Editors,** Sinai's Moscow Seminar on Dynamical Systems